T0371861

Design Optimization using MATLAB® and SOLIDWORKS®

A unique text integrating numerics, mathematics and applications to provide a hands-on approach to learning optimization techniques, this mathematically accessible textbook emphasizes conceptual understanding and importance of theorems rather than elaborate proofs. It allows students to develop fundamental optimization methods before delving into MATLAB®'s optimization toolbox, and to link MATLAB's results with the results from their own code. Following a practical approach, the text demonstrates several applications, from error-free analytic examples to truss (size) optimization, and 2D and 3D shape optimization, where numerical errors are inevitable. The principle of minimum potential energy is discussed to highlight the deep relationship between engineering and optimization. MATLAB code in every chapter illustrates key concepts and the text demonstrates the coupling between MATLAB and SOLIDWORKS® for design optimization. A wide variety of optimization problems are covered including constrained non-linear, linear-programming, least-squares, multi-objective, and global optimization problems.

Krishnan Suresh is the Philip and Jean Myers Professor of Mechanical Engineering at the University of Wisconsin–Madison, and a Fellow of the American Society of Mechanical Engineers. He is also the co-founder of SciArt Software (www.sciartsoft.com), a UW–Madison spinoff that creates and supports high-performance topology optimization software solutions.

Design Optimization using MATLAB® and SOLIDWORKS®

Krishnan Suresh
University of Wisconsin–Madison

CAMBRIDGE
UNIVERSITY PRESS

University Printing House, Cambridge CB2 8BS, United Kingdom

One Liberty Plaza, 20th Floor, New York, NY 10006, USA

477 Williamstown Road, Port Melbourne, VIC 3207, Australia

314–321, 3rd Floor, Plot 3, Splendor Forum, Jasola District Centre, New Delhi – 110025, India

79 Anson Road, #06–04/06, Singapore 079906

Cambridge University Press is part of the University of Cambridge.

It furthers the University's mission by disseminating knowledge in the pursuit of education, learning, and research at the highest international levels of excellence.

www.cambridge.org
Information on this title: www.cambridge.org/highereducation/suresh
DOI: 10.1017/9781108869027

First published 2021

Printed in the United Kingdom by TJ Books Limited, Padstow Cornwall

A catalogue record for this publication is available from the British Library.

Library of Congress Cataloging-in-Publication Data
Names: Suresh, Krishnan, 1970– author.
Title: Design optimization using MATLAB and SOLIDWORKS / Krishnan Suresh.
Description: Cambridge, United Kingdom ; New York, NY : Cambridge University Press, 2021. | Series: Cambridge texts in biomedical engineering | Includes bibliographical references and index.
Identifiers: LCCN 2020052134 | ISBN 9781108491600 (hardback) | ISBN 9781108869027 (ebook)
Subjects: LCSH: MATLAB. | SolidWorks. | Engineering design – Data processing – Textbooks. | Mathematical optimization – Textbooks.
Classification: LCC TA174 .S925 2021 | DDC 620/.00420285536–dc23
LC record available at https://lccn.loc.gov/2020052134

ISBN 978-1-108-49160-0 Hardback

Additional resources for this publication at www.cambridge.org/highereducation/suresh

Dedicated to my family.

They mean the world to me.

Contents

Preface

Origins of This Text

As is of true of many textbooks, this text has evolved from teaching a formal course, specifically, "Optimum Design of Mechanical Elements and Systems," at the University of Wisconsin–Madison. It has undergone several revisions over the last decade.

Target Audience

The primary audience for this text includes senior undergraduate students, junior graduate students and practicing engineers. Given this wide audience, prerequisites are kept to a minimum. Basic undergraduate-level mathematics is assumed, but no prior background in optimization is required. Prior experience in programming, especially MATLAB, is helpful.

Topics Covered

This text covers three complementary topics in optimization: (1) the underlying mathematics, (2) numerical methods and nuances and (3) engineering applications.

On the first topic, we will recall basic results from single- and multi-variable calculus. Emphasis is given to conceptual understanding of theorems rather than to elaborate proofs (which can be found in mathematically oriented texts).

The second topic, numerical methods, lies at the heart of optimization. It is impossible to understand optimization without a good grasp of numerical methods. Therefore, this text encourages the reader to implement simple optimization methods "from scratch," before delving into MATLAB's optimization toolbox. This will provide a good understanding of how optimization algorithms work, or sometimes fail! Besides the generic class of constrained non-linear optimization problems, we will also discuss special types of problems, including linear-programming problems, least-squares problems and multi-objective problems, that require special treatment. Finally, we will briefly explore global optimization methods that are becoming increasingly important in engineering.

It is the author's strong opinion that engineers should learn optimization within the context of applications. This text largely focuses on geometric and structural applications. For example, we will study how truss systems can be optimized, and some of the pitfalls. Lessons learned from this exercise will be applied toward the structural shape optimization of 2D and 3D designs. The reader is encouraged to apply underlying optimization principles and methods to his/her field of study.

Topics Not Covered

Engineering optimization is too broad to be covered in a single text. Consequently, a few topics have been omitted; these include surrogate modeling, stochastic optimization and optimization under uncertainty.

Software Resources

This text assumes that the reader has access to MATLAB® (www.mathworks.com). The MATLAB code accompanying this text is an integral part of student learning, and can be downloaded from the author's website at www.ersl.wisc.edu under the "Research" tab. The MATLAB code is object-oriented; it teaches the basic concepts of data encapsulation, inheritance and code reuse. Through exercises, the reader will learn to extend the code to address various optimization problems.

In the last couple of chapters, we will create and analyze 3D designs using SOLIDWORKS® (www.solidworks.com). It is assumed that the reader is familiar with creating simple models in SOLIDWORKS. Basic finite element analysis using SOLIDWORKS is covered in the text. Then a toolbox, namely SOLIDLAB, which serves as an interface between SOLIDWORKS and MATLAB, is presented; SOLIDLAB was developed by the author's research group. Through SOLIDLAB, one can, for example, modify feature dimensions of SOLIDWORKS models, query mass properties, carry out a finite element analysis and optimize, all from within the comfort of MATLAB.

Acknowledgments

Textbook writing inevitably takes time away from family, and I would like to acknowledge the support of my dear wife, Vanitha, for her constant encouragement to finish this text. Meanwhile, our two sons, Sanjay and Arjun, have smartly learned to work around my busy schedule. I would also like to thank my graduate students:

Aaditya Chandrasekhar, Josh Danczyk, Tej Kumar, Amir M. Mirzendehdel, Bhagyashree Prabhune, Saketh Sridhara and Subodh Subedi, who have contributed both directly and indirectly to this text. I would also like to acknowledge the support of the National Science Foundation and the Graduate School at the University of Wisconsin–Madison; they have helped nurture this text.

1 Introduction

Optimization is an integral part of engineering today. Engineers use optimization techniques to design civil structures, machine components, electrical circuits, plant layouts, chemical processes and so on. Indeed, one simply cannot make technological advances without optimization.

1.1 An Example of Optimization

While there are numerous examples of optimization, we will consider here a design problem posed in Figure 1.1 where a plate is subject to a uniform pressure loading on one face, and is fixed at the other. The plate can be modeled and analyzed using any of the popular finite element packages; we will discuss one such package, namely SOLIDWORKS ([1]), later in the text.

Based on finite element analysis (FEA), one can determine the stress distribution within the plate, as illustrated in Figure 1.2; the maximum stress happens to be around 515 MPa, and occurs on the periphery of the large hole, as expected.

A typical design objective now is to reduce the maximum stress, without increasing the mass of the plate. Further, the overall plate dimensions and the diameter of the larger hole cannot be modified. The location and size of the two smaller holes can, however, be modified. In this example, the engineer finds out, through trial-and-error, that the stress can be reduced by simply enlarging the two smaller holes. This is illustrated in Figure 1.3, where the maximum stress is reduced to 466 MPa; this is done by increasing the diameter of the two holes from 15 mm to 25 mm. The maximum stress now occurs at the periphery of one of the smaller holes. *Observe that, in the process of reducing the stress, the mass has also been reduced!*

One can now ask if the stress can be further reduced, i.e., *is there an optimal location and optimal diameter for the two smaller holes such that the maximum stress is minimized?* This is an example of shape optimization that we shall study later in the text.

1.2 Challenges

There are numerous such "simple" optimization examples in engineering. However, in the author's experience, while solving such problems engineers often run into several challenges: (1) *How does one translate the above problem*

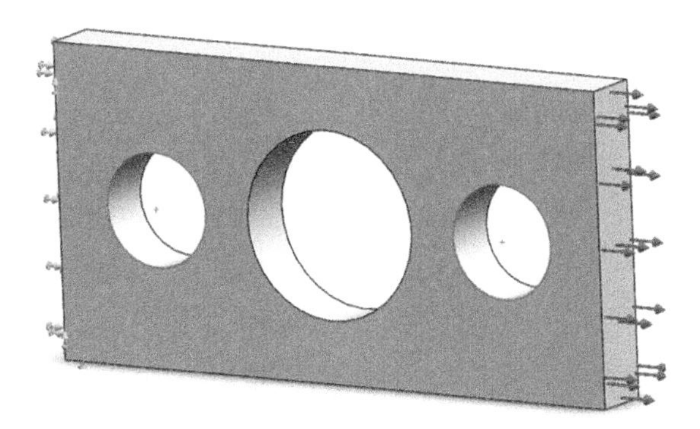

Figure 1.1 Design problem.

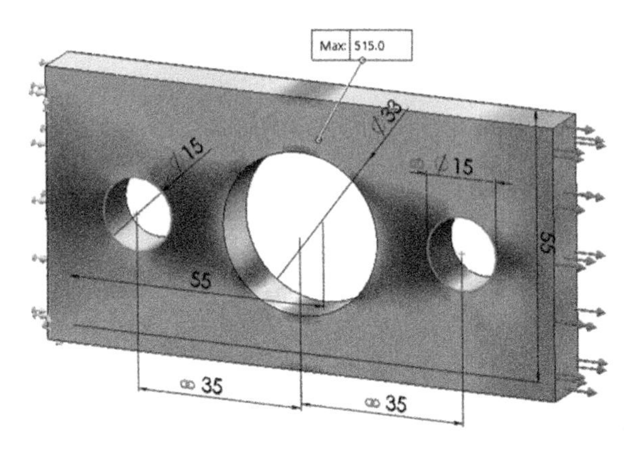

Figure 1.2 Stress plot based on finite element analysis (FEA).

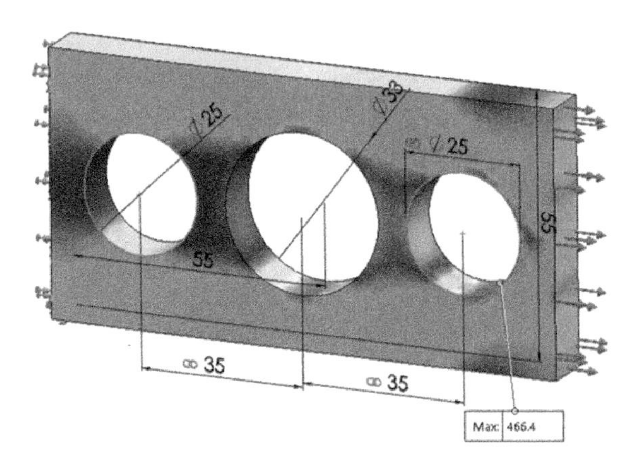

Figure 1.3 Reduced stress for a modified design.

into a formal optimization statement (with objective, constraints, feasible space, design variables, etc.)? (2) *What optimization method should one use and why?* (3) *What if the optimization method does not converge?* And so on.

After reading through this text, and completing the exercises, the author sincerely believes that the reader will be able to answer such questions confidently, and also extend the concepts to his/her field of study.

Study of optimization can be fun and enriching. However, for most effective learning, engineers should study optimization within the context of a relevant and familiar application. Learning to code sophisticated optimization methods, without understanding why these different methods exist in the first place, is both meaningless and counterproductive. Therefore, this text introduces fundamental optimization concepts through concrete applications.

For example, to introduce the important concept of *numerical scaling*, we will consider optimizing a truss system. We will observe that, without numerical scaling, even the best optimization method will not converge correctly. By considering a variety of such applications, fundamental optimization concepts can be assimilated easily.

No prior background in optimization is assumed in this text; basic undergraduate-level mathematics is sufficient. Prior experience in programming, especially MATLAB programming [2], is helpful. For this text, we will rely entirely on MATLAB programming, basics of which are covered in Chapters 3, 6 and 7.

1.3 MATLAB Code

The MATLAB code accompanying this text is an integral part of student learning, and can be downloaded as a zip-file from the author's website at www.ersl.wisc.edu under the "Research" tab. The code is organized chapter-wise; the use of this code is discussed in Chapter 3, where MATLAB is introduced.

1.4 Organization of Text

This text covers three complementary topics in engineering optimization: (1) the underlying mathematics, (2) numerical methods and nuances and (3) engineering applications; see Figure 1.4.

The text is organized as in Table 1.1; each chapter will introduce a critical mathematical concept/numerical method (identified by rows in the table) by considering a specific engineering application (identified by columns in the table).

Chapter 2 introduces the critical concept of *modeling*, i.e., the art of translating a loosely worded optimization problem into a formal mathematical statement. The notions of objective and constraint are introduced by considering various applications. This chapter will serve as an overview for the remainder of this text.

Chapter 3 is a short introduction to MATLAB. It is by no means an exhaustive review. The reader will be introduced to basic computing and plotting

Table 1.1 Text organization across various chapters.

Applications → *Concepts/Tools*↓	Analytical optimization	Truss optimization	Shape optimization
Modeling	Chapter 2	Chapter 2	Chapter 2
Basic MATLAB programming	Chapter 3		
Unconstrained theory	Chapter 4		
Basic algorithms	Chapter 5		
MATLAB optimization toolbox	Chapter 6		
Constrained theory	Chapter 7		
Specialized problems	Chapter 8		
Structural analysis		Chapter 9	
Numerical scaling		Chapter 10	
Gradient computation		Chapter 11	
Finite element analysis (2D)			Chapters 12, 13
Finite element analysis (3D)			Chapters 14, 15, 16

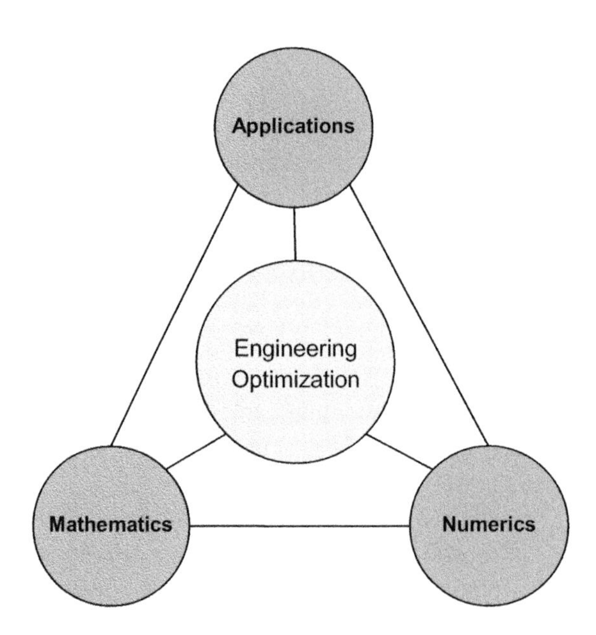

Figure 1.4 Engineering optimization encompasses three topics: mathematics, numerics and applications.

routines available within MATLAB. Various programming constructs such as the for-loop and if-then-else will be discussed. The concept of object-oriented programming will be reviewed through examples.

In Chapter 4, we will address optimization theory, focusing on the *unconstrained optimization* problems discussed in Chapter 2. Some of the questions raised in Chapter 2 will be answered. Critical concepts such as global/local minima and stationary points will be introduced.

Chapter 5 complements Chapter 4 in that we will *implement a few basic algorithms* to appreciate the nuances of numerical optimization. While engineers do not typically implement optimization algorithms "from scratch," numerical implementation will provide a good understanding of how algorithms work, and sometimes fail!

In Chapter 6, we will delve into MATLAB's *optimization toolbox* and study some of the algorithms for solving unconstrained problems. Using several test cases, we will observe the behavior of these methods, and correlate them to the observations made in the previous chapter.

Chapter 7 addresses the basic theory behind *constrained optimization*. Critical concepts such as Lagrange multipliers will be introduced within the context of engineering applications. Physical and mathematical interpretations of Lagrange multipliers will be discussed. We will also study MATLAB-supported algorithms for solving constrained problems.

Chapter 8 covers certain special types of optimization problems, including linear-programming problems, least-squares problems and multi-objective problems, that are not addressed in the previous chapters. One of the primary reasons for treating them separately is that they often require the use of specialized numerical methods.

In Chapter 9, trusses are analyzed through force balance and potential energy principles. The main concept emphasized is that "elastic structures, such as truss systems, when subject to an external load, reach a stable configuration when their potential energy reaches a local minimum." In other words, *physical systems behave in an optimal fashion*! This observation is both fascinating and important in engineering.

Chapter 10 will build on Chapter 9 by considering the *size optimization of truss systems*, where the goal is to find the optimal size (diameter) of truss members such that a given objective (say, the mass of the truss) is minimized, subject to certain constraints (such as deflection and stresses). MATLAB optimization methods introduced in Chapter 7 will be deployed to find optimal solutions to such problems. We will find that the algorithms do not always converge to the correct answer (even for simple problems). This will motivate the need for *numerical scaling* – a critical concept in numerical optimization! This concept will be covered and illustrated through several examples.

Chapter 11 will cover an equally important concept of gradient (i.e., sensitivity) computation. Gradient computation is critical if first-order methods of optimization are employed. This chapter covers the pitfalls of finite-difference-based gradient computation and provides alternative methods.

In Chapters 12 and 13, we consider FEA and optimization of 2D elastic problems. Conceptually, this is a direct extension of truss analysis and optimization, discussed in Chapters 9 and 10 respectively. However, new concepts such as shape parameters and finite element discretization enter the picture.

Chapter 14 is an extension of Chapter 12 to 3D FEA. Here, we will rely on SOLIDWORKS for modeling and analysis. It is assumed that the reader is familiar with the process of creating 3D models and carrying out basic FEA within SOLIDWORKS. The objective of this chapter is to develop a foundation for parametric study and optimization to be pursued in the next chapter.

Chapter 15 introduces SOLIDLAB, an interface between SOLIDWORKS and MATLAB. SOLIDLAB was developed by the author and his graduate students. This chapter will illustrate the use of SOLIDLAB to query and analyze SOLIDWORKS models, from within the comfort of MATLAB.

Chapter 16 addresses 3D shape optimization by combining SOLIDWORKS, MATLAB and SOLIDLAB. The fundamental difference between compliance and stress minimization is highlighted.

The Appendix covers additional mathematical concepts and proofs.

Finally, it goes without saying that no textbook is ever complete or comprehensive. There are several excellent textbooks on engineering optimization (see references [3], [4], [5], [6], [7]) and numerical optimization (see references [8], [9]) that complement this text. The reader is strongly encouraged to consult these for topics not covered here.

2 Modeling

Highlights

1. This chapter discusses *modeling*, the first step in optimization. Modeling is the process of converting a loosely worded optimization problem into a formal mathematical statement.
2. Several modeling examples from geometry and structural mechanics are considered, and each example is converted into the standard formulation.
3. Through these examples, relevant optimization terminology is also explained.
4. Finally, important observations are made about modeling; these include: (1) introducing appropriate design variables, (2) exploiting optimality criteria for simplification and (3) the iterative nature of modeling.

Modeling is the process of converting a loosely worded "optimization" problem into a mathematically precise and standard formulation. For example, converting the stress minimization problem discussed in Section 1.1 into a formal optimization statement would be a modeling effort. Accurate modeling is crucial; an inaccurate model will lead to erroneous conclusions. To illustrate the concept of modeling, we consider several examples in this chapter. However, first, the standard optimization formulation is presented, together with an explanation of the terminology.

2.1 Standard Optimization Formulation

Almost all optimization problems considered in this text will be posed in the following standard form ("s.t." is short for "such that"):

$$\begin{aligned} &\underset{\mathbf{x}}{\text{minimize}}\ f(\mathbf{x}) \\ &\text{s.t.}\quad h_i(\mathbf{x}) = 0;\ i = 1, 2, \ldots \\ &\qquad\ g_j(\mathbf{x}) \le 0;\ j = 1, 2, \ldots \\ &\qquad\ \mathbf{x}^{\min} \le \mathbf{x} \le \mathbf{x}^{\max} \\ &\qquad\ \mathbf{x} = \{x_1 \quad x_2 \quad \ldots \quad x_N\} \end{aligned} \tag{2.1}$$

where

- $f(\mathbf{x})$ is the single *objective* that we are trying to minimize. Given a loosely worded problem statement, the first task is to identify and *quantify* the objective. In structural design problems, the objective is typically the volume or weight of a part, or the maximum deflection, or the maximum stress, and so on. Two points are worth noting here: (1) Often, an engineer may be interested in multiple objectives (for example, minimize weight and minimize stress); such *multi-objective* problems are briefly considered in Section 6.7, but are not the main focus of this text; further, it is often possible to interpret one or more of these objectives as constraints. (2) If we desire to *maximize* an objective (for example, maximize the stiffness of a part), it is easy to convert this into a minimization problem in multiple ways, as discussed later on.
- $\mathbf{x}$ are the *optimization or design variables*. These are the free parameters that can be modified to meet the objective. In structural design problems, these could be geometric parameters (thickness of a truss member, the location and size of a hole, topology), or material properties (Young's modulus, yield strength), and so on. In this text, we will assume that the optimization variables are *continuously varying* (such as the radius of a hole). Discrete variables, such as the number of holes in a design, are not explicitly treated in this text; however, references and examples are provided on how such integer problems can be handled.
- $\mathbf{x}^{\min}$ and $\mathbf{x}^{\max}$ are the lower and upper bounds on the optimization variables; for example, a lower bound and/or upper bound on a hole radius. It is not essential for optimization variables to exhibit lower and upper bounds. However, if such bounds exist, it is important to include them in the problem statement.
- $h_i(\mathbf{x})$ are the *equality constraints*; for example, "the diameter of truss member A must be exactly equal to three times the diameter of truss member B." While we are not distinguishing here between linear and non-linear constraints in the formulation, such differences can be important in numerical analysis. They are highlighted later on.
- $g_j(\mathbf{x})$ are the *inequality constraints*; for example, "the diameter of truss member A must be less than three times the diameter of truss member B." Again, we are not distinguishing here between linear and non-linear inequality constraints, but such differences can be important in numerical analysis.

In the remainder of this chapter, we shall study several optimization examples and convert them into the standard form shown in Equation (2.1). We will observe that there are several special cases of the standard form; for example, when the constraints are absent, we obtain an *unconstrained minimization* problem, which is mathematically and numerically easy to analyze:

$$\underset{\mathbf{x}}{\text{minimize}}\ f(\mathbf{x}) \tag{2.2}$$

where $\mathbf{x} = \{x_1 \quad x_2 \quad \dots \quad x_N\}$. Further, when there is precisely one optimization variable, this reduces to a *single-variable unconstrained minimization*, which is one of the simplest optimization problems that one can pose:

$$\underset{x}{\text{minimize}}\ f(x) \tag{2.3}$$

2.2 Illustrative Examples

Before moving to modeling, we consider here several examples of the standard form. Consider

$$\underset{x}{\text{minimize}}\ x^2 \tag{2.4}$$

This is a single-variable unconstrained minimization problem, with a trivial solution of $x = 0$. On the other hand, the problem

$$\underset{x}{\text{minimize}}\ 3\sin x - x + 0.1x^2 + 0.1\cos(2x) \tag{2.5}$$

is also a single-variable unconstrained minimization problem, but its solution will require numerical analysis. The problem

$$\underset{\{u,v\}}{\text{minimize}}\ \left\{\begin{aligned}&\frac{1}{2}100\left(\sqrt{u^2+(1+v)^2}-1\right)^2+\frac{1}{2}50\left(\sqrt{u^2+(1-v)^2}-1\right)^2\\&-(10u+8v)\end{aligned}\right\} \tag{2.6}$$

is a two-variable unconstrained minimization problem. The physical interpretation of the above problem is discussed later in this chapter. One can add upper and lower bounds to the above problem, leading to

$$\begin{aligned}&\underset{\{u,v\}}{\text{minimize}} && \left\{\begin{aligned}&\frac{1}{2}100\left(\sqrt{u^2+(1+v)^2}-1\right)^2+\frac{1}{2}50\left(\sqrt{u^2+(1-v)^2}-1\right)^2\\&-(10u+8v)\end{aligned}\right\}\\&\text{s.t.} && 0 \le u \le 1.0\\& && 0 \le v\end{aligned} \tag{2.7}$$

One can also impose an equality constraint linking the two variables:

$$\begin{aligned}&\underset{\{u,v\}}{\text{minimize}} && \left\{\begin{aligned}&\frac{1}{2}100\left(\sqrt{u^2+(1+v)^2}-1\right)^2+\frac{1}{2}50\left(\sqrt{u^2+(1-v)^2}-1\right)^2\\&-(10u+8v)\end{aligned}\right\}\\&\text{s.t.} && v - u^3 = 0\end{aligned} \tag{2.8}$$

The physical meaning behind such constraints is discussed later in this chapter, and such constraints can completely change the complexity of the problem.

2.3 Geometry Problems

We now consider problems in geometry; the primary goal is to translate each of these problems into the standard form presented in Equation (2.1).

Example 2.1 Closest-Point Computation

Problem: Given a point p on a plane, and a straight line passing through the origin, find the closest point on the straight line to the given point p.

Modeling: Let the straight line passing through the origin be denoted by $y = mx$ and the given point be $p = (x_0, y_0)$. The task is to find the point q on the straight line that is closest to p; see Figure 2.1.
Let $q = (\alpha, m\alpha)$ be the point on the straight line. Observe that q is defined such that it always lies on the straight line $y = mx$. Introducing such intermediate variables that satisfy the problem constraints is one of the most important steps in modeling. Observe that the only unknown now is α.

Finding a point q closest to p is equivalent to minimizing the distance between them, where the distance is given by

$$D(\alpha) = \sqrt{(x_0 - \alpha)^2 + (y_0 - m\alpha)^2} \tag{2.9}$$

Thus, the optimization problem is posed as

$$\underset{\alpha}{\text{minimize}} \; D(\alpha) = \sqrt{(x_0 - \alpha)^2 + (y_0 - m\alpha)^2} \tag{2.10}$$

Equation (2.10) is interpreted as "Find α that minimizes $D(\alpha)$."

Classification: The problem posed in Equation (2.10) is *a single-objective, single-variable, unconstrained* optimization problem since: (1) there is only one objective namely, the distance D, that must be minimized, (2) there is only one continuous (unknown) variable α and (3) there are no constraints.

Solution: In the chapters that follow, we shall discuss various numerical methods to solve optimization problems. However, for now, recall from basic calculus that, if a continuous *differentiable* function takes a minimum at a point, then the derivative of

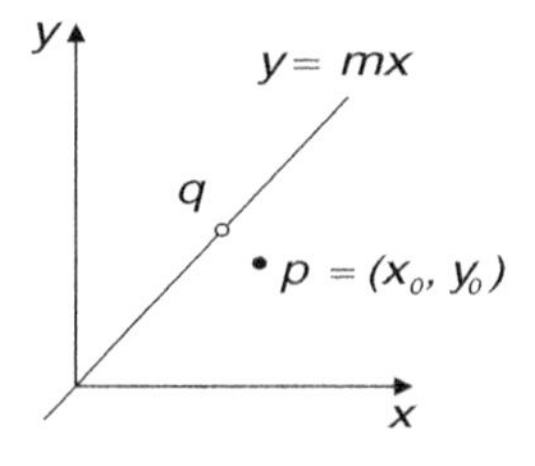

Figure 2.1 Closest point on a straight line.

(cont.)

that function vanishes at that point. Thus, differentiating the objective $D(\alpha)$ with respect to α and setting it equal to zero will yield the value of α:

$$\frac{dD}{d\alpha} = \frac{-2(x_0 - \alpha) - 2m(y_0 - m\alpha)}{2\sqrt{(x_0 - \alpha)^2 + (y_0 - m\alpha)^2}} = 0 \tag{2.11}$$

i.e.,

$$\alpha = \frac{x_0 + my_0}{(1 + m^2)} \tag{2.12}$$

Thus, the closest point is

$$q = \left(\frac{x_0 + my_0}{(1 + m^2)}, \frac{mx_0 + m^2 y_0}{(1 + m^2)} \right) \tag{2.13}$$

and the shortest distance is

$$D = \frac{|(x_0 m - y_0)|}{\sqrt{(1 + m^2)}} \tag{2.14}$$

Caveat: We have not formally established that Equation (2.11) yields a minimum; we shall address this in later chapters.

Verification: Verify whether the above solution yields the "expected answer." For example, suppose the straight line is the x axis, and $p = (0, 1)$: do you recover the expected solution? Verification, if possible, is highly recommended in order to catch modeling errors.

Example 2.2 Fitting a Straight Line

Problem: Yet another commonly occurring optimization problem is "data-fitting," where, given a set of points, the task is to find a straight line that best fits the set of points.

Modeling: Let the set of 2D data points be (x_i, y_i), $i = 1, 2, \ldots, N$; one must find the straight line that best fits the data-set (see Figure 2.2).

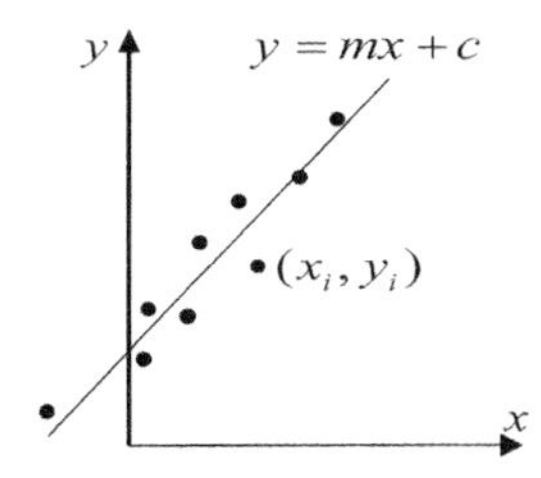

Figure 2.2 Find the best-fitting straight line.

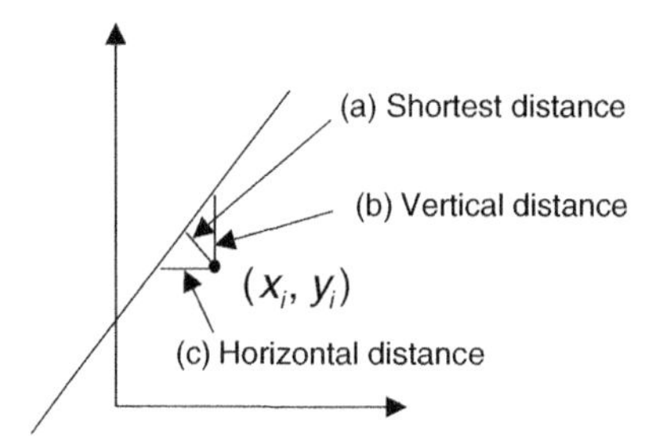

Figure 2.3 Various interpretations of deviation.

The overall strategy is to construct a straight line that minimizes the total deviation from the data points. There are, however, various definitions for "deviation" (see Figure 2.3): (a) the *shortest distance* between the data point and straight line, (b) the *vertical distance* between the data point and straight line and (c) the *horizontal distance* between the data point and straight line. Each of these will yield a different mathematical problem and solution!

This highlights a second important aspect of modeling: *a problem description can often be interpreted in multiple ways, yielding different solutions.* Here, we choose to minimize the vertical distance.

Let the straight line be of the form $y = mx + c$. The task then is to find the optimal values of m and c. Since deviation is defined as the vertical distance between a data point and the straight line, for a given point (x_i, y_i) the deviation (i.e., error) is given by

$$E_i = |mx_i + c - y_i| \tag{2.15}$$

The "best-fitting" straight line can be defined as the one that minimizes the sum of squares of all errors, i.e.,

$$\underset{m,c}{\text{minimize}} \; E = \sum_{i=1,2,\ldots}^{N} (mx_i + c - y_i)^2 \tag{2.16}$$

Equation (2.16) is interpreted as "find m and c that minimize the sum of squares of individual deviations." The reason for squaring the error measure is to make the objective differentiable, an important consideration in optimization.

Classification: The problem posed in Equation (2.16) is a *single-objective, two-variable, unconstrained* optimization problem since: (1) there is only one objective, namely the total error E, that needs to be minimized, (2) there are two continuous (unknown) variables, m and c, and (3) there are no constraints.

Solution: For multi-variable unconstrained problems, we set the partial derivative of the objective with respect to each variable to zero (this is discussed formally in later chapters), i.e.,

$$\frac{\partial E}{\partial m} = 0 \Rightarrow \sum_{i=1,2,\ldots}^{N} 2(mx_i + c - y_i)x_i = 0 \tag{2.17}$$

and

$$\frac{\partial E}{\partial c} = 0 \Rightarrow \sum_{i=1,2,\ldots}^{N} 2(mx_i + c - y_i) = 0 \tag{2.18}$$

Thus we have two linear equations involving two unknowns:

$$m \sum_{i=1,2,\ldots}^{N} (x_i)^2 + c \sum_{i=1,2,\ldots}^{N} x_i - \sum_{i=1,2,\ldots}^{N} x_i y_i = 0 \tag{2.19}$$

and

$$m \sum_{i=1,2,\ldots}^{N} x_i + c \sum_{i=1,2,\ldots}^{N} 1 - \sum_{i=1,2,\ldots}^{N} y_i = 0 \tag{2.20}$$

One can express the two equations in a standard linear algebraic form:

$$\begin{bmatrix} \sum_{i=1,2,\ldots}^{N} (x_i)^2 & \sum_{i=1,2,\ldots}^{N} x_i \\ \sum_{i=1,2,\ldots}^{N} x_i & \sum_{i=1,2,\ldots}^{N} 1 \end{bmatrix} \begin{Bmatrix} m \\ c \end{Bmatrix} = \begin{Bmatrix} \sum_{i=1,2,\ldots}^{N} x_i y_i \\ \sum_{i=1,2,\ldots}^{N} y_i \end{Bmatrix} \tag{2.21}$$

Many optimization problems reduce to linear algebra problems involving symmetric matrices. Equation (2.21) can be solved provided the 2×2 matrix is invertible.

Verification: As a specific example, consider finding the best-fitting straight line to the two data points (0, 0) and (1, 1). Equation (2.21) reduces to

$$\begin{bmatrix} 1 & 1 \\ 1 & 2 \end{bmatrix} \begin{Bmatrix} m \\ c \end{Bmatrix} = \begin{Bmatrix} 1 \\ 1 \end{Bmatrix} \tag{2.22}$$

One can verify that the solution is $m = 1$ and $c = 0$ (as expected).

Example 2.3 Area Maximization

Problem: Construct the largest rectangle within a circle of radius R. Figure 2.4 illustrates three sub-optimal rectangles within a circle.

(cont.)

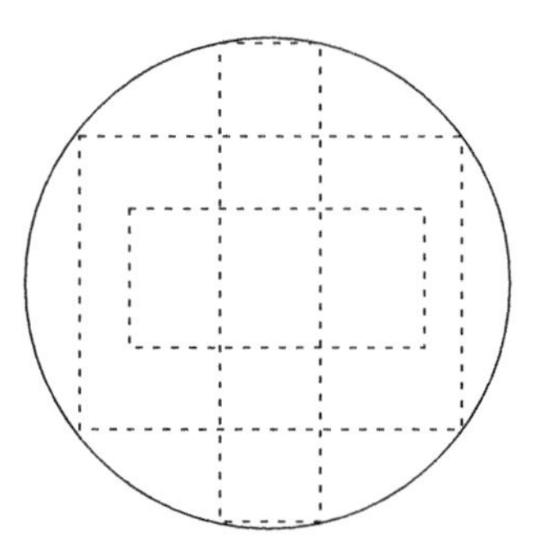

Figure 2.4 Find the largest rectangle within a circle.

Modeling: We leave it as an exercise for the reader to show that, for a rectangle within a circle to have maximum area, all four corners of the rectangle must touch the circle (a necessary condition). This *optimality condition* (a condition that must be satisfied when the solution is optimal) simplifies the problem considerably.

Thus, one of the rectangles in Figure 2.4 is clearly non-optimal. Now let an optimal rectangle (with corners touching the circle) be of width W and height H. Observe that since the corners touch the circle, W and H are not independent. Indeed, we must have the following constraint:

$$\left(\frac{W}{2}\right)^2 + \left(\frac{H}{2}\right)^2 = R^2 \tag{2.23}$$

Further, since the area is given by WH, the optimization problem reads

$$\begin{aligned} &\underset{\{W,H\}}{\text{maximize}} \; (WH) \\ &\text{s.t.} \quad \left(\frac{W}{2}\right)^2 + \left(\frac{H}{2}\right)^2 = R^2 \end{aligned} \tag{2.24}$$

Observe that this is a *maximization* problem with one objective, two variables and one equality constraint. One can easily transform this into the standard minimization problem by introducing a negative sign to the objective:

$$\begin{aligned} &\underset{\{W,H\}}{\text{minimize}} \; (-WH) \\ &\text{s.t.} \quad \left(\frac{W}{2}\right)^2 + \left(\frac{H}{2}\right)^2 = R^2 \end{aligned} \tag{2.25}$$

Solution: Later in the text, we shall consider formal methods for solving optimization problems with constraints. But, for now, we will introduce an auxiliary variable θ such that the constraint is automatically satisfied and can therefore be eliminated. In particular, let

$$\begin{aligned} W &= 2R\cos\theta \\ H &= 2R\sin\theta \end{aligned} \tag{2.26}$$

Observe that this ensures that the equality constraint is always satisfied. Thus, one can pose Equation (2.24) as

$$\underset{\{\theta\}}{\text{minimize}} \quad -4R^2\sin\theta\cos\theta \tag{2.27}$$

i.e., we have reduced the problem to a one-variable unconstrained problem. As the reader can verify, the solution is $\theta = \pi/4$, i.e.,

$$\begin{aligned} W &= R\sqrt{2} \\ H &= R\sqrt{2} \end{aligned} \tag{2.28}$$

and the maximum area is $2R^2$. Is the answer reasonable?

Example 2.4 Container Optimization

Problem: Consider an empty cylinder of radius R and height H that is closed at both ends. The objective is to minimize the surface area of the cylinder while ensuring that it contains a certain volume. This has practical implications – for example, minimizing the material usage of a soda-can of a given volume.

Modeling: The total surface area of the cylinder is $A = 2\pi R^2 + 2\pi RH$. Further, the volume of the cylinder is $V = \pi R^2 H$. Thus, one can state the optimization problem as follows:

$$\begin{aligned} &\underset{\{R,H\}}{\text{minimize}} \quad (2\pi R^2 + 2\pi RH) \\ &\text{s.t.} \quad \pi R^2 H - V_0 = 0 \end{aligned} \tag{2.29}$$

Equation (2.29) represents a *minimization problem involving two variables, with an equality constraint*. Observe that one can eliminate the constraint in Equation (2.29) by substituting $H = V_0/(\pi R^2)$ in the objective (we shall discuss pitfalls of constraint elimination later on), simplifying the problem to

$$\underset{\{R\}}{\text{minimize}} \left(2\pi R^2 + \frac{2V_0}{R}\right) \tag{2.30}$$

Observe that we have reduced the problem to a single-variable, unconstrained minimization problem.

Solution: Once again we set the derivative of the objective to zero:

$$\begin{aligned} &\frac{d\left(2\pi R^2 + \frac{2V_0}{R}\right)}{dR} = 0 \Rightarrow 4\pi R - \frac{2V_0}{R^2} = 0 \\ &\Rightarrow R = \sqrt[3]{\frac{V_0}{2\pi}} \\ &H = V_0/(\pi R^2) \end{aligned} \tag{2.31}$$

(cont.)

Instances: Now, suppose $V_0 = 16$oz (0.00047 m^3). We have

$$R = \sqrt[3]{\frac{V_0}{2\pi}} = 0.042 \text{ m} = 1.65''$$
$$H = V_0/(\pi R^2) = 0.084 \text{ m} = 3.3''$$

Observation: If the goal is to design a hand-held soda-can, clearly the large radius is inappropriate. Proper ergonomic constraints must be placed; for example, a reasonable modification to Equation (2.29) is

$$\begin{aligned} \underset{\{R,H\}}{\text{minimize}} \quad & (2\pi R^2 + 2\pi RH) \\ \text{s.t.} \quad & \pi R^2 H - V_0 = 0 \\ & R - R_{\max} \leq 0 \end{aligned} \tag{2.32}$$

It is indeed common to iterate on problem formulation until the solution is "satisfactory."

2.4 Analytical Design Problems

Next, we consider simple structural and engineering design problems where the objective and/or constraints can be expressed analytically.

Example 2.5 Point of Maximum Deflection

Problem: Consider a beam that is pinned at both ends, with an asymmetric load applied as illustrated in Figure 2.5; the Young's modulus is E and the moment of inertia is I. Find the point of maximum deflection; assume that $b < L/2$.

Modeling: This is, of course, a trivial problem in strength of materials where the beam deflection is known analytically; the deflection in the left segment is given by (see reference [10])

$$y(x) = \frac{Pb}{6EIL}[x^3 - (L^2 - b^2)x] \tag{2.33}$$

i.e.,

$$\underset{x}{\text{minimize}} \left(\frac{Pb}{6EIL}[x^3 - (L^2 - b^2)x] \right) \tag{2.34}$$

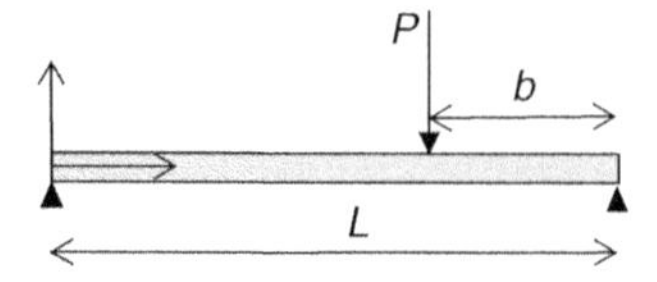

Figure 2.5 Beam deflection.

(cont.)

This is a single-variable, unconstrained minimization problem. Setting the derivative to zero, we have

$$\frac{dy}{dx} = \frac{Pb}{6EIL}[3x^2 - (L^2 - b^2)] = 0 \tag{2.35}$$

Thus,

$$x = \sqrt{\frac{(L^2 - b^2)}{3}} \tag{2.36}$$

Example 2.6 Spring Design

Problem: Next consider the classic spring design problem where the objective is to minimize the volume of a helical spring (see Figure 2.6) subject to deflection and stress constraints. The primary design variables are the number of coils (N), the outer diameter (D) and the wire diameter (d).

Modeling: Spring analysis is addressed in machine design handbooks such as reference [11]. The example is used here merely to highlight a typical design optimization problem.

The volume of a helical spring, after accounting for end-effects, can be approximated via

$$V = (N+2)\frac{\pi D}{2}\frac{\pi d^2}{4} = \pi^2\frac{(N+2)}{8}Dd^2 \tag{2.37}$$

Further, given an external load F, the deflection is given ([11]) by

$$\delta = \frac{8FD^3N}{d^4G} \tag{2.38}$$

where G is the shear modulus. Finally, the maximum shear stress can be approximated ([11]) by

$$\tau = \frac{8F(D/d)(1 + 0.5d/D)}{\pi d^2} \tag{2.39}$$

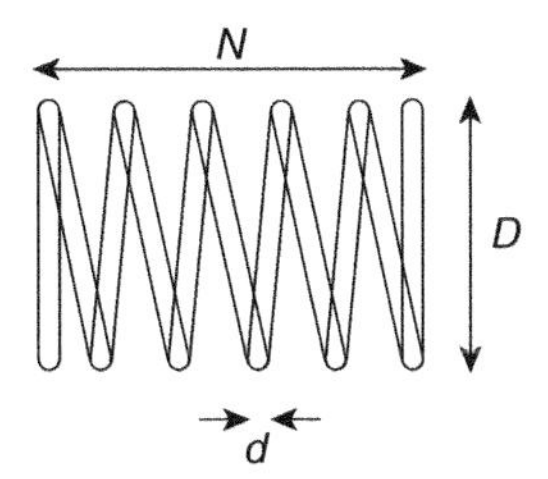

Figure 2.6 Helical spring.

(cont.)

Thus, the optimization problem can be posed as

$$\begin{aligned} \underset{\{N,D,d\}}{\text{minimize}} \quad & \pi^2 \frac{(N+2)}{8} D d^2 \\ \text{s.t.} \quad & \frac{8FD^3N}{d^4 G} \leq \delta_{\max} \\ & \frac{8F(D/d)(1+0.5d/D)}{\pi d^2} \leq \tau_{\max} \end{aligned} \tag{2.40}$$

where the constraint limits are provided by the user. This is a *multi-variable, multi-constrained minimization* problem. In practice, additional constraints on the ratio D/d and the number of coils will be needed.

Solution: Since the objective and constraints are analytical functions of the design variables, such problems are computationally easy to solve using numerical algorithms to be discussed later.

2.5 Structural Analysis

Next, consider the problem illustrated in Figure 2.7, where a simple truss bridge is pinned at two ends, and a load is applied in the middle. Assume that all material and cross-sectional properties are given. The objective is to find the displacement at the point of load application.

At first glance, this does not appear to be an optimization problem, i.e., we are not trying to optimize any objective function. It is a structural *analysis* problem. However, there is an intricate relationship between structural analysis and optimization. This is captured by the *principle of minimum potential energy.*

The principle of minimum potential energy states that *a structural system* (such as the one in Figure 2.7) *subject to an external force will come to rest when its potential energy is a minimum* [12]. The potential energy is defined as

$$\Pi = U - 2W \tag{2.41}$$

where U is the internal elastic energy and W is the *quasi-static* work done by the force.

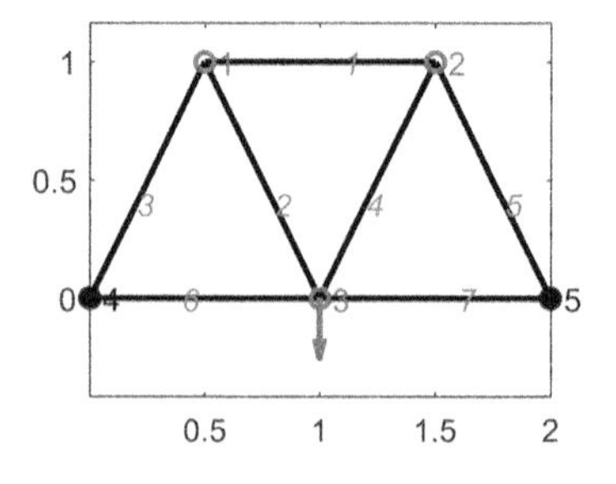

Figure 2.7 A truss analysis problem.

In other words, structural analysis is really an optimization problem. This duality between structural analysis and optimization is fascinating and important. Using the above principle, one can pose the problem of computing the displacement as a minimization problem. This is illustrated below through a few examples.

Example 2.7 Tensile Bar Analysis

Problem: Consider a tensile bar subject to an external force on one end and pinned at the other end (Figure 2.8). The properties of the tensile bar are as follows: E is the Young's modulus, A is the cross-sectional area and L is the length. Compute the deformation of the bar by minimizing its potential energy.

Modeling: Once again, observe that this is an analysis problem. To pose this as an optimization problem, we will appeal to the principle of minimum potential energy. The potential energy consists of two terms: elastic energy and work done. If the tensile bar undergoes a deformation δ, then its elastic energy is given by [10]

$$U = \frac{1}{2}k\delta^2 \tag{2.42}$$

where

$$k = \frac{EA}{L} \tag{2.43}$$

is the stiffness of the bar. On the other hand, the quasi-static work done is (see reference [10])

$$W = \frac{1}{2}P\delta \tag{2.44}$$

Observe the "1/2" in Equation (2.44); this arises because of the quasi-static nature of the force, i.e., the force is gradually increased from zero to the maximum value, rather a full force applied instantaneously. Thus, the potential energy is given by

$$\Pi(\delta) = U - 2W = \frac{1}{2}k\delta^2 - P\delta \tag{2.45}$$

Observe that the potential energy is a function of the unknown deformation δ. Since, at equilibrium, the potential energy takes a minimum, we differentiate Equation (2.45) with respect to δ, and set it equal to zero, leading to

Figure 2.8 A tensile bar problem.

(cont.)

$$\frac{d\Pi}{d\delta} = k\delta - P = 0 \tag{2.46}$$

i.e.,

$$\delta = P/k \tag{2.47}$$

Observe that Equation (2.46) is essentially the force balance equation, i.e., we have arrived at the force balance equation by differentiating the potential energy equation and setting it equal to zero! This *fundamental duality* is explored further in later chapters.

Example 2.8 Truss Analysis

Problem: Consider the truss system in Figure 2.9, subject to an external force. Find the deflection of the free node by minimizing its potential energy (the deflection is illustrated schematically by dashed lines). A and l are respectively the cross-sectional area and length of each bar.

Modeling: Let the displacement of the free node in the horizontal and vertical directions be u and v respectively. Assuming small displacements, using simple vector calculus one can show that the two bars undergo deformations of

$$\delta_1 \approx u \cos\theta - v \sin\theta \tag{2.48}$$

$$\delta_2 \approx -u \cos\theta - v \sin\theta \tag{2.49}$$

(see Exercise 2.11 at the end of this chapter). Thus, the total elastic energy is

$$U = \frac{1}{2}k(\delta_1)^2 + \frac{1}{2}k(\delta_2)^2 \tag{2.50}$$

where k captures the stiffness of the truss bar and is given by

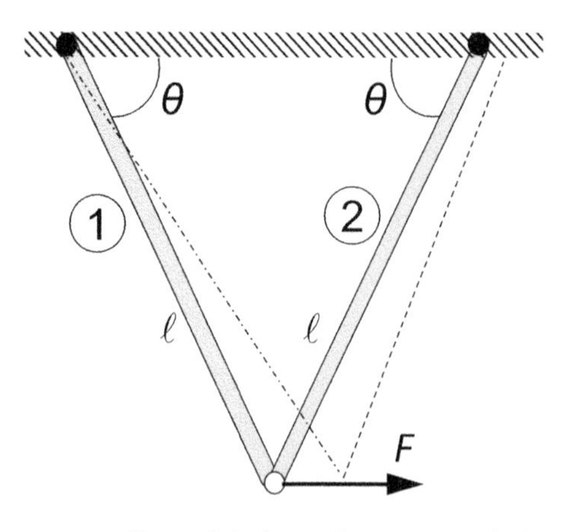

Figure 2.9 A two-bar truss analysis.

(cont.)

$$k = \frac{EA}{l} \tag{2.51}$$

From Equations (2.48) and (2.49), we have

$$U = \frac{1}{2}k(u\cos\theta - v\sin\theta)^2 + \frac{1}{2}k(-u\cos\theta - v\sin\theta)^2 \tag{2.52}$$

Simplifying,

$$U = k(u^2\cos^2\theta + v^2\sin^2\theta) \tag{2.53}$$

Further, since a horizontal force is applied, only the horizontal displacement plays a role in the quasi-static work done:

$$W = \frac{1}{2}Fu \tag{2.54}$$

Once again, the factor of "1/2" is due to the quasi-static nature of the force. Finally, the potential energy is given by

$$\Pi = U - 2W = k(u^2\cos^2\theta + v^2\sin^2\theta) - Fu \tag{2.55}$$

As stated earlier, the structural system will come to rest when the potential energy is a minimum. Thus, we pose the optimization problem as

$$\underset{\{u,v\}}{\text{minimize}}\ \Pi = k(u^2\cos^2\theta + v^2\sin^2\theta) - Fu \tag{2.56}$$

Solution: This is a single-objective, two-variable, unconstrained minimization problem. For multi-variable, unconstrained problems, we set the partial derivative of the objective with respect to each variable to zero, i.e.,

$$\frac{\partial[k(u^2\cos^2\theta + v^2\sin^2\theta) - Fu]}{\partial u} = 0 \tag{2.57}$$

$$\frac{\partial[k(u^2\cos^2\theta + v^2\sin^2\theta) - Fu]}{\partial v} = 0 \tag{2.58}$$

This results in

$$\begin{aligned} u &= F/(2k\cos^2\theta) = Fl/(2EA\cos^2\theta) \\ v &= 0 \end{aligned} \tag{2.59}$$

Again, the reader can verify that we will arrive at the same result via force balance.

Example 2.9 Spring System Analysis

Problem: Consider the spring system in Figure 2.10, consisting of two springs (marked as 1 and 2) that are fixed at top and bottom (nodes 2 and 3). An external force is applied at the middle node (node 1). The lengths of the two springs are 1 unit each (in the undeformed state), while their stiffnesses are 100 and 50 units respectively. The force components are 10 units in the x-direction and 8 units in the y-direction. Find the deflection of node 1 by minimizing the potential energy of the spring system.

(cont.)

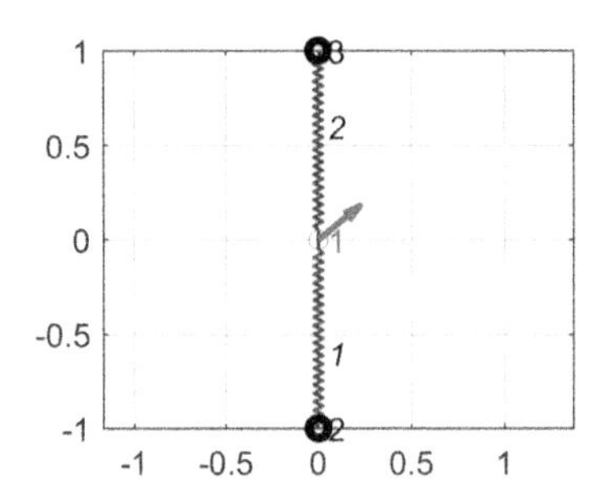

Figure 2.10 A spring system.

Modeling: Let the displacements of node 1 in the horizontal and vertical directions be u and v respectively. Observe that the undeformed length of each spring is 1 unit. Further, given an arbitrary displacement (u, v) of node 1, the deformed lengths of the two springs are

$$\begin{aligned} L_1 &= \sqrt{(0-u)^2 + (-1-v)^2} = \sqrt{u^2 + (1+v)^2} \\ L_2 &= \sqrt{(0-u)^2 + (1-v)^2} = \sqrt{u^2 + (1-v)^2} \end{aligned} \tag{2.60}$$

Since the deformation can be fairly large, one cannot make the truss-based simplification. Therefore, the increase in length is given by

$$\begin{aligned} \Delta L_1 &= \sqrt{u^2 + (1+v)^2} - 1 \\ \Delta L_2 &= \sqrt{u^2 + (1-v)^2} - 1 \end{aligned} \tag{2.61}$$

This results in an elastic energy of

$$\begin{aligned} U_1 &= 0.5k_1(\Delta L_1)^2 = 0.5k_1\left(\sqrt{u^2 + (1+v)^2} - 1\right)^2 \\ U_2 &= 0.5k_2(\Delta L_2)^2 = 0.5k_2\left(\sqrt{u^2 + (1-v)^2} - 1\right)^2 \end{aligned} \tag{2.62}$$

Given the spring stiffnesses of 100 and 50 units, the total elastic energy is given by

$$U = 0.5\begin{bmatrix} 100\left(\sqrt{u^2 + (1+v)^2} - 1\right)^2 \\ + 50\left(\sqrt{u^2 + (1-v)^2} - 1\right)^2 \end{bmatrix} \tag{2.63}$$

The external work done is

$$W = (1/2)(f_x u + f_y v) = (1/2)(10u + 8v) \tag{2.64}$$

Finally, the potential energy is given by

$$\Pi = 0.5\begin{bmatrix} 100\left(\sqrt{u^2 + (1+v)^2} - 1\right)^2 \\ + 50\left(\sqrt{u^2 + (1-v)^2} - 1\right)^2 \end{bmatrix} - (10u + 8v) \tag{2.65}$$

To find the displacements, we pose the following optimization problem:

$$\underset{\{u,v\}}{\text{minimize}} \begin{Bmatrix} \frac{1}{2}100\left(\sqrt{u^2+(1+v)^2}-1\right)^2+\frac{1}{2}50\left(\sqrt{u^2+(1-v)^2}-1\right)^2 \\ -(10u+8v) \end{Bmatrix} \tag{2.66}$$

This is a single-objective, two-variable, unconstrained minimization problem that is difficult to solve analytically. We will need numerical methods to solve such problems.

Example 2.10 Constrained Spring System Analysis

Problem: We consider again the problem in Example 2.9, but impose an additional constraint: that the free node must slide (without friction) along a curve $y = x^3$ (see Figure 2.11).

Modeling: The modeling is exactly as before except that the two deformations are coupled by $v = u^3$. This leads to the following equality-constrained minimization problem:

$$\begin{aligned} \underset{\{u,v\}}{\text{minimize}} \quad & \begin{Bmatrix} \frac{1}{2}100\left(\sqrt{u^2+(1+v)^2}-1\right)^2+\frac{1}{2}50\left(\sqrt{u^2+(1-v)^2}-1\right)^2 \\ -(10u+8v) \end{Bmatrix} \\ \text{s.t.} \quad & v-u^3=0 \end{aligned} \tag{2.67}$$

We will need numerical methods to solve such problems.

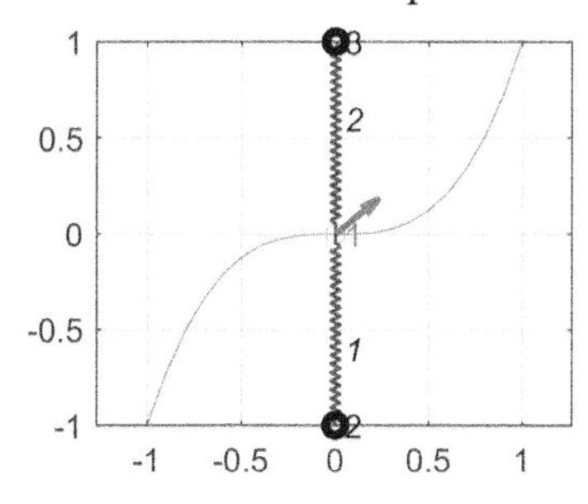

Figure 2.11 A constrained spring system.

2.6 Structural Optimization

Finally, we will consider problems from structural (design) optimization, one of the primary focus areas of this text.

Example 2.11 Determinate Truss Optimization

Problem: We once again consider the truss system in Figure 2.12, but now we will assume that the two tensile bars are of equal but *unknown* cross-sectional area. The objective is to minimize the total volume of the structure, subject to a constraint on the horizontal deflection of the free node.

Modeling: Let the cross-sectional areas of the two bars be A and the length of each bar be l; therefore the volume of the truss system is $V = 2lA$. Recall that the horizontal deflection of the node is given by

$$u = Fl/(2EA\cos^2\theta) \tag{2.68}$$

Thus, the design optimization problem is posed as follows:

$$\begin{aligned} \underset{A}{\text{minimize}} \quad & 2lA \\ \text{s.t.} \quad & Fl/(2EA\cos^2\theta) \leq \delta_{\max} \end{aligned} \tag{2.69}$$

Solution: Note that the problem can be posed as

$$\begin{aligned} \underset{A}{\text{minimize}} \quad & 2lA \\ \text{s.t.} \quad & A \geq Fl/(\delta_{\max} 2E\cos^2\theta) \end{aligned} \tag{2.70}$$

Clearly, given the lower bound on the area, the objective is minimized when

$$A^* = Fl/(\delta_{\max} 2E\cos^2\theta) \tag{2.71}$$

In this example, we were "lucky" in that the deflection could be determined analytically. Most engineering problems do not admit such easy solution, as the next example illustrates.

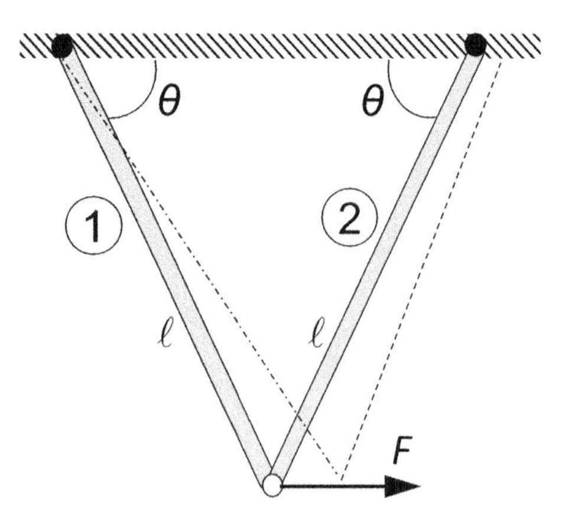

Figure 2.12 A structural design problem.

Example 2.12 Indeterminate Truss Optimization

Problem: Consider the truss system illustrated in Figure 2.13 (all length units in meters), consisting of five bars and six nodes. Four of the nodes are fixed, and a force of (100, −20) N is applied at node 1. The Young's modulus of all bars is 2×10^{11} N/m^2, but the cross-sectional areas of each of them can be different. Minimize the volume of truss subject to a deflection constraint on node 1.

Modeling: The volume of the truss can be expressed in an analytical form as a function of the cross-sectional areas A_i, $i = 1, 2, \ldots, 5$. However, the deflection at node 1 cannot be obtained in closed form. We will see in Chapter 9 that one needs to solve a system of equations to determine the deflection. We will see in Chapter 10 that this problem can be posed as

$$\begin{aligned}
&\underset{\{A_1, A_2, \ldots, A_5\}}{\text{minimize}} && \sum_{i=1}^{5} A_i l_i \\
&\text{s.t.} && \delta_1 \leq \delta_{\max} \\
& && A_i > 0 \\
& && \mathbf{Ku} = \mathbf{f}
\end{aligned} \tag{2.72}$$

Observe that, to evaluate the constraint, one must solve the linear system of equations $\mathbf{Ku} = \mathbf{f}$. Indeed, almost all optimization problems in engineering will exhibit this property, i.e., they will require the solution of linear systems of equations to evaluate the objective and/or the constraint.

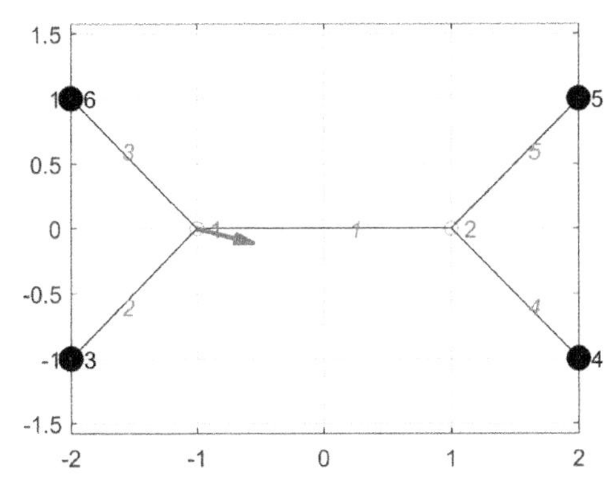

Figure 2.13 Design of a five-bar truss system.

2.7 Conclusions and Observations

We considered various examples from geometry and structural mechanics to illustrate the modeling process. Modeling is perhaps one of the most important (but often neglected) topics in engineering optimization. By considering several examples, we observed that:

- Introducing intermediate variables (as in the case of the closest point on a straight line) that satisfy the problem constraints is an important aspect of modeling.
- A problem description can often be interpreted, i.e., modeled, in multiple ways (as in the case of the best-fitting straight line), yielding different solutions.
- Exploiting an optimality condition (as in the case of the largest rectangle within a circle) can simplify the problem statement considerably.
- Modeling is often an iterative process (as in the case of the soda-can problem).
- Modeling can exploit fundamental principles of nature, as in structural mechanics problems.
- Almost all optimization problems in engineering will require the solution of linear system of equations to evaluate the objective and/or the constraint.

2.8 EXERCISES

Exercise 2.1 Let p be a 2D point and $y = mx + c$ be a straight line; further, let q be the closest point on the straight line to p. Assuming p does not lie on the straight line, show that the line joining p and q is perpendicular to the straight line. This result has important implications in approximation theory.

Exercise 2.2 Given three points $(0,0)$, $(1,1)$ and $(2,2.1)$, using the results derived in this chapter, find the best-fitting line that minimizes the vertical distance deviation. Where does the line intersect the x-axis?

Exercise 2.3 Given a set of points (x_i, y_i), $i = 1, 2, \ldots, N$, find the best-fitting line if one chooses the horizontal distance as a measure of deviation. Your solution should be reduced to solving a linear algebra problem. Hint: You might find it more convenient to define the straight line as $x = ay + b$. Use your result to best-fit a line through $(0,0)$, $(1,1)$ and $(2,2.1)$. Where does the line intersect the x-axis? Compare this result against the one in Exercise 2.2.

Exercise 2.4 Given an ellipse $x^2/a^2 + y^2/b^2 = 1$ and a point $p = (x_0, y_0)$, pose the problem of finding the closest point q on the ellipse to p as a two-variable constrained optimization problem. Then reduce it to a single-variable unconstrained problem by introducing an intermediate variable; you do not need to solve the problem.

Exercise 2.5 Suppose you are given a set of data points (x_i, y_i), $i = 1, 2, \ldots, N$, $N \geq 3$. The objective is to find a circle of the form $(x - x_c)^2 + (y - y_c)^2 = r^2$ that best fits these data points, using the shortest distance as a metric. Pose this as an optimization problem; you do not need to solve the problem.

Exercise 2.6 Find the minimum value of $f(x) = 1 - xe^{-x^2}$ analytically.

Exercise 2.7 Sometimes the global minimum of an analytical function can be found by inspection. For example, consider the 2D Rosenbrock function (which we will encounter again in Chapter 5):

$$f(x, y) = 100(y - x^2)^2 + (1 - x)^2$$

Find the minimum value of this function, and the corresponding (x, y).

Exercise 2.8 Consider the N-dimensional Rosenbrock function:

$$f(\mathbf{x}) = \sum_{i=1}^{N} [100(x_{i+1} - x_i^2)^2 + (1 - x_i)^2]$$

By inspection, find the minimum value of this function, and the corresponding point $\mathbf{x}$.

Exercise 2.9 Consider the function

$$f(x, y) = 100\sqrt{|y - 0.01x^2|} + 0.01|x + 10|$$

By inspection, find the minimum value of this function, and the corresponding point (x, y).

Exercise 2.10 Consider the beam bending problem illustrated in Figure 2.14, where one end is fixed and the other end is pinned. A moment M is applied at the

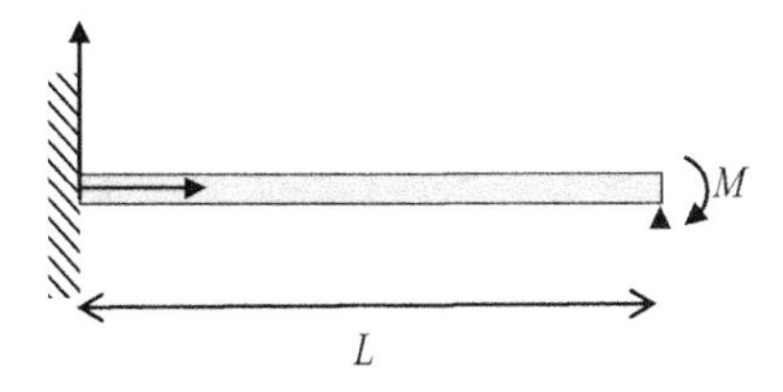

Figure 2.14 A beam bending problem.

pinned end. Find the location and magnitude of the maximum deflection. Note that the deflection is given by (reference [10])

$$y(x) = -\frac{M}{2EI}x^2 + \frac{3M}{2EI}\left(\frac{x^2}{2} - \frac{x^3}{6L}\right)$$

Exercise 2.11 In the truss analysis problem discussed in Example 2.8, we stated that, as a first-order approximation, we have

$$\Delta\delta_1 \approx u\cos\theta - v\sin\theta$$
$$\Delta\delta_2 \approx -u\cos\theta - v\sin\theta$$

Prove this result.

Exercise 2.12 Consider the structural analysis problem in Figure 2.15, where the stiffness of the right bar is k, and the stiffness of the left-bar is 2k. Find the displacement of the free node by minimizing the potential energy.

Exercise 2.13 Consider the structural analysis problem shown in Figure 2.16. If the force acts at 45° to horizontal, find the displacement of the free node via potential energy minimization. Assume the stiffness of each bar is k.

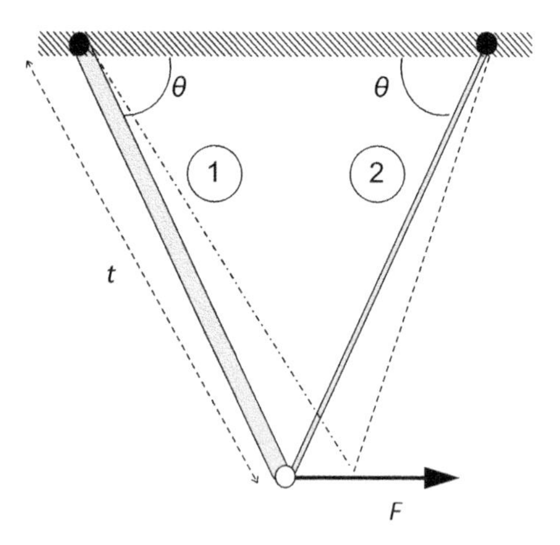

Figure 2.15 An asymmetric structural analysis problem.

Exercise 2.14 Consider the three-spring system in Figure 2.17. Each of the three springs is of length 1 and has a spring constant of 100 units, and a force of (10, 8) acts on node 1. Formulate the minimization problem governing the displacements (u, v) of node 1. You do not need to solve the optimization problem.

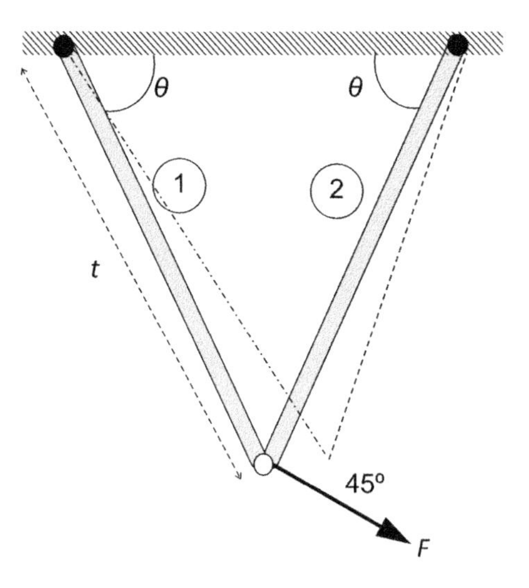

Figure 2.16 A structural analysis problem.

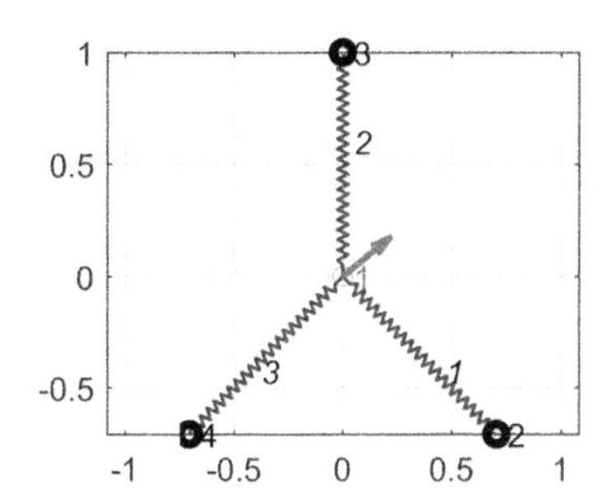

Figure 2.17 A three-spring system.

3 Introduction to MATLAB

Highlights

1. This chapter is a brief introduction to MATLAB; the discussion here is by no means exhaustive, but is sufficient for the remainder of this text.
2. Besides focusing on the programming fundamentals, this chapter also introduces the reader to object-oriented programming using MATLAB.
3. Object-oriented programming has numerous benefits over traditional functional programming, perhaps the most important ones being data sharing within a class, and inheritance.

Chapter 2 provided an overview of modeling and optimization through illustrative problems. Many of the problems were intentionally simple, and could therefore be solved analytically. In practice, most optimization problems entail numerical analysis.

In this chapter, we discuss a popular numerical analysis package, namely MATLAB. The MATLAB topics discussed in this chapter are sufficient for the remainder of this text. However, the reader is encouraged to consult reference [2] and other resources, such as references [13] and [14].

3.1 Basics

A typical MATLAB integrated development environment (IDE) is illustrated in Figure 3.1, with three windows displayed: (1) command window at the bottom right, (2) editor window at the top right and (3) current folder window at the left.

In MATLAB's command window, one can create and manipulate variables as shown in Figure 3.2.

Henceforth, for clarity, all interactive sessions will be highlighted, as shown below:

```
>> a = 2;
>> b = 2*a -1
b =
     3
```

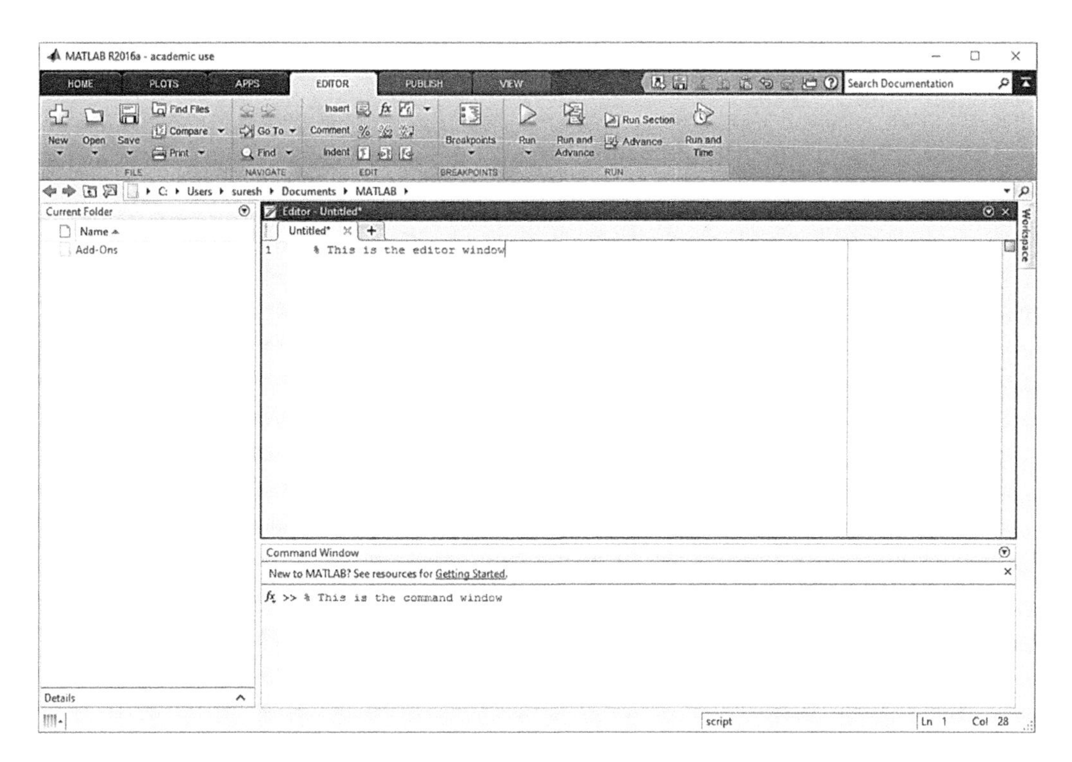

Figure 3.1 MATLAB integrated development environment (IDE).

Various constants, such as π, and operations such as sin() and cos(), are predefined in MATLAB:

```
>> x = pi
x =
   3.141592653589793
>> y = sin(x)
y =
    1.224646799147353e-016
```

In MATLAB, one often creates and manipulates vectors and matrices. For example, one can construct two vectors and carry out the dot product, as shown below. Observe that `a` is a row vector, while `b` is a column vector (indicated by the apostrophe); consequently, `a*b` results in a scalar, while b*a results in a matrix.

```
>> a = [3 2 4.2]
a =
    3.0000    2.0000    4.2000
>> b = [1  0 -1]'
```

Command Window

New to MATLAB? Watch this Video, see Demos, or read Getting Started.

```
>> a = 2;
>> b = 2*a-1

b =

     3

fx >>
```

Figure 3.2 MATLAB interactive command window.

```
b =
     1
     0
    -1
>> a*b
ans =
   -1.2000
>> b*a
ans =
    3.0000    2.0000    4.2000
         0         0         0
   -3.0000   -2.0000   -4.2000
```

One can create matrices and carry out matrix-vector multiplication as shown below. The semicolon (;) serves to demarcate rows.

```
>> A = [2 -1 0; -1 2 -1; 0 -1 2]
A =
     2    -1     0
    -1     2    -1
     0    -1     2
>> x = A*b
x =
     2
     0
    -2
```

One of the easiest means of creating vectors is through the colon (:) operator:

```
>> x = 0:0.25:1
x =
         0    0.2500    0.5000    0.7500    1.0000
>> y = 0:.1:pi
y =
  Columns 1 through 6
         0    0.1000    0.2000    0.3000    0.4000    0.5000
  Columns 7 through 12
    0.6000    0.7000    0.8000    0.9000    1.0000    1.1000
  Columns 13 through 18
    1.2000    1.3000    1.4000    1.5000    1.6000    1.7000
  Columns 19 through 24
    1.8000    1.9000    2.0000    2.1000    2.2000    2.3000
  Columns 25 through 30
    2.4000    2.5000    2.6000    2.7000    2.8000    2.9000
  Columns 31 through 32
    3.0000    3.1000
```

In the first example `0.25` is the step size.

3.2 MATLAB Code Resource

The MATLAB code accompanying this text is available as a zip-file (DesignOptimizationSoftware.zip) from the author's website at www.ersl.wisc.edu under the "Research" tab. Readers can download this file to their home directory and unzip the contents; see Figure 3.3.

Name	Date modified	Type
Chapter1-Overview	6/22/2020 12:16 PM	File folder
Chapter2-Modeling	6/22/2020 12:16 PM	File folder
Chapter3-MatlabIntro	6/22/2020 9:53 AM	File folder
Chapter4-UnconstrainedOpt-Theory	6/22/2020 12:16 PM	File folder
Chapter5-UnconstrainedOptimization	2/24/2020 8:38 AM	File folder
Chapter6-MatlabOptimizationToolbox	6/22/2020 9:53 AM	File folder
Chapter7-ConstrainedOptimization	6/22/2020 9:53 AM	File folder
Chapter8-SpecializedProblems	6/22/2020 12:15 PM	File folder
Chapter9-TrussAnalysis	6/22/2020 9:53 AM	File folder
Chapter10-TrussOptimization	6/22/2020 9:53 AM	File folder
Chapter11-GradientComputation	6/22/2020 9:53 AM	File folder
Chapter12-FEA2D	6/22/2020 9:53 AM	File folder
Chapter13-ShapeOptimization2D	6/22/2020 9:53 AM	File folder
Chapter14-SolidWorksFEA	6/22/2020 9:53 AM	File folder
Chapter15-SolidLab	6/22/2020 9:53 AM	File folder
Chapter16-ShapeOptimization3D	6/22/2020 9:53 AM	File folder
runMe	5/18/2020 9:51 AM	MATLAB Code

Figure 3.3 Contents of the Design Optimization Software accompanying this text.

Then, from within MATLAB, they may navigate to this directory, and type `runMe` from MATLAB's command window.

```
>> runMe
Design Optimization Software Initialized
Version 2019.03
```

Then all directories listed in Figure 3.3 will be included in MATLAB's search path. Readers will find most of the examples discussed in each chapter under the corresponding directory.

3.3 Memory Pre-allocation and Vectorization

Consider the simple mathematical operation

$$y_i = x_i^2 + 1; \; i = 1, 2, \ldots, N \tag{3.1}$$

In MATLAB, one can implement this via a `for` loop; for example:

```
>> x = 0:0.25:1;
>> for i = 1:numel(x)
y(i) = x(i)*x(i) +1;
end
```

However, in this implementation, the memory for the `y` variable is allocated "on-the-fly," and this can slow down the code drastically. In MATLAB, it is highly recommended that memory for arrays be pre-allocated. For example:

```
>> x = 0:0.25:1;
>> y = zeros(1,numel(x));
>> for i = 1:numel(x)
y(i) = x(i)^2 + 1;
end
```

Further, in this example, the `for` loop can be replaced by vectorization as follows:

```
>> x = 0:0.25:1;
>> y = x.^2 + 1;
```

Observe the use of the full stop (.), which implies vectorization. *Whenever possible,* `for` *loops must be replaced by vectorization, and memory must always be pre-allocated.*

3.4 MATLAB Script Files

One can create, save and execute a series of MATLAB commands through a "script" file. For example, the script file in Figure 3.4, created in the editor window named `flowControl.m`, illustrates the use of `for`, `if` and `while`

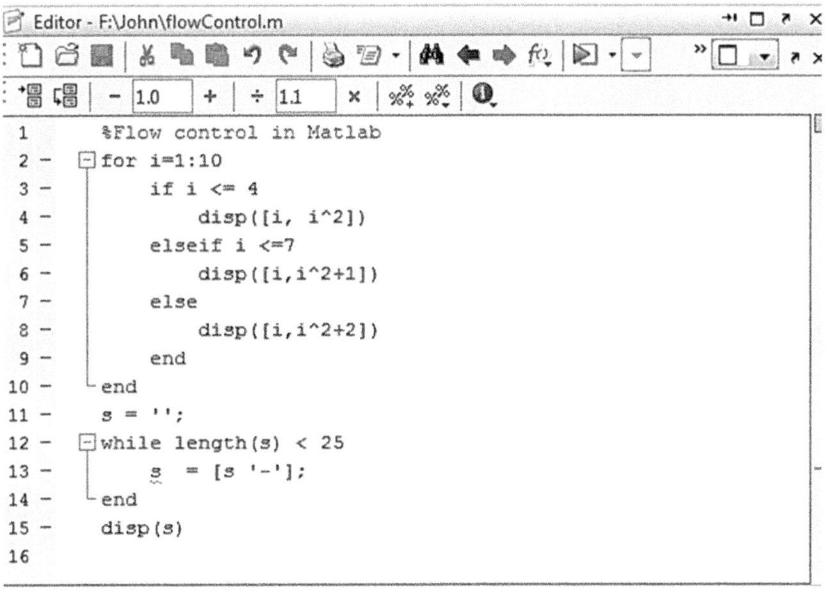

```
Editor - F:\John\flowControl.m
1      %Flow control in Matlab
2 -    for i=1:10
3 -        if i <= 4
4 -            disp([i, i^2])
5 -        elseif i <=7
6 -            disp([i,i^2+1])
7 -        else
8 -            disp([i,i^2+2])
9 -        end
10 -   end
11 -   s = '';
12 -   while length(s) < 25
13 -       s  = [s '-'];
14 -   end
15 -   disp(s)
16
```

Figure 3.4 MATLAB editor window.

commands in MATLAB. Note that any line that starts with the percentage sign (`%`) is treated as a comment in MATLAB.

For clarity, script files will be displayed as below:

```
%Flow control in MATLAB
for i=1:10
    if i <= 4
        disp([i, i^2])
    elseif i <=7
        disp([i,i^2+1])
    else
        disp([i,i^2+2])
    end
end
s = '';
while length(s) < 25
    s  = [s '-'];
end
disp(s)
```

This script file, saved as `flowControl.m`, can be executed from the command window, as shown below. Readers are encouraged to “play around” with the script file, for example, by modifying the `if` conditions.

```
>> flowControl
     1     1
     2     4
     3     9
     4    16
     5    26
     6    37
     7    50
     8    66
     9    83
    10   102
--------------------------
>>
```

3.5 Linear Algebra

MATLAB supports numerous matrix operations; here are a few examples:

```
% Linear Algebra
A = [2 -1; -1 2];
b = [1;1];
x = A\b
[Lambda, V] = eig(A)
A = rand(50,50);
x = rand(50,1);
b = A*x;
res = A\b;
err = norm(x-res);
disp(['Error for a random matrix solve is ' num2str(err)])
```

Observe, in the above script, that:

- backslash (\) is an efficient implementation for solving a linear system of equations
- the `rand` function creates a uniformly distributed random vector or matrix of specified size, where each element is between 0 and 1
- the eigenvalues of a matrix are computed via the `eig` function
- `norm` is the usual 2-norm of a vector, defined as

$$\|\mathbf{x}\| = \sqrt{(x_1)^2 + (x_2)^2 + \ldots + (x_N)^2}$$

The output of this script, saved as `linearAlgebra.m`, is shown below.

```
>> linearAlgebra
x =
   1.000000000000000
   1.000000000000000
Lambda =
  -0.707106781186547  -0.707106781186547
  -0.707106781186547   0.707106781186547
V =
     1     0
     0     3
Error for a random matrix solve is 5.7673e-014
>>
```

3.6 Complex Numbers

MATLAB also supports the use of complex numbers; below are a few examples.

```
>> i = sqrt(-1);
>> a = 3+2i
a =
  3.000000000000000 + 2.000000000000000i
>> sqrt(a)
ans =
  1.817354021023971 + 0.550250522700337i
```

Complex numbers will be used in Chapter 11 for numerical differentiation.

3.7 Plots

The ability to plot and visualize functions is extremely valuable in optimization. MATLAB supports 2D and 3D plotting; below is a simple example of 2D plotting.

```
%2-D plots
x = 0:pi/50:2*pi;
y = sin(x);
z = cos(x);
plot(x,y,'-','Linewidth',2.0);hold on;
plot(x,z,':','Linewidth',2.0);
xlabel('x')
ylabel('sin(x) & cos(x)')
legend('sin(x)','cos(x)')
```

The results are shown in Figure 3.5.

MATLAB offers various 3D plotting routines as well; below is an example of plotting the function

$$f(x,y) = (x-1)^2 + (y-4)^4 \tag{3.2}$$

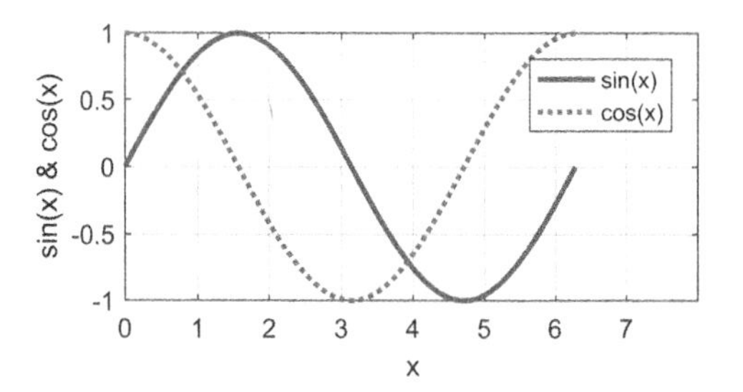

Figure 3.5 Plot of trigonometric functions.

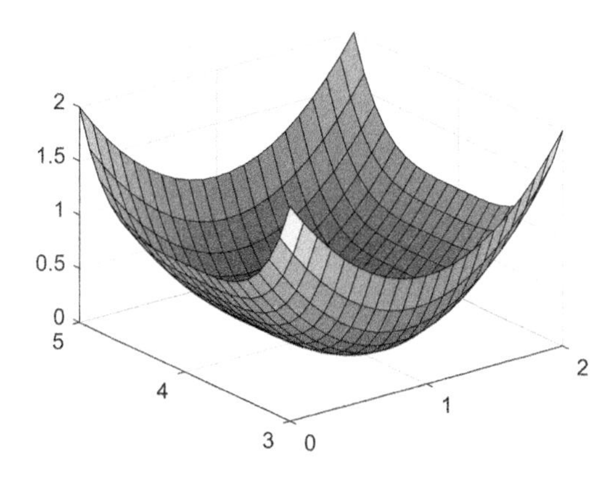

Figure 3.6 Surface plot of Equation (3.2)

The `meshgrid` function creates two matrices over a grid-space, while the `surf` command plots a function over a grid-space.

```
% 3-D plotting
[x,y] = meshgrid(0:.1:2, 3:.1:5);
f = (x-1).^2 + (y-4).^4;
surf(x,y,f);
```

The result is illustrated in Figure 3.6.

An alternative means of visualizing 2D functions is through a contour plot. The contour of a function $f(x,y)$ is the set (x, y) of points at which the function takes a prescribed contour value c_0, i.e.,

$$C(f, c_0) = \{(x,y)|f(x,y) = c_0\} \tag{3.3}$$

Observe that by varying the contour value, different contours can be generated. In other words, there are an infinite number of contours associated with a function, but two different contours can never intersect! We will study their importance in later chapters. Continuing with the previous script, Figure 3.7

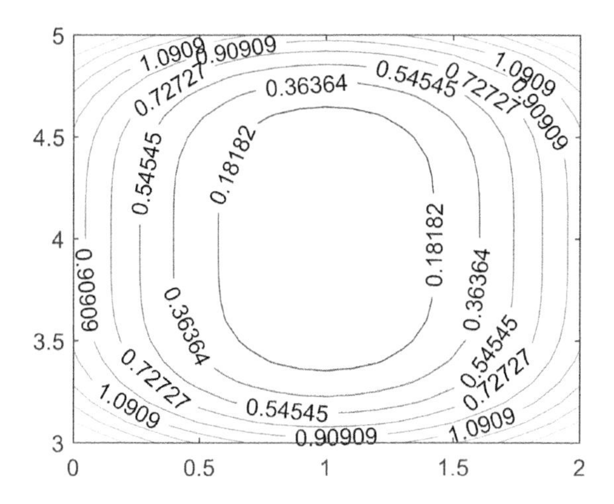

Figure 3.7 Contour plot of Equation (3.2).

illustrates contour plotting with 10 contour values (generated internally by MATLAB).

```
[c, h] = contour(x,y,f,10);
clabel(c, h);
```

Observe in the contour plot that the contour values continue to decrease until they reach a value of zero at (1, 4) as expected. Further, note the following:

1. While it is possible to generate an infinite number of contours by varying c_0, there may be values of c_0 for which a contour may not exist since the function may not be able to reach the value of c_0.
2. For every point (x_0, y_0) there exists a unique contour passing through that point.
3. At every point (x_0, y_0), the *gradient* of the function is perpendicular to the contour passing through that point (we will learn more about gradients in Chapter 4).

3.8 Symbolic Operations

MATLAB also provides a convenient symbolic toolbox that we shall exploit later in the text. For example, consider the function

$$f(x) = x \sin x^2 \tag{3.4}$$

The command `syms x real` below informs MATLAB that we are creating a symbolic variable x that is guaranteed to take only real values. After defining a function f symbolically, one can differentiate it as shown below.

```
>> syms x real
>> f = x*sin(x^2);
>> diff(f,x)
ans =
sin(x^2) + 2*x^2*cos(x^2)
```

This can be particularly helpful in differentiating complicated functions. *Note that symbolic operations are expensive and should be used sparingly, and never within an optimization routine.*

3.9 User-Defined Functions

Consider the function

$$f(x) = 1 - xe^{-x^2} \tag{3.5}$$

An easy way to create such a function is through MATLAB's *anonymous function*. Observe the use of vectorization.

```
>> f = @(x) (1-x.*exp(-x.^2));
```

Then,

```
>>  f(3)
ans =
   0.999629770587740
>>  f([1 2 3])
ans =
  0.632120558828558   0.963368722222532   0.999629770587740
```

Alternatively, one can create a MATLAB file `f1.m` with the contents shown below; observe the keyword `function`.

```
function f = f1(x)
f = 1-x.*exp(-x.^2);
```

Once the file has been created, it can be used as follows:

```
>> f1(3.0)
ans =
   0.999629770587740
```

One can now also combine this with the plot function; observe the vectorized function call:

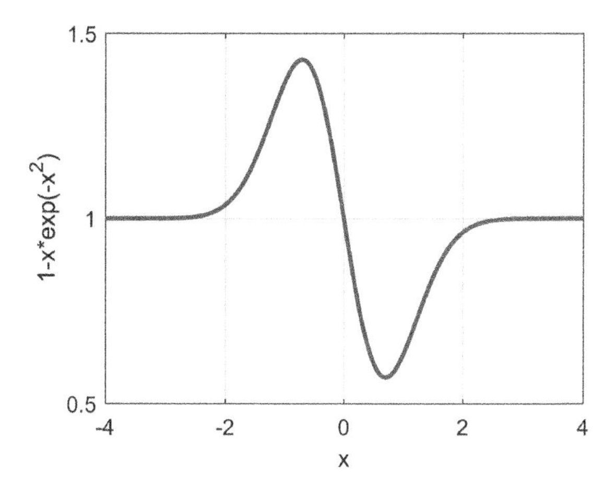

Figure 3.8 Plot of Equation (3.5).

```
x = -4:0.01:4;
y = f1(x);
plot(x,y,'Linewidth',2.0)
xlabel('x');
ylabel('1-x*exp(-x^2)');
```

The resulting plot is shown in Figure 3.8.

3.10 Variable Arguments

It is often convenient to define functions that can accept one or more inputs and return one or more outputs. For example, suppose we wish to generalize Equation (3.5) to

$$f(x) = 1 - xe^{-x^2/c} \tag{3.6}$$

where the value of c is user-defined, with a default value of $c = 1$. One can achieve this as follows:

```
function f = f1(x,c)
if (nargin == 1), c=1;end
f = 1-x.*exp(-x.^2/c);
```

Observe the use of the keyword `nargin` (number of arguments in): if only one argument is passed by the user, then the default value for "c" is used.

The following examples illustrate this concept.

```
>> f1(3)
ans =
   0.999629770587740
>> f1(3,1)
ans =
   0.999629770587740
>> f1(3,-1)
ans =
  -2.430825178272615e+004
```

Similarly, one can control the output variables using the keyword `nargout` (number of arguments out), where one can optionally return one or more outputs depending on the user request. In the example below, the function value is always returned; then, depending on the user request, the first derivative and second derivative are also computed. This concept is often used to improve the efficiency of MATLAB code.

```
function [f,g,h] = f1(x,c)
if (nargin == 1), c =1;end
f = 1-x.*exp(-x.^2/c);
if (nargout > 1), g = -exp(-x.^2) + 2*(x.^2).*exp(-x.^2); end
if (nargout > 2), h = 6*x.*exp(-x.*x)- 4*x.^3.*exp(-x.*x); end
```

The following examples illustrate this concept.

```
>> f = f1(3)
f =
   0.999629770587740
>> [f,g] = f1(3)
f =
   0.999629770587740
g =
   0.002097966669474
>> [f,g,h] = f1(3)
f =
   0.999629770587740
g =
   0.002097966669474
h =
  -0.011106882367801
```

3.11 Solving Non-linear Equations

Next, we briefly consider the solution of non-linear equations since they often arise in optimization. In particular, we will consider the use of MATLAB's `fsolve` method to solve such equations. In Chapter 5 we will revisit this method, and study how it is implemented.

Consider a single non-linear equation

$$\sin x - x = 0 \tag{3.7}$$

Clearly $x = 0$ is a solution; to find other possible solutions, we create an anonymous function:

```
>> r = @(x) sin(x)-x;
```

Then we pass this function to `fsolve` as follows, with an initial guess of 1:

```
>> fsolve(r,1)
ans =
    0.0852
```

There are many solutions to this non-linear equation; depending on the initial guess, we can find various solutions:

```
>> fsolve(r,10)
ans =
 0.0949
```

When there are multiple non-linear equations to solve simultaneously, it is more convenient to create a script file to capture these equations. For example, consider the pair of non-linear equations involving two variables:

$$\begin{aligned} \sin x - 3y &= 0 \\ \cos y - 2x &= 0 \end{aligned} \tag{3.8}$$

We first create a function that takes a 2D input and returns the "residuals" corresponding to the two equations:

```
function [r] = fsolveTestFunction(xVec)
x = xVec(1);
y = xVec(2);
r(1) = sin(x)-3*y;
r(2) = cos(y)-2*x;
```

Now we can use `fsolve` to drive the residuals to zero, i.e., essentially solving the non-linear equations, with an initial guess of (0, 0):

```
>> fsolve(@(xVec)fsolveTestFunction(xVec),[0 0])
    0.4938    0.1580
```

We will come across many such examples of non-linear equations later in the text.

3.12 Modules

One can include multiple functions within a single MATLAB file and access each one of them individually. For example, the file below contains multiple single-variable functions. The keyword `classdef` states that `testFunctions1D` is a class containing methods and possibly shared data. The keyword `Static` states that the functions within do not share data (data sharing is explained in Section 3.14). Collecting similar functions into a single class module can simplify file management considerably.

```
classdef testFunctions1D
    % This class encapsulates numerous 1D test functions
    methods(Static)
        function [f,g,h] = f1(x)
            % returns (x-1)^2 and its derivatives
            f = (x-1).^2;
            if (nargout > 1), g = 2*(x-1);end
            if (nargout > 2), h = 2;end
        end
        function [f,g,h] = f2(x)
            % returns (x-1)^3 and its derivatives
            f = (x-1).^3;
            if (nargout > 1), g = 3*(x-1).^2;end
            if (nargout > 2), h = 6*(x-1);end
        end
        function [f,g,h] = f3(x)
            % returns (x-1)^4 and its derivatives
            f = (x-1).^4;
            if (nargout > 1), g = 4*(x-1).^3;end
            if (nargout > 2), h = 12*(x-1).^2;end
        end
        function [f] = f4(x)
            % returns abs(x-1)
            f = abs(x-1);
        end
    ...
end
```

Each of these functions is accessible as illustrated below.

```
>> test = testFunctions1D;
>> test.f1(2)
ans =
     1
>> x = -1:0.1:3;
>> [f,g] = test.f1(x);
```

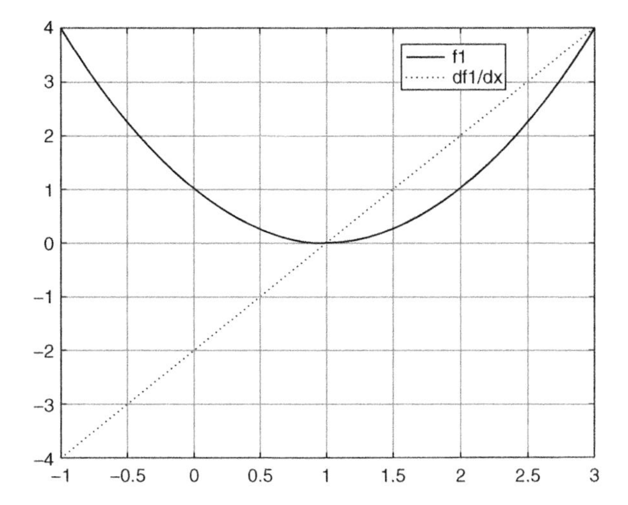

Figure 3.9 Plot of $(x - 1)^2$ and its gradient.

```
>> plot(x,f,'r-'); hold on;
>> plot(x,g,'k:');
>> legend('f1','df1/dx')
```

The resulting plot is shown in Figure 3.9.

3.13 Sampling Algorithm

Consider now a simple algorithm that finds the "global" minimum of a 1D function $f(x)$ within a given interval $[a, b]$ through uniform sampling. The MATLAB script below takes four parameters as arguments: (1) a function name that must be minimized, (2) the start point of the interval to search within, (3) the end point of the interval and (4) the number of sampling points.

```
function [fMin,xMin] = fminSampling(funToMinimize,a,b,nSamples)
% Find the global minimum of a function
% in an interval [a,b] through uniform sampling
% Usage: Sampling(@(x)cos(x),0,2*pi,1000)
if (nargin < 4), nSamples = 100; end
if (nargin == 1), a = 0; b = 1; end
xSamples = linspace(a,b,nSamples);
fSamples = feval(funToMinimize,xSamples); % evaluate function
[fMin, index] = min(fSamples);
xMin = xSamples(index);
```

Observe that if the user does not specify some of the input arguments, default values are assigned. The function `linspace` creates a uniform sample in the specified interval. The function `min` finds the smallest value and its location in the specified vector. Its usage is illustrated as follows.

```
>> [xMin,fMin] = fminSampling(@(x)cos(x),0,2*pi,100)
xMin =
    3.109859394462623
fMin =
  -0.999496542383185
>> [xMin,fMin] = fminSampling (@(x)cos(x),0,2*pi,1000)
xMin =
    3.138447916198813
fMin =
  -0.999995055317446
>> [xMin,fMin] = fminSampling (@(x)cos(x),0,2*pi,10000)
xMin =
    3.141278462905366
fMin =
  -0.999999950642107
```

Observe that as the number of samples increases, the accuracy of the estimated minimum increases. One can now find the minimum of the function `f1` defined in Section 3.9. The notation `@(x)` can be confusing, but it simply states that the function that follows has x as the primary argument.

```
>> test = testFunctions1D;
>> [xMin,fMin] = fminSampling(@(x)test.f1(x),0,2*pi,1000)
xMin =
    1.0000
fMin =
     7.0174e-010
```

The sampling algorithm is rather naïve in that the function must be sampled at thousands of points to get an accurate estimate for the minimum. While this is not a serious issue for simple functions, in practical scenarios each function call can take minutes or even hours to execute. Thus the naïve sampling method can be prohibitively expensive. *A key goal in optimization is to reduce the number of function calls.*

In other words, we will need smarter algorithms to find the minimum. One such function is the built-in `fminbnd`. The underlying theory behind `fminbnd` is discussed in Section 5.5.

```
>> help fminbnd
 fminbnd Single-variable bounded nonlinear function
minimization.
```

`X = fminbnd(FUN, x1, x2)` attempts to find a local minimizer `X` of the function `FUN` in the interval `x1 < X < x2`. `FUN` is a function handle which accepts scalar input `X` and returns a scalar function value `F` evaluated at `X`.

Observe in the following example that, using the `fminbnd` function (rather than the naïve sampling algorithm), the minimum has been found, to within machine precision, in about six function calls! We shall study the theory behind `fminbnd` in Chapter 5, since it forms the foundation of more complex optimization routines.

```
>> [xMin,fMin,exitflag,output]
=fminbnd(@(x)test.f1(x),0,2*pi)
xMin =

   1.0000
fMin =

   4.9304e-032
exitflag =

     1
output =

    iterations: 5
     funcCount: 6
     algorithm: 'golden section search, parabolic
interpolation'
       message: [1x112 char]
```

3.14 Polynomial Class

In an earlier example in Section 3.12, we collected numerous 1D functions into a single "class" without data sharing. Here, we develop a class with data sharing.

As a vehicle for our discussion, consider the polynomial

$$p(x) = a_0 + a_1x + a_2x^2 + \cdots + a_{N-1}x^{N-1} \tag{3.9}$$

Observe that the polynomial is fully defined by the array of coefficients

$$\mathbf{a} = \{a_0 \quad a_1 \quad a_2 \quad \cdots \quad a_{N-1}\} \tag{3.10}$$

For example, the polynomial

$$p(x) = 1 - x + 3.2x^3 \tag{3.11}$$

is captured by the coefficients

$$\mathbf{a} = \{1, -1, 0.0, 3.2\} \tag{3.12}$$

This array will be treated as the "data" shared by multiple functions within the class for manipulating, evaluating and printing polynomials. In the language of object-oriented programming (OOP), polynomials are treated here as objects with intrinsic methods.

Specifically, observe that, given two polynomials of the same order, i.e., given their coefficients, one can add the two polynomials by simply adding their coefficients. For example:

$$\begin{aligned} &p(x) = 1 - x + 2x^2 \equiv \{1, -1,\ 2\} \\ &q(x) = 0 + 2x + 3x^2 \equiv \{0,\ 2,\ 3\} \\ &\Rightarrow \\ &r(x) = p(x) + q(x) = (1+0) + (-1+2)x + (2+3)x^2 \\ &r(x) = \{1,\ 1,\ 5\} \end{aligned} \tag{3.13}$$

With these observations, a polynomial class may be created in MATLAB as illustrated below. Observe that the keyword `properties` specifies the data shared by all member functions within the class. A `'public' GetAccess` states the data can be read, for example, from the command line. A `'private' SetAccess` states that the data cannot be overwritten except through a member function. This is typically recommended for the classes that you may develop using MATLAB.

The special member function `PolynomialClass` is the constructor for the class and must be present if object instances must be created. A polynomial object is created as follows:

```
>> p = PolynomialClass([1 -1 3.2]);
```

This code creates the polynomial object representing

$$p(x) = 1 - x + 3.2x^2 \tag{3.14}$$

```
classdef PolynomialClass
    properties(GetAccess = 'public', SetAccess = 'private')
        % public read access, but private write access.
        myCoefficients;
    end
    methods
        function obj = PolynomialClass(a)
            % constructor for a PolynomialClass
            % a is a vector
            obj.myCoefficients = a;
        end
        function string = char(obj)
            % return a string representation of polynomial
            string = '';
```

```
            for i = 1:length(obj.myCoefficients)
                coeff = obj.myCoefficients(i);
                if coeff == 0
                    continue
                elseif (coeff < 0)
                    sign = '-';
                else
                    sign = '+';
                end
                % handle first two coefficients
                if (i == 1)
                    string = [string sign
num2str(abs(coeff))];
                elseif (i == 2)
                    string = [string sign
num2str(abs(coeff)) '*x'];
                else
                    string = [string sign
num2str(abs(coeff)) '*x^' num2str(i-1)];
                end
            end
            if (string(1) == '+') % remove initial '+' sign
                string = string(2:end);
            end
        end
        function disp(obj)
            % Pretty print of a Polynomial object
            disp(char(obj))
        end
        function y = evaluate(obj,x)
            % evaluate the Polynomial at x
            % y = a1 + a2*x + a3*x^2 + ...
            y = 0;
            temp = 1;
            for i = 1:length(obj.myCoefficients)
                y = y + obj.myCoefficients(i)*temp;
                temp = temp*x;
            end
        end
```

```
        function r = plus(obj1,obj2)
            % add two Polynomials;
            % the two Polynomials may be of different order
            % Find the difference in order
            n1 = length(obj1.myCoefficients);
            n2 = length(obj2.myCoefficients);
            if (n1 > n2) % append zeros to p2, and then add
                r = PolynomialClass(obj1.myCoefficients +
[obj2.myCoefficients zeros(1,n1-n2)]);
            elseif (n1 < n2) % append zeros to p1, and then
add
                r = PolynomialClass([obj1.myCoefficients
zeros(1,n2-n1)] + obj2.myCoefficients);
            else % just add coefficients
                r = PolynomialClass(obj1.myCoefficients +
obj2.myCoefficients);
            end
        end
    end
end
```

We can now use other functions within the class. The `disp` method prints a human-readable form:

```
>> p.disp();
1-1*x^1+3.2*x^2
```

This code essentially retrieves the coefficients and creates a string for display; the reader is encouraged to understand the logic behind the code. In fact, you can simply type the name of the object to call the `disp` function. MATLAB looks to see if there is a special function named `disp` within the class, and calls this function.

```
>> p
1-1*x^1+3.2*x^2
```

To evaluate the polynomial at a particular value of `x`, we have

```
>> p.evaluate(2.0)
ans =
   11.8000
```

The reader can verify that the reported value is indeed correct.

Next, we consider the addition of polynomials. Recall that the addition operator takes on different interpretations depending on the object associated with it. For integers and floats, it is interpreted as the usual addition, while for strings and lists, it is interpreted as concatenation. We now overload the addition operator for the polynomial class, with the natural interpretation:

$$
\begin{aligned}
&p(x) = a_0 + a_1 x + a_2 x^2 + \cdots + a_{N-1} x^{N-1} \\
&q(x) = b_0 + b_1 x + b_2 x^2 + \cdots + b_{N-1} x^{N-1} \\
&\Rightarrow \\
&p(x) + q(x) = (a_0 + b_0) + (a_1 + b_1)x + (a_2 + b_2)x^2 + \cdots + (a_{N-1} + b_{N-1})x^{N-1}
\end{aligned} \tag{3.15}
$$

In `PolynomialClass`, we have created a method with the `add` operator. Its usage is illustrated as follows:

```
>> p = PolynomialClass([1 -1 3.2])
p =
1-1*x+3.2*x^2
>> q = PolynomialClass([0 2 -1 5.5])
q =
2*x-1*x^2+5.5*x^3
>> r = p + q
r =
1+1*x+2.2*x^2+5.5*x^3
```

Once again, when the "+" sign is used between two objects, MATLAB checks to see if there is a function called `add` that takes another object of the same type. We can continue to insert other methods such as subtraction of polynomials, multiplication of a polynomial by a scalar, etc.

3.15 Extending the Polynomial Class

Now consider extending `PolynomialClass` so that one can find the roots of the polynomial. Classes can be extended as shown below, using the "<" operator: the `PolynomialClass` in Section 3.14 is the parent, while the new `PolynomialClassRoots` is the child. The child inherits all the member functions of the parent. Additional member functions can be added to the child. In the present context, by relying on the `roots` function we have added an additional member function to the `PolynomialClassRoots` class.

```
classdef PolynomialClassRoots < PolynomialClass
  methods
function obj = PolynomialClassRoots(a)
      obj = obj@PolynomialClass(a); % call Parent class
      % can add other function calls here
end
function r = findRoots(obj)
            r = roots(obj.myCoefficients); % calls MATLAB's
roots
      end
  end
end
```

Its usage is illustrated as follows:

```
>> p = PolynomialClassRoots([1 -1 3.2])
p =
1-1*x+3.2*x^2
>> p.findRoots()
ans =
   0.5000 + 1.7176i
   0.5000 - 1.7176i
```

3.16 EXERCISES

Exercise 3.1 The importance of memory pre-allocation and vectorization can be understood through the following exercise. Create and run the following script file; observe that the memory for the y-vector has not been pre-allocated. Note down the time taken.

```
clear x y;
tic
N = 10^7;
x = 0:2*pi/N:2*pi;
for i = 1:numel(x)
      y(i) = sin(x(i));
end
timeTaken = toc;
```

Now modify the script file by inserting memory pre-allocation for y as follows:

```
clear x y;
tic
N = 10^7;
x = 0:2*pi/N:2*pi;
y = zeros(1,N);
for i = 1:numel(x)
      y(i) = sin(x(i));
end
timeTaken = toc;
```

Next vectorize the loop as follows:

```
clear x y;
tic
N = 10^7;
x = 0:2*pi/N:2*pi;
y = sin(x);
timeTaken = toc;
```

Create a single semi-log plot, with log(N) as x-axis and time taken on the y-axis, showing the time taken for all three scripts for N: 10^3, 10^4, 10^5, 10^6, 10^7. Use the `legend` function in MATLAB to identify the three different results as `Naive`, `Memory Allocate` or `Vectorized`.

Exercise 3.2 Under the MATLAB `command` prompt, evaluate the following expressions: (a) $10^3 + 1 - 10^3$, (b) $10^{16} + 1 - 10^{16}$, (c) $1 + 10^{-5} - 1$ and (d) $1 + 10^{-16} - 1$. Explain the apparent discrepancies.

Exercise 3.3 Numerically solve the following three linear algebra problems:

$$\left\{\begin{array}{l} 2x + 3y + 4z = 5 \\ x - 6y - 2z = 2 \\ 5x - 2y + 6z = -1 \end{array}\right\}, \left\{\begin{array}{l} 2x + 3y + 4z = 5 \\ x - 6y - 2z = 2 \\ 5x - 15y - 2z = 11 \end{array}\right\}, \left\{\begin{array}{l} 2x + 3y + 4z = 5 \\ x - 6y - 2z = 2 \\ 5x - 15y - 2z = 15 \end{array}\right\} \quad (3.16)$$

Observe that the matrix is singular in the second and third problems. However, one of them yields a finite solution, while the other "blows up"; explain.

Exercise 3.4 Find the eigenvalues and eigenvectors of the matrices in Exercise 3.3. Now explain the conclusions via eigenvalues and eigenvectors.

Exercise 3.5 Using the `fminSampling` code provided, find the minimum x-location of the function

$$f(x) = 1 - xe^{-x^2}$$

using 10 000 samples, within the range [0, 2]. Verify that the gradient of the function is nearly zero at that location, and that the second derivative is positive.

Exercise 3.6 Write a MATLAB function `fzeroSampling.m` that finds the approximate zero location of a 1D function via sampling. The format of your function should be

```
function [xZero,fZero] = fzeroSampling(g,a,b,nSamples)
```

The first argument is typically the gradient of a function you are trying to minimize. Use this function to find the minimum x-location of the function

$$f(x) = 1 - xe^{-x^2}$$

using 10 000 samples in the range [0, 2]. Note that you must pass the gradient of $f(x)$ into your code. Compare the result with the previous exercise.

Exercise 3.7 Write a MATLAB function `binSearch.m` that asks the user to think of an integer between 0 and 1000 (inclusive). Then guess this number by repeatedly using the prompt: "Is your number (e)qual to, (l)ess than or (g)reater than x?"

Exercise 3.8 Write a MATLAB function `coins.m` that takes as input an integer between 1 and 100 (inclusive). Then the function should return *the smallest number of coins* required to reconstruct the input integer. Allowed coins are: penny (1 cent), nickel (5 cents), dime (10 cents) and quarter (25 cents). For example, if the user specifies 38 cents, the output should be 1 quarter, 1 dime, 3 pennies. The format of your function should be

```
function [pennies,nickels,dimes,quarters] = coins(value)
```

Exercise 3.9 Write a MATLAB function `quadfit.m` that takes as input three points described via a (2, 3) array, and finds the quadratic curve $y = a + bx + cx^2$ passing through these points. For example, given $(-1, 1)$, $(0, 0)$ and $(1, 1)$, described via [−1 0 1; 1 0 1], the function should return (0, 0, 1) since the unique curve passing through these points is $y = x^2$. There is no need to plot the curve (do not do more computation than asked for). What does your function return if the points array is [−1 0 1; 1 0 −1]? The format of your function should be

```
function [a,b,c] = quadfit(pts)
```

Exercise 3.10 We will use contour plots and visual inspection to approximately find the minimum location of the function

$$f(u,v) = \left\{ \begin{array}{l} \frac{1}{2}100\left(\sqrt{u^2+(v+1)^2}-1\right)^2 \\ +\frac{1}{2}500\left(\sqrt{u^2+(v-1)^2}-1\right)^2 - (10u+8v) \end{array} \right\}$$

Start with a range of $u \in [0, 1]$ and $v \in [-0.5, 0.5]$, plot ten contour lines and visually guess the minimum. Using this guess, narrow down the range for u and v, plot the contours again and visually guess the minimum until the confidence interval for u and v is within ± 0.05. Verify that the gradient of the function is nearly zero at that location.

Exercise 3.11 Consider the following Holder table function that exhibits numerous minima:

$$f(x,y) = -\left|\sin x \cos y \exp\left(\left|1 - \frac{\sqrt{x^2+y^2}}{\pi}\right|\right)\right|$$

Plot this function, using MATLAB's `meshgrid` and `surf` commands, in the range $-10 \leq x, y \leq 10$ with a sampling of 0.25 along both axes.

Exercise 3.12 Numerically find solutions to the following polynomial equation using the MATLAB's function `fsolve` with three different initial guesses, −10, 0 and 10:

$$x^3 - 4x^2 - 7x + 10 = 0$$

Exercise 3.13 One can carry out both definite and indefinite integrals using the MATLAB symbolic function `int`. Integrate the following equations using `int`:

$$\int \frac{1}{(x-2)(x-6)} dx$$

$$\int_{10}^{20} \frac{1}{(x-2)(x-6)} dx$$

Exercise 3.14 Implement an `IntegerSet` class to carry out the following operations:

1. You should be able to create instances as follows by passing positive integers:

```
>> A = IntegerSet([1 2 4]);
>> B = IntegerSet([2 4 9 10]);
```

2. If the user passes real numbers, or duplicates, return an error:

```
>> C = IntegerSet([1.3 pi]);
Error: input numbers must be positive integers
>> D = IntegerSet([1 1 3]);
Error: duplicates are not allowed
```

3. For display, simply print the contents of the set as follows:

```
>> A
[1 2 4]
```

4. Given two instances, implement the union of sets (without any duplicates); note that `C` should be an `IntegerSet` as well:

```
>> C = A.Union(B);
>> C
[1 2 4 9 10]
```

5. Given two instances, implement the intersection of sets; note that `C` should be an `IntegerSet` as well:

```
>> C = A.Intersection(B);
>> C
[2 4]
```

6. Given two instances, implement the difference of sets; note that `C` should be an `IntegerSet` as well:

```
>> C = A.Difference(B);
>> C
[1]
```

7. You should be able combine these operations:

```
>> C = A.Difference(B);
>> D = C.Union(B);
>> E = D.Intersection(A)
[1 2 4]
```

Exercise 3.15 Expand `PolynomialClass` to include subtraction of polynomials of different orders; for example:

$$\begin{aligned} p(x) &= 1 - x + 3.2x^2 \\ q(x) &= 1 + x \\ r(x) &= p(x) - q(x) = ? \\ r(x) &= q(x) - p(x) = ? \end{aligned} \tag{3.17}$$

The format of your class, containing two functions, should be

```
classdef PolynomialClassSubtract < PolynomialClass
  methods
function obj = PolynomialClassSubtract (a)
      obj = obj@PolynomialClass(a); % call Super class
end
function obj3 = minus(obj1,obj2)
            %implement the subtraction here
      end
  end
end
```

Exercise 3.16 Expand `PolynomialClass` to include multiplication of polynomials of different orders; for example:

$$\begin{aligned} p(x) &= 1 - x + 3.2x^2 \\ q(x) &= 1 + 2x \\ r(x) &= p(x)q(x) = ? \\ r(x) &= q(x)p(x) = ? \end{aligned} \tag{3.18}$$

The format of your class, containing two functions, should be

```
classdef PolynomialClassMultiply < PolynomialClass
  methods
function obj = PolynomialClassMultiply (a)
      obj = obj@PolynomialClass(a); % call Super class
end
function obj3 = mtimes(obj1,obj2)
            %implement the multiplication here
      end
  end
end
```

An elegant way of computing the coefficients of r is to construct a matrix R as follows, for the above example:

```
>> R = [1 -1 3.2]'*[1 2]
R =
    1.0000    2.0000
   -1.0000   -2.0000
    3.2000    6.4000
```

Note that the coefficients of r can be gathered by adding all cross-diagonal coefficients of R as follows: [1 (2−1) (−2+3.2) 6.4]. Test your code using the following example (and other examples as well):

```
>> p = PolynomialClassMultiply([1 -1 3.2]);
>> q = PolynomialClassMultiply([1 2]);
>> r = q*p
```

Exercise 3.17 Write a MATLAB function `myFactorialForLoop.m` that takes as input an integer N (assume N is less than 50), and returns $N!$ using a `for` loop. Check your solution against those obtained through MATLAB built-in function `factorial` for $N = 1, 2, \ldots, 10$.

Exercise 3.18 Write a MATLAB function `myFactorialRecursive.m` that takes as input an integer N (assume N is less than 50), and returns $N!$ using recursive calls. Recall that the factorial can be computed recursively as follows:

$$N! = N\,(N-1)! \tag{3.19}$$

$$1! = 1 \tag{3.20}$$

Check your solution against those obtained through MATLAB built-in function `factorial`, for $N = 1, 2, \ldots, 10$.

Exercise 3.19 Write a MATLAB function `myDeterminant.m` that takes as input an $N \times N$ (square) matrix, and returns the determinant value of the matrix using recursive calls. Specifically, suppose the matrix is

$$A = \begin{bmatrix} a_{11} & a_{12} & \dots & a_{1N} \\ a_{21} & a_{22} & \dots & a_{2N} \\ \dots & \dots & \dots & \dots \\ a_{N1} & a_{N2} & \dots & a_{NN} \end{bmatrix} \tag{3.21}$$

Then recall that the determinant of A can be computed recursively as follows:

$$\det(A) = \sum_{i=1}^{N} a_{1i}(-1)^{i+1}\det(M) \tag{3.22}$$

where the (minor) matrix M is obtained by deleting the first row and ith column. You can assume that only square matrices will be provided as input. As a test case, create a tridiagonal matrix as follows, for $N = 2, 3, \dots, 10$.

```
A = zeros(N,N);
A(1:1+N:N*N) = 2;
A(N+1:1+N:N*N) = -1;
A(2:1+N:N*N-N) = -1;
```

Compute the determinant of this matrix using your code, and compare your solutions against those obtained through MATLAB's built-in function `det`.

4 Unconstrained Optimization: Theory

Highlights

1. In this chapter, we study the mathematics behind *unconstrained* minimization problems of the form

$$\underset{\mathbf{x}}{\text{minimize}}\ f(\mathbf{x})$$

2. Such unconstrained problems may exhibit no solution, one solution or multiple solutions.
3. If a minimization problem exhibits multiple solutions, each of the solutions is referred to as a *local minimum*.
4. We will summarize and illustrate necessary and sufficient conditions for a chosen point to be a local minimum.

In this chapter, we turn our attention to mathematical aspects of optimization. The reader may be tempted to skip the mathematics and plunge directly into numerical optimization and algorithm development. This is not advisable! Optimization is a non-trivial discipline; a poorly designed algorithm can lead to erroneous results if care is not taken. A strong mathematical foundation (at least the basics) is essential.

The focus of this chapter is on *unconstrained minimization* problems of the form

$$\underset{\mathbf{x}}{\text{minimize}}\ f(\mathbf{x}) \tag{4.1}$$

where $\mathbf{x} = \{x_1 \quad x_2 \quad \ldots \quad x_N\}$ are the design variables and $f(\mathbf{x})$ is the objective function. An example of such a problem is

$$\underset{\mathbf{x}=\{x,y\}}{\text{minimize}} f(\mathbf{x}) = x^2 - xy + y^2 - x - 2y \tag{4.2}$$

Observe that this optimization problem is constraint-free; constraints will be introduced in Chapter 7. We now review the conditions under which the function $f(\mathbf{x})$ in Equation (4.1) has a minimum, and the mathematical properties of the function at the minimum.

4.1 Local versus Global Minimum

A concept of utmost significance in optimization is that of a *local minimum*. A function $f(\mathbf{x})$ is said to exhibit a local minimum at a point $\mathbf{x}^*$ if its value at that point is less than or equal to all other values of the function within some neighborhood of $\mathbf{x}^*$.

For example, consider the 1D function $f(x) = 3\sin x - x + 0.1x^2 + 0.1\cos(2x)$ in Figure 4.1. Observe that it "dips" at multiple points: $x \approx -8, x \approx -2, x \approx 4.5, x \approx 11$, etc. All such points are local minima of that function, since the values of the function at these points are smaller than all other values in a small neighborhood. Formally:

Definition: A function $f(\mathbf{x})$ exhibits a *local minimum* at $\mathbf{x}^*$ if there exists a δ such that $f(\mathbf{x}^*) \leq f(\mathbf{x})$ for all $\mathbf{x}$ such that $\|\mathbf{x} - \mathbf{x}^*\| < \delta$.

Further, among all the local minima, observe that the function in Figure 4.1 appears to take a *global minimum* at $x \approx 4.5$. Formally:

Definition: $f(\mathbf{x})$ exhibits a *global minimum* at $\mathbf{x}^*$ if $f(\mathbf{x}^*) \leq f(\mathbf{x})$ for all $\mathbf{x}$.

Most numerical methods of optimization inevitably yield only a local minimum. (We will briefly study global optimization methods in Section 6.7.) The particular minimum they will yield depends on the initial guess that the user provides.

The next concept of importance is *differentiability*. Most mathematical theorems and many numerical methods assume that the underlying function is differentiable. For example, consider the two functions x^2 and $|x|$ in Figure 4.2.

While x^2 is differentiable (i.e., its derivative is well defined), $|x|$ is not differentiable at $x = 0$. The theorems to be discussed only apply to differentiable functions. Observe that both x^2 and $|x|$ exhibit a global minimum at $x = 0$. Thus, a function need not be differentiable to exhibit a global or local minimum at a point.

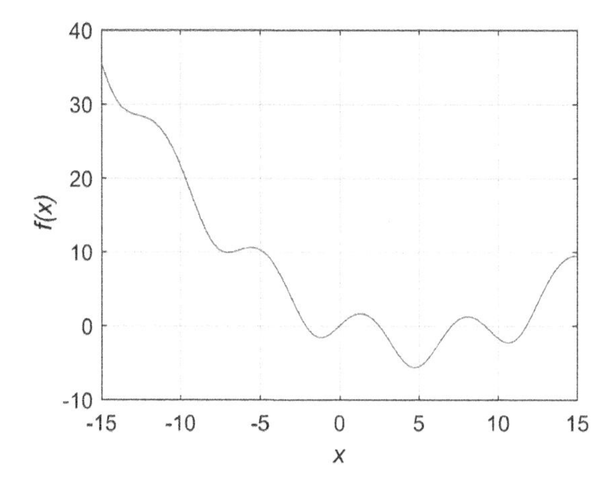

Figure 4.1 Plot of $f(x) = 3\sin x - x + 0.1x^2 + 0.1\cos(2x)$.

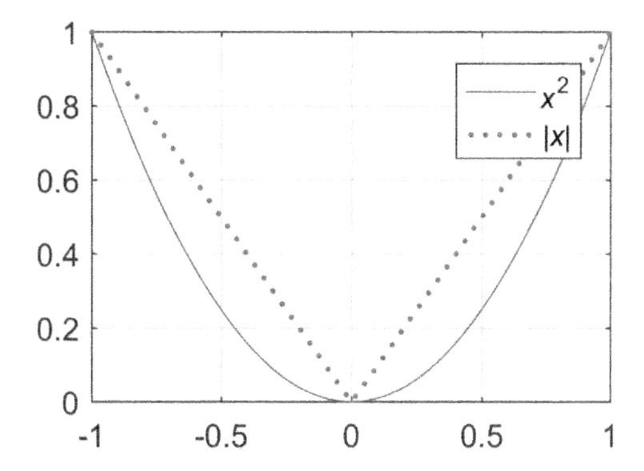

Figure 4.2 Differentiable and non-differentiable functions.

The concept of differentiability extends to higher dimensions. For example, the function $f(x,y) = x^2 - xy + y^2 - x - 2y$ is differentiable everywhere, but the function $f(x,y) = \sin(x)|x+y-3|$ is not differentiable with respect to x or y over the straight line $x+y-3=0$ but infinitely differentiable elsewhere. On the other hand, the (continuous) function

$$f(\mathbf{x}) = \{x,y\} = \begin{cases} (x-1)^2 y; & x < 2 \\ (x-1)^3 y; & x \geq 2 \end{cases} \tag{4.3}$$

is infinitely differentiable with respect to y, but not differentiable with respect to x at $x = 2$.

For the remainder of this chapter, we shall assume that the functions being investigated are infinitely differentiable.

4.2 First-Order Condition

From basic mathematics ([15]) recall that, for a single-variable differentiable function $f(x)$, the derivative vanishes at a local minimum. As a specific example, consider the function

$$f(x) = e^{-x^2} \tag{4.4}$$

The reader can verify that its derivative is given by

$$\frac{df(x)}{dx} = -e^{-x^2} + 2x^2 e^{-x^2} \tag{4.5}$$

The function and its derivative are plotted in Figure 4.3.

Theorem 4.1 *If x^* is a local minimum of a differentiable function $f(x)$, then the gradient of the function vanishes at that point, i.e., the necessary condition for a local minimum is that*

$$\left.\frac{df}{dx}\right|_{x^*} = 0 \tag{4.6}$$

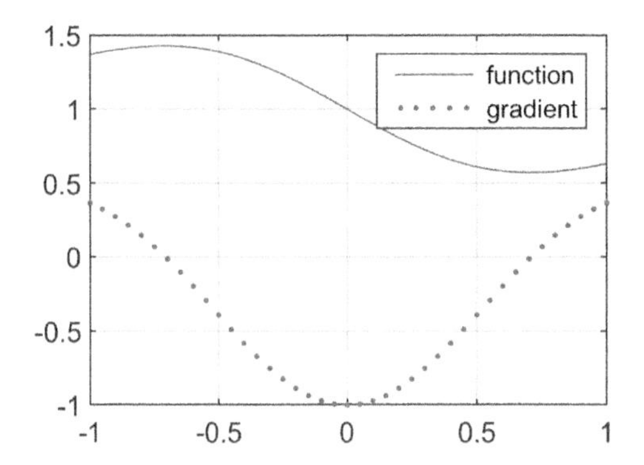

Figure 4.3 The function $f(x)$ and its derivative.

Proof: Consider the Taylor series expansion of $f(x)$ at a point x^*:

$$f(x^* + \Delta x) = f(x^*) + \left.\frac{df}{dx}\right|_{x^*} \Delta x + o(\Delta x)^2 \tag{4.7}$$

where Δx is sufficiently small. If x^* is a local minimum, then by definition $f(x^*)$ must be less than both $f(x^* + \Delta x)$ and $f(x^* - \Delta x)$ for sufficiently small Δx (see the definition of local minimum), i.e.,

$$\begin{aligned} f(x^*) &\leq f(x^* \pm \Delta x) = f(x^*) \pm \left.\frac{df}{dx}\right|_{x^*} \Delta x + o(\Delta x)^2 \\ \Rightarrow \quad 0 &\leq \pm \left.\frac{df}{dx}\right|_{x^*} \Delta x + o(\Delta x)^2 \end{aligned} \tag{4.8}$$

In the limit $\Delta x \to 0$, we have

$$\begin{gathered} 0 \leq \pm \left.\frac{df}{dx}\right|_{x^*} \Delta x \\ \Rightarrow \\ 0 \leq \left.\frac{df}{dx}\right|_{x^*} \Delta x \leq 0 \end{gathered} \tag{4.9}$$

The only way this is possible is if

$$\left.\frac{df}{dx}\right|_{x^*} = 0 \tag{4.10}$$

We will now generalize this concept to functions of arbitrary dimension $f(\mathbf{x})$ via their gradient. The gradient of a function is nothing but the derivative generalized to higher dimensions; for example, the gradient of a 2D function $f(x, y)$ is defined as follows:

$$\nabla f(\mathbf{x}) \equiv \begin{Bmatrix} \dfrac{\partial f}{\partial x} \\ \dfrac{\partial f}{\partial y} \end{Bmatrix} \tag{4.11}$$

For an N-dimensional function, the gradient is defined as

$$\nabla f(\mathbf{x}) \equiv \begin{Bmatrix} \frac{\partial f}{\partial x_1} \\ \frac{\partial f}{\partial x_2} \\ \vdots \\ \frac{\partial f}{\partial x_N} \end{Bmatrix} \tag{4.12}$$

Observe that the gradient is a vector.

Example 4.1 Find the gradient of $f(x, y) = x^2 - 2xy + 4y^2 - 2x - 4y - 5$ at (0, 0) and at (2, 1).

Solution: Observe that the gradient is

$$\nabla f = \begin{Bmatrix} 2x - 2y - 2 \\ -2x + 8y - 4 \end{Bmatrix}$$

At (0, 0), the gradient is

$$\nabla f = \begin{Bmatrix} -2 \\ -4 \end{Bmatrix}$$

At (2, 1), the gradient is

$$\nabla f = \begin{Bmatrix} 0 \\ 0 \end{Bmatrix}$$

For convenience, we summarize here a few gradient identities; these will come in handy in the remainder of the text. It is recommended that the reader prove each of the following statements, where $\mathbf{A}$ is a symmetric matrix, $\mathbf{b}$ is a vector and c is a scalar, while $\mathbf{g}(\mathbf{x})$ is a vector function of $\mathbf{x}$:

$$\nabla(c) = \mathbf{0} \tag{4.13}$$

$$\nabla(\mathbf{x}^T\mathbf{b}) = \nabla(\mathbf{b}^T\mathbf{x}) = \mathbf{b} \tag{4.14}$$

$$\nabla(\mathbf{x}^T\mathbf{A}\mathbf{x}) = 2\mathbf{A}\mathbf{x} \tag{4.15}$$

$$\nabla\left(\mathbf{g}^T(\mathbf{x})\mathbf{g}(\mathbf{x})\right) = 2\nabla\mathbf{g}(\mathbf{x})\mathbf{g}^T(\mathbf{x}) \tag{4.16}$$

We can now state the *necessary condition* for a local minimum in higher dimension.

Theorem 4.2 *If $\mathbf{x}^*$ is a local minimum of a differentiable function $f(\mathbf{x})$, then the gradient of the function vanishes at that point, i.e., the necessary condition for a local minimum is that*

$$\nabla f|_{\mathbf{x}^*} = \begin{Bmatrix} \frac{\partial f}{\partial x_1} \\ \frac{\partial f}{\partial x_2} \\ \vdots \\ \frac{\partial f}{\partial x_N} \end{Bmatrix}_{\mathbf{x}^*} = \begin{Bmatrix} 0 \\ 0 \\ \vdots \\ 0 \end{Bmatrix} \tag{4.17}$$

Proof: See the Appendix, Section A.2.

Observe the following:

1. The theorem is valid only for differentiable functions. Thus, although the function $f(x) = |x|$ exhibits a local minimum at $x = 0$, its gradient is not defined at that point, and therefore the theorem does not apply.
2. The theorem only states that if x^* is a local minimum then $\nabla f(\mathbf{x}^*) = 0$. It does *not* state the reverse, i.e., $\nabla f(\mathbf{x}^*) = 0$ *does not imply* that $\mathbf{x}^*$ is a local minimum.

To illustrate, consider the four functions x^2, $-x^2$, x^3 and x^4, illustrated in Figure 4.4. Observe that $f(x) = x^2$ and $f(x) = x^4$ exhibit local minima at $x = 0$. As per Theorem 4.2, the gradient of x^2 and x^4 must vanish at $x = 0$. Indeed,

$$\nabla(x^2)|_{x=0} = 2x|_{x=0} = 0$$

$$\nabla(x^4)|_{x=0} = 4x^3|_{x=0} = 0$$

On the other hand, although the gradient of x^3 vanishes at $x = 0$ (as the reader can verify), the function does *not* take a local minimum or a local maximum at that point, as Figure 4.4 illustrates. This motivates the following definition.

Definition: Points where the gradient of a function vanishes are called *stationary points*. Thus $x = 0$ is a stationary point of x^2, $-x^2$, x^3 and x^4; it is a local minimum of x^2 and x^4, and a local maximum of $-x^2$.

As an application of Theorem 4.2 in multiple dimensions, consider the function $f(x, y) = x^2 - 2xy + 4y^2 - 2x - 4y - 5$, discussed earlier. Recall from Example 4.1 that its gradient vanishes at (2, 1). Thus (2, 1) is a stationary point

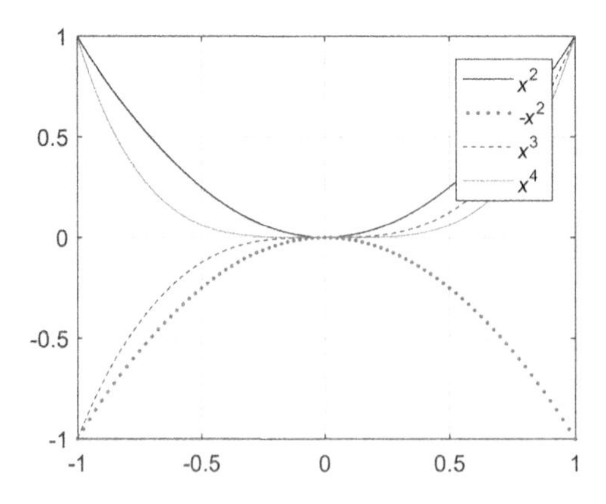

Figure 4.4 Plot of x^2, $-x^2$, x^3 and x^4.

of the function. But is it a local minimum? To answer this question, we turn to the second-order sufficient condition.

4.3 Second-Order Condition

Given a stationary point, we now study the curvature, i.e., the second derivative, of the function at the stationary point. Consider again the functions plotted in Figure 4.4. Observe that the second derivative of the function x^2 is positive at the stationary point $x = 0$, while the second derivative of the function $-x^2$ is negative at $x = 0$. This suggests that one can classify stationary points as local minima or maxima by considering the second derivative. This is indeed true (with some qualifiers).

Theorem 4.3 *Let x^* be a stationary point of $f(x)$. Further, let $f(x)$ be twice differentiable at x^*. Then*

(a) If $\left.\frac{d^2f}{dx^2}\right|_{x^*} > 0$ *then x^* is a* local minimum *of $f(x)$.*

(b) If $\left.\frac{d^2f}{dx^2}\right|_{x^*} < 0$ *then x^* is a* local maximum *of $f(x)$.*

(c) If $\left.\frac{d^2f}{dx^2}\right|_{x^*} = 0$ *then the test is* inconclusive.

Proof: Note that when the first derivative vanishes, the Taylor series reduces to

$$f(x^* \pm \Delta x) = f(x^*) + \frac{1}{2}\left.\frac{d^2f}{dx^2}\right|_{x^*} \Delta x^2 \pm o(\Delta x)^3 \tag{4.18}$$

Suppose

$$\left.\frac{d^2f}{dx^2}\right|_{x^*} > 0 \tag{4.19}$$

Then, clearly, for sufficiently small Δx,

$$f(x^*) + \frac{1}{2}\left.\frac{d^2f}{dx^2}\right|_{x^*} \Delta x^2 > f(x^*) \tag{4.20}$$

i.e., x^* is a local minimum. Similarly, one can establish that

If $\left.\frac{d^2f}{dx^2}\right|_{x^*} < 0$ then x^* is a *local maximum* of $f(x)$.

The second derivative is sometimes denoted by $f''(x)$.

Example 4.2 Use Theorem 4.3 to classify the four functions x^2, $-x^2$, x^3 and x^4 at the stationary point $x^* = 0$.

Solution: The second derivatives of the four functions are

$$\begin{aligned}
(x^2)'' &= 2 \\
(-x^2)'' &= -2 \\
(x^3)'' &= 6x \\
(x^4)'' &= 12x^2
\end{aligned}$$

At the stationary point $x = 0$, we have

$$\begin{aligned}
(x^2)'' \,|_{x=0} &= 2 \\
(-x^2)'' \,|_{x=0} &= -2 \\
(x^3)'' \,|_{x=0} &= 0 \\
(x^4)'' \,|_{x=0} &= 0
\end{aligned}$$

Thus, applying Theorem 4.3, we conclude that

- The function x^2 takes a minimum at $x = 0$, while the function $-x^2$ takes a maximum at $x = 0$.
- On the other hand, the second derivatives of x^3 and x^4 at $x = 0$ are both zero. Therefore, the theorem is inconclusive.

The generalization of the second derivative to higher dimension is the *Hessian matrix*. In 2D, the Hessian matrix is defined as

$$\mathbf{H}_f(\mathbf{x}) = \begin{bmatrix} \dfrac{\partial^2 f}{\partial x^2} & \dfrac{\partial^2 f}{\partial x \partial y} \\ \dfrac{\partial^2 f}{\partial x \partial y} & \dfrac{\partial^2 f}{\partial y^2} \end{bmatrix} \tag{4.21}$$

For an N-dimensional function, it is defined as

$$\mathbf{H}_f(\mathbf{x}) = \begin{bmatrix} \dfrac{\partial^2 f}{\partial x_1^2} & \dfrac{\partial^2 f}{\partial x_1 \partial x_2} & \cdots & \dfrac{\partial^2 f}{\partial x_1 \partial x_N} \\ \dfrac{\partial^2 f}{\partial x_2 \partial x_1} & \dfrac{\partial^2 f}{\partial x_2^2} & \cdots & \dfrac{\partial^2 f}{\partial x_2 \partial x_N} \\ \cdots & \cdots & \cdots & \cdots \\ \dfrac{\partial^2 f}{\partial x_N \partial x_1} & \dfrac{\partial^2 f}{\partial x_N \partial x_2} & \cdots & \dfrac{\partial^2 f}{\partial x_N^2} \end{bmatrix} \tag{4.22}$$

When there is no ambiguity, we will drop the subscript and simply denote the Hessian as $\mathbf{H}(\mathbf{x})$ or $\mathbf{H}$. The following theorem states the *sufficient condition* for a stationary point to be a local minimum for higher-dimensional functions.

Theorem 4.4 *Let* $\mathbf{x}^*$ *be a stationary point of* $f(\mathbf{x})$*. Further, let* $f(\mathbf{x})$ *be twice differentiable at* $\mathbf{x}^*$*, and let* $\mathbf{H}(\mathbf{x}^*)$ *be the Hessian matrix of the function at that point, defined per Equation (4.22).*

(a) If all eigenvalues of $\mathbf{H}(\mathbf{x}^*)$ *are positive, then* $\mathbf{x}^*$ *is a* local minimum.
(b) If all eigenvalues of $\mathbf{H}(\mathbf{x}^*)$ *are negative, then* $\mathbf{x}^*$ *is a* local maximum.
(c) If some eigenvalues of $\mathbf{H}(\mathbf{x}^*)$ *are positive, while others are negative, then* $\mathbf{x}^*$ *is a* saddle point, *i.e., neither a local minimum nor a local maximum.*
(d) If some of the eigenvalues of $\mathbf{H}(\mathbf{x}^*)$ *are zero while the remaining eigenvalues are of the same sign, then the test is* inconclusive.

Proof: See the Appendix, Section A.2.

Example 4.3 Find all stationary points of $f(x, y) = x^2 - 2xy + 4y^2 - 2x - 4y - 5$ and classify these points.

Solution: First compute the gradient:

$$\nabla f = \begin{Bmatrix} \dfrac{\partial f}{\partial x} \\ \dfrac{\partial f}{\partial y} \end{Bmatrix} = \begin{Bmatrix} 2x - 2y - 2 \\ -2x + 8y - 4 \end{Bmatrix}$$

Set both terms equal to zero and solve for the stationary points:

$$\begin{Bmatrix} 2x - 2y - 2 \\ -2x + 8y - 4 \end{Bmatrix} = \begin{Bmatrix} 0 \\ 0 \end{Bmatrix}$$
$$\Rightarrow \quad \begin{bmatrix} 2 & -2 \\ -2 & 8 \end{bmatrix} \begin{Bmatrix} x \\ y \end{Bmatrix} = \begin{Bmatrix} 2 \\ 4 \end{Bmatrix}$$

This is a linear system of equations and can be solved in MATLAB, as discussed in Chapter 3, leading to the stationary point $\mathbf{x}^* = (x^*, y^*) = (2, 1)$. Further, the Hessian of the function is

$$\mathbf{H} = \begin{bmatrix} \dfrac{\partial^2 f}{\partial x^2} & \dfrac{\partial^2 f}{\partial x \partial y} \\ \dfrac{\partial^2 f}{\partial x \partial y} & \dfrac{\partial^2 f}{\partial y^2} \end{bmatrix} = \begin{bmatrix} 2 & -2 \\ -2 & 8 \end{bmatrix}$$

The Hessian happens to be a constant, and the eigenvalues of this matrix are 1.39 and 8.6 (computed via MATLAB). Since both eigenvalues are positive, we conclude that the function exhibits a local minimum at $\mathbf{x}^* = (2, 1)$. This can be confirmed by studying the contour plot of the function in Figure 4.5. The contour values decrease toward the stationary point, i.e., the function takes a local minimum.

(cont.)

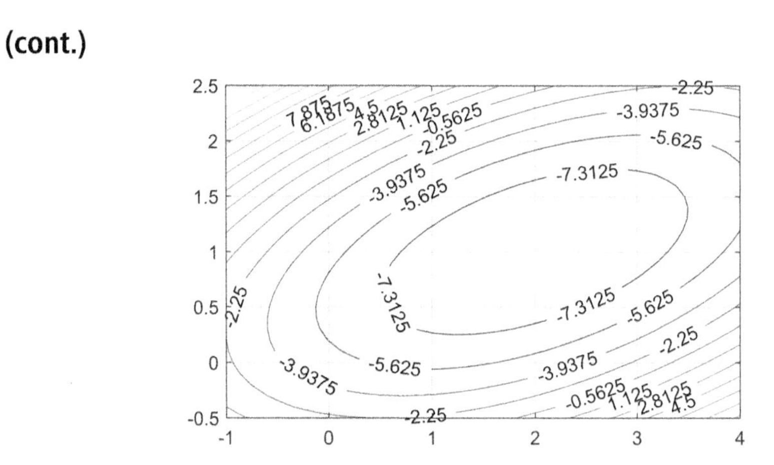

Figure 4.5 Contour plot of $f(x, y) = x^2 - 2xy + 4y^2 - 2x - 4y - 5$.

4.4 Sylvester's Criterion

An alternative, and often simpler, approach for verifying second-order conditions is via Sylvester's criterion. Given a square symmetric matrix H, *if all its eigenvalues are positive, then* H *is said to be positive definite* ([15]). Further, the kth *leading principal minor* of the matrix H is defined to be the determinant of its upper-left $k \times k$ sub-matrix. Sylvester's criterion states: *A symmetric matrix* H *is positive definite if and only if all its principal minors are positive.*

Example 4.4 For the function $f(x, y) = x^2 - 2xy + 4y^2 - 2x - 4y - 5$, using Sylvester's criterion, determine whether the stationary point (2, 1) is a local minimum.

Solution: From Example 4.3, the Hessian at the stationary point is

$$\mathbf{H} = \begin{bmatrix} 2 & -2 \\ -2 & 8 \end{bmatrix}$$

There are only two principal minors. The first principal minor is

$$m_1 = det[2] = 2$$

The second principal minor is

$$m_2 = det\begin{bmatrix} 2 & -2 \\ -2 & 8 \end{bmatrix} = 12$$

Since both are positive, the matrix is positive definite.

4.5 Quadratic and Convex Functions

Theorems 4.2 and 4.4 form the mathematical foundation of continuous optimization, and apply to any differentiable function. We now discuss two special classes of functions for which additional theorems apply. The first is the class of quadratic functions, defined below.

Definition: A function $f(\mathbf{x})$ is *quadratic* if it can be expressed as

$$f(\mathbf{x}) = c + \mathbf{x}^T\mathbf{g} + \frac{1}{2}\mathbf{x}^T\mathbf{H}\mathbf{x} \tag{4.23}$$

where $\mathbf{g}$ is a vector of size N, and $\mathbf{H}$ is a symmetric matrix of size (N, N).

In other words, quadratic functions are polynomial functions of degree 2. Examples include

$$f(x) = x^2 + 3x + 1 \tag{4.24}$$

$$f(x,y) = x^2 - 2xy + 4y^2 - 2x - 4y - 5 \tag{4.25}$$

$$f(x_1,x_2,x_3) = 4x_1^2 + 4x_2^2 + 4x_3^2 - x_1x_2 - x_2x_3 + 10x_3 \tag{4.26}$$

Observe that these functions can be represented respectively as

$$f(x) = 1 + \{x\}\{1\} + \frac{1}{2}\{x\}[4]\{x\} \tag{4.27}$$

$$f(x,y) = -5 + \{x \quad y\}\begin{Bmatrix} -2 \\ -4 \end{Bmatrix} + \frac{1}{2}\{x \quad y\}\begin{bmatrix} 2 & -2 \\ -2 & 8 \end{bmatrix}\begin{Bmatrix} x \\ y \end{Bmatrix} \tag{4.28}$$

$$f(x_1,x_2,x_3) = 0 + \{x_1 \quad x_2 \quad x_3\}\begin{Bmatrix} 0 \\ 0 \\ 10 \end{Bmatrix} + \frac{1}{2}\{x_1 \quad x_2 \quad x_3\}\begin{bmatrix} 8 & -1 & 0 \\ -1 & 8 & -1 \\ 0 & -1 & 8 \end{bmatrix}\begin{Bmatrix} x_1 \\ x_2 \\ x_3 \end{Bmatrix} \tag{4.29}$$

For quadratic functions, we now state the following theorem.

Theorem 4.5 *The stationary points* $\mathbf{x}^*$ *of a quadratic function satisfy the linear equation*

$$\mathrm{H}\mathbf{x}^* = -x \tag{4.30}$$

The stationary point is a local minimum (or local maximum) if all the eigenvalues of H *are positive (or negative).*

Proof: This is a special case of Theorem 4.4.

Thus, quadratic functions with positive definite Hessians have one and only one local minimum! *Positive definite quadratic functions* are important in optimization for many reasons:

1. Such functions are easy to analyze from a theoretical perspective.
2. For the same reason, it is extremely easy to numerically find the minima of such functions.
3. In Chapter 10, we shall see that many optimization problems in engineering reduce to minimizing quadratic functions.
4. Even when engineering problems lead to non-quadratic functions, these can be approximated locally by quadratic functions.

A generalization of positive definite quadratic functions is the class of *convex functions*. Formally, convex functions are defined as follows.

Definition: A function f is *convex* if, for all points p and q in the domain of f, the following property holds true:

$$f\Big((1-t)p + tq\Big) \leq (1-t)f(p) + tf(q); \;\; 0 \leq t \leq 1 \tag{4.31}$$

Convex functions are also common in engineering. Examples of analytic convex functions include $f(x) = x^2$, $f(x) = |x|$ (see Figure 4.2) and $f(x) = x^4$. For such convex functions, we now state the following theorem.

Theorem 4.6 *A convex function has one and only one local minimum.*

Proof: See reference [8].

Note that functions such as $f(x) = 1 - xe^{-x^2}$ are neither quadratic *nor* convex.

4.6 Illustrative Examples

In this section, we illustrate the theorems discussed in this chapter through additional examples.

Example 4.5 Find the stationary points of the function $f(x) = x^3 - 3x + 5$, and classify.

Solution: The gradient of the function is $\nabla f = 3x^2 - 3$. Thus the stationary points are obtained by setting the gradient to zero:

$$3x^2 - 3 = 0 \Rightarrow \quad x^* = \pm 1$$

(cont.)

We have two stationary points. The Hessian is given by $H_f = 6x$. At the two stationary points, we have

$$H_f|_{x^*=1} = 6 > 0$$
$$H_f|_{x^*=-1} = -6 < 0$$

Thus

$x^* = +1$ is a local minimum of the function, while

$x^* = -1$ is a local maximum of the function.

We confirm these findings in Figure 4.6.

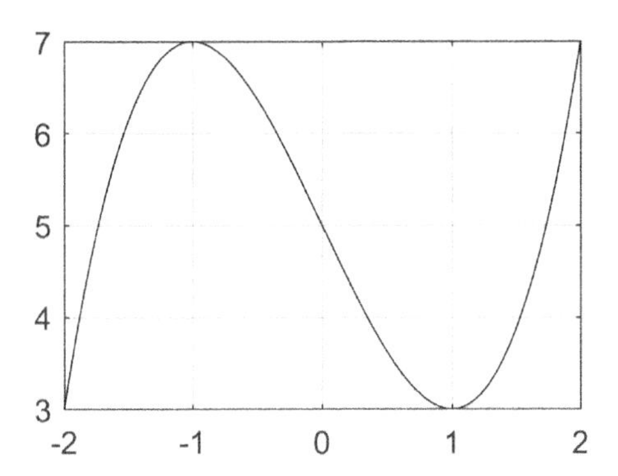

Figure 4.6 Plot of $f(x) = x^3 - 3x + 5$.

Example 4.6 Find the stationary points of $f(x) = \sin x - x/\sqrt{2}$, and classify.

Solution: Setting the gradient of the function to zero, we have

$$\nabla f = \cos x - 1/\sqrt{2} = 0 \Rightarrow \quad \cos x^* = \frac{1}{\sqrt{2}}$$

There are an infinite number of stationary points captured by the following set:

$$x^* = \{\pm\pi/4 \quad \pm 7\pi/4 \quad \pm 9\pi/4 \quad \ldots \quad \pm(2n\pi + \pi/4)\}$$

The Hessian is $H = -\sin x$. Observe that the Hessian alternates in sign:

$$\sin(\pi/4) < 0$$
$$\sin(7\pi/4) > 0$$
$$\sin(9\pi/4) < 0$$
$$\ldots$$

Thus, there are an infinite number of local minima and local maxima for this function, and they alternate, as confirmed in Figure 4.7.

(cont.)

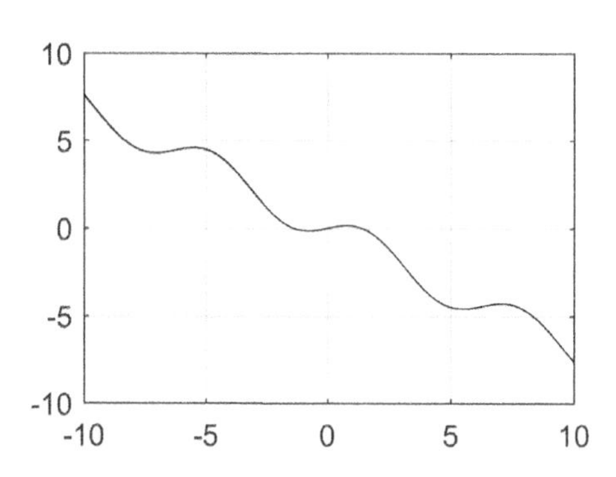

Figure 4.7 Plot of $f(x) = \sin x - x/\sqrt{2}$.

Observe that although the function has an infinite number of local minima and local maxima, it has no global minimum or maximum.

Example 4.7 Find the stationary point and determine its type for the function $f(x, y) = -4x^2 - 6y^2$.

Solution: One can verify that the stationary point is $(0, 0)$ and it is a *maximum* point, since the eigenvalues of the Hessian are -8 and -12. The 2D contour is shown in Figure 4.8; one can observe that the contour values increase toward the origin, i.e., the function takes a maximum.

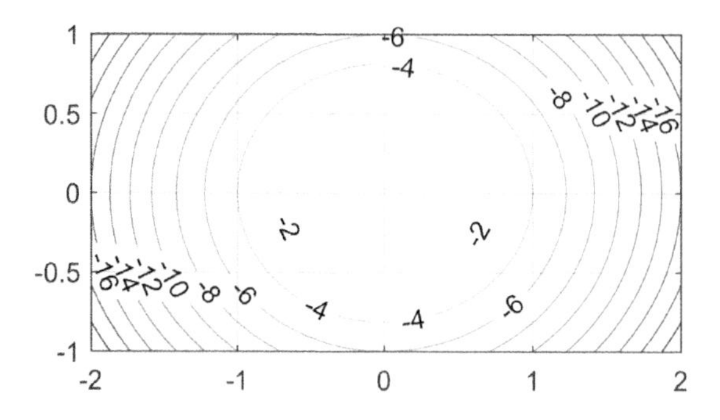

Figure 4.8 Contour plot of $f(x, y) = -4x^2 - 6y^2$.

Example 4.8 Find the stationary point and determine its type for the function

$$f(x, y) = -x^2 - 3y^2 + 9xy + 5x + 10y.$$

Solution: The stationary point is $(-1.7391, -0.9420)$, and it is a *saddle point* since the eigenvalues are 5.21 and -13.21, i.e., of opposite signs. It is a minimum along x and a maximum along y, therefore a saddle point, as illustrated in Figure 4.9.

(cont.)

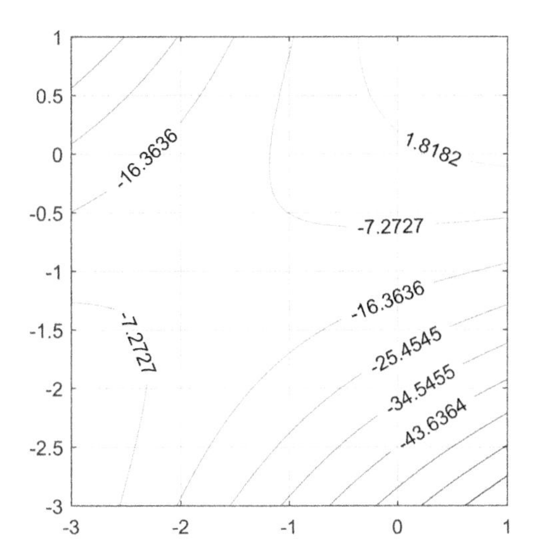

Figure 4.9 Contour plot of $f(x, y) = -x^2 - 3y^2 + 9xy + 5x + 10y$.

Example 4.9 Consider the function

$$f = 4x_1^2 + 4x_2^2 + 4x_3^2 + 4x_3^2 - x_1x_2 - x_2x_3 - x_3x_4 + 10x_3 + 8x_4.$$

Find the stationary values and the nature of these stationary values.

Solution: Find the gradient of the function and set it equal to zero to find the stationary points:

$$\nabla f = \begin{Bmatrix} \dfrac{\partial f}{\partial x_1} \\ \dfrac{\partial f}{\partial x_2} \\ \dfrac{\partial f}{\partial x_3} \\ \dfrac{\partial f}{\partial x_4} \end{Bmatrix} = \begin{Bmatrix} 8x_1 - x_2 \\ 8x_2 - x_1 - x_3 \\ 8x_3 - x_2 - x_4 + 10 \\ 8x_4 - x_3 + 8 \end{Bmatrix} = 0$$

The four equations can be written in a matrix form:

$$\begin{bmatrix} 8 & -1 & 0 & 0 \\ -1 & 8 & -1 & 0 \\ 0 & -1 & 8 & -1 \\ 0 & 0 & -1 & 8 \end{bmatrix} \begin{Bmatrix} x_1 \\ x_2 \\ x_3 \\ x_4 \end{Bmatrix} = \begin{Bmatrix} 0 \\ 0 \\ -10 \\ -8 \end{Bmatrix}$$

(cont.)

Solving the linear system, the stationary point is (0.0225, −0.180, −1.42, 1.18), and the Hessian is

$$\mathbf{H} = \begin{bmatrix} \frac{\partial^2 f}{\partial x_1^2} & \frac{\partial^2 f}{\partial x_1 \partial x_2} & \cdots & \frac{\partial^2 f}{\partial x_1 \partial x_4} \\ \frac{\partial^2 f}{\partial x_1 \partial x_2} & \frac{\partial^2 f}{\partial x_2^2} & \cdots & \frac{\partial^2 f}{\partial x_2 \partial x_4} \\ \cdots & \cdots & \cdots & \cdots \\ \frac{\partial^2 f}{\partial x_4 \partial x_1} & \frac{\partial^2 f}{\partial x_4 \partial x_2} & \cdots & \frac{\partial^2 f}{\partial x_4^2} \end{bmatrix} = \begin{bmatrix} 8 & -1 & 0 & 0 \\ -1 & 8 & -1 & 0 \\ 0 & -1 & 8 & -1 \\ 0 & 0 & -1 & 8 \end{bmatrix}$$

Note that the Hessian is a constant since the function is quadratic. The eigenvalues of this matrix are $(6.38, 7.38, 8.62, 9.62)$. Since all the eigenvalues are positive, the stationary point is a local minimum.

We can also check for positive definiteness by appealing to Sylvester's criterion. The four sub-matrices that we need to consider are

$$\mathbf{H}^1 = [8]$$
$$\mathbf{H}^2 = \begin{bmatrix} 8 & -1 \\ -1 & 8 \end{bmatrix}$$
$$\mathbf{H}^3 = \begin{bmatrix} 8 & -1 & 0 \\ -1 & 8 & -1 \\ 0 & -1 & 8 \end{bmatrix}$$
$$\mathbf{H}^4 = \begin{bmatrix} 8 & -1 & 0 & 0 \\ -1 & 8 & -1 & 0 \\ 0 & -1 & 8 & -1 \\ 0 & 0 & -1 & 8 \end{bmatrix}$$

The corresponding determinants are 8, 63, 496 and 3905. Since all of these are positive, the point is a local minimum.

It is not always possible to find the stationary points analytically. The next two examples illustrate this important point.

Example 4.10 Find stationary points of $f(x) = 3\sin x - x + 0.1x^2 + 0.1\cos(2x)$

Solution: To find the stationary points, we must solve

$$\nabla f = 3\cos x - 1 + 0.2x - 0.2\sin(2x) = 0$$

Observe that this is non-trivial. Thus, optimality criteria can rarely be used to find the local minimum analytically. For most optimization problems, one must devise numerical methods for finding stationary points. This is the subject matter of Chapter 5.

Example 4.11 Consider the function

$$f(u,v) = \begin{bmatrix} \frac{1}{2}100\left(\sqrt{u^2+(v+1)^2}-1\right)^2 \\ +\frac{1}{2}500\left(\sqrt{u^2+(v-1)^2}-1\right)^2 - (10u+8v) \end{bmatrix} \tag{4.32}$$

Find the approximate local minimum using contour plots.

Solution: Recall from Chapter 2 that this function represents the potential energy of a two-spring system. Figure 4.10 plots this function as a function of (u, v). Observe that the function exhibits both a local minimum close to the origin and a local maximum further away from it.

Indeed, from the contour plot in Figure 4.11, observe that the minimum occurs at approximately (0.25, 0.1) and a maximum occurs at approximately (0, 1). We must rely on numerical methods to find both these stationary points.

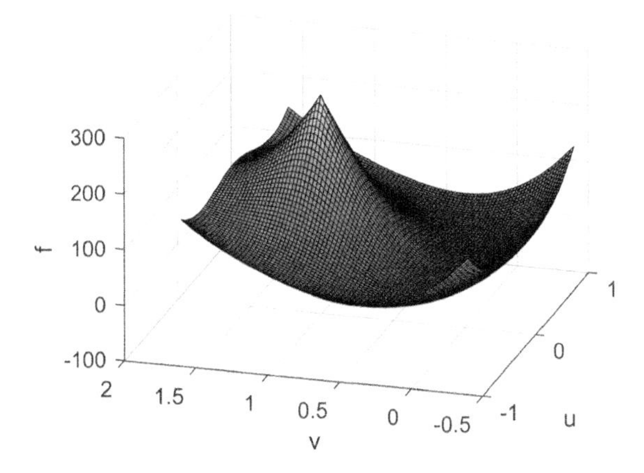

Figure 4.10 Surface plot of Equation (4.32).

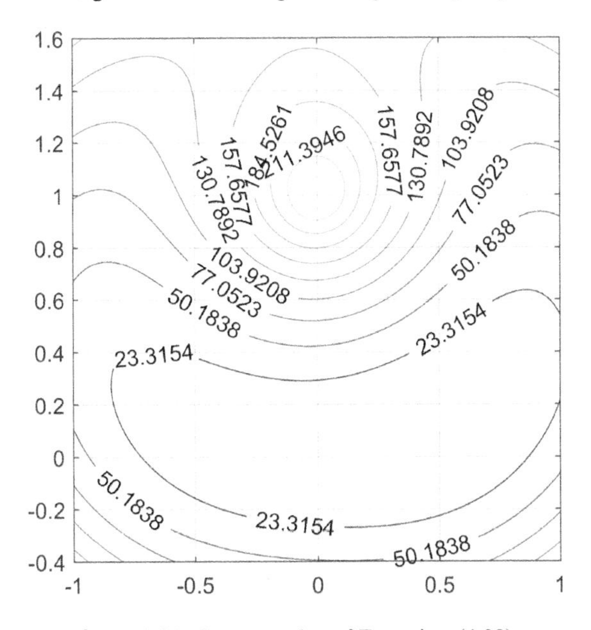

Figure 4.11 Contour plot of Equation (4.32).

4.7 EXERCISES

Exercise 4.1 Give an example of a 1D function whose first derivative is well defined everywhere, but not its second derivative at the origin.

Exercise 4.2 Consider the 2D Rosenbrock function

$$f(x,y) = 100(y - x^2)^2 + (1 - x)^2$$

Recall that the global minimum occurs at (1, 1). Are the first- and second-order conditions satisfied at the minimum?

Exercise 4.3 The following N-dimensional Rosenbrock function, which is often used as a test case for optimization algorithms, exhibits a minimum at (1, 1, . . ., 1).

$$f(\mathbf{x}) = \sum_{i=1}^{N-1} [100(x_{i+1} - x_i^2)^2 + (1 - x_i)^2]$$

For $N = 4$, are the first- and second-order conditions satisfied at the minimum?

Exercise 4.4 Find and classify the stationary points for:

(a) $f(x) = x^3 + 6x^2 - 15x + 2$
(b) $f(x) = x^2 e^x$
(c) $f(x) = x + x^{-1}$

Exercise 4.5 Suppose the first and second derivatives of a 1D function are zero at a particular point. What are the necessary and sufficient conditions for that point to be a local minimum? Assume the function is infinitely differentiable. Justify your answer and give an example of such a function.

Exercise 4.6 Find the point on the parabola $y = (1/5)(x - 1)^2$ that is closest to (2, 3), in the shortest-distance sense. Through elimination, reduce the problem to an unconstrained minimization problem. Then use MATLAB's `fsolve` on the first-order condition to find the stationary point(s). Use the second-order condition to identify the local minimum.

Exercise 4.7 Using the definition of the gradient, prove the gradient identities in Equations (4.13) through (4.16).

Exercise 4.8 Given an ellipse $(x/2)^2 + (y/1)^2 = 1$ and a point $p = (5, 5)$, reduce the problem of finding the closest point q on the ellipse to p to a single-variable unconstrained problem by introducing an intermediate variable. Then, by exploiting the first-order condition, find the point q using `fsolve`.

Exercise 4.9 Suppose you are given the data points $(1, 0)$, $(0, 0.99)$, $(-1.01, 0)$ and $(0, -0.98)$. The objective is to find a circle of the form

$$(x - x_c)^2 + (y - y_c)^2 = r^2$$

that best fits these data points, using the shortest distance as a metric. Pose this as an optimization problem. Next, using the first-order condition, find the three non-linear equations governing the center and radius. Finally, find the solution using MATLAB's `fsolve`.

Exercise 4.10 Find and classify the stationary point(s) of the function

$$f(x, y) = (x - 2)^2 + (4y + 4)^2 - xy$$

Exercise 4.11 Recall the N-dimensional Taylor series (see the Appendix, Section A.1).

$$f(\mathbf{x} + \Delta\mathbf{x}) = f(\mathbf{x}) + \frac{1}{1!}\Delta\mathbf{x}^T\nabla f(\mathbf{x}) + \frac{1}{2!}\Delta\mathbf{x}^T(\mathbf{H}_f)\Delta\mathbf{x} + o(\|\Delta\mathbf{x}\|)^3$$

Find the Taylor series of $f(x, y) = x^2 - 2xy + 4y^2 - 2x - 4y - 5$ about:

(a) (0, 0)
(b) (2, 1)

with $\Delta\mathbf{x} = \{\Delta x \quad \Delta y\}$.

Exercise 4.12 For $f = x^4 - xy + y^3$, find the gradient and Hessian at (1, 1). Then, using Taylor series, estimate the function value at (1.1, 0.8). Compare this with the exact value.

Exercise 4.13 Find expressions for the gradient and the Hessian for the following function:

$$f = x_1 - \frac{10}{x_1 x_2} + 5x_2$$

Find the stationary point, and check if it is a local minimum.

Exercise 4.14 For the generic quadratic function

$$f = Ax^2 + Bxy + Cy^2 + Dx + Ey + F$$

using Sylvester's criterion, determine the analytical condition under which a stationary point is a local minimum. Apply this result to each of the following problems:

(a) $f = x^2 - 5xy + y^2$
(b) $f = x^2 + 5xy + y^2$
(c) $f = x^2 - xy + y^2$
(d) $f = -x^2 + 5xy + y^2$

Exercise 4.15 Prove the following:

(a) There are an infinite number of contours for a function, but no two contours may cross each other.
(b) For every point (x, y) there exists a unique contour passing through that point; however, for a given contour value c_0, a contour may not exist.
(c) At every point (x, y), the gradient of a function, i.e., ∇f (a 2D vector), is perpendicular to the contour passing through that point, and points in the direction of increasing values of f.

5 Unconstrained Optimization: Algorithms

Highlights

1. The objective of this chapter is to introduce readers to basic numerical methods for solving an unconstrained minimization problem:

$$\underset{\mathbf{x}}{\text{minimize}}\ f(\mathbf{x})$$

2. We will discuss the implementation of these algorithms. This will prepare the reader for Chapter 6, where in-built MATLAB algorithms will be discussed.
3. The numerical methods discussed here can be classified into three types: (1) *zeroth-order* methods that only entail function evaluations, (2) *first-order* methods that entail function and gradient evaluations and (3) *second-order* methods that entail function, gradient and Hessian evaluations.
4. Zeroth-order methods are robust but slow. First-order methods are the most popular; examples include the non-linear conjugate gradient. Second-order methods converge rapidly, and are used when the Hessian of the function is readily available.
5. A popular 1D algorithm that we will discuss at length is the golden-section method.
6. Multi-dimensional algorithms rely heavily on such 1D algorithms; thus, a good understanding of 1D algorithms is critical.

In this chapter, we consider numerical algorithms to solve *unconstrained* minimization problems of the form

$$\underset{\mathbf{x}}{\text{minimize}}\ f(\mathbf{x}) \tag{5.1}$$

The objective is to cover the basic ideas and concepts. This will prepare the reader to understand and exploit advanced algorithms supported by MATLAB.

5.1 Test Functions

We will test the optimization methods developed here on a variety of functions designed to identify the strengths and weaknesses of these methods. For example, consider the suite of test functions in Equation (5.2). Most of these functions are sufficiently simple that we can easily find the minimum analytically. Thus, one can verify whether the numerical algorithms yield the correct solutions.

$$\begin{aligned} f_1(x) &= (x-1)^2 \\ f_2(x) &= (x-1)^3 \\ f_3(x) &= (x-1)^4 \\ f_4(x) &= |x-1| \\ f_5(x) &= \begin{Bmatrix} 2(x-1)^2; x<1 \\ 4(x-1)^2; x\geq 1 \end{Bmatrix} \\ f_6(x) &= \cos^2 x \\ f_7(x) &= 1 - xe^{-x^2} \\ f_8(x) &= 3\sin x - x + 0.1x^2 + 0.1\cos(2x) \\ f_9(x) &= (x-1)^2 + 0.025\sin^2\left(\frac{1000(x-1)}{\pi}\right) \end{aligned} \tag{5.2}$$

These test functions are captured in the MATLAB class `testFunctions1D.m` within the software package accompanying this text. For example, the function

$$f(x) = (x-1)^2 \tag{5.3}$$

is implemented as $f_1(x)$ in `testFunctions1D.m`, while the function

$$f(x) = |x-1| \tag{5.4}$$

is implemented as $f_4(x)$ in `testFunctions1D.m`.

The function

$$f(x) = (x-1)^2 + 0.025\sin^2(1000(x-1)/\pi) \tag{5.5}$$

is differentiable but has numerous local minima, and can therefore pose challenges to optimization routines; see Figure 5.1. This function is captured as $f_9(x)$, and so on. To access these 1D functions, we create an instance

```
>> test1D = testFunctions1D;
```

We can then create a plot similar to the one in Figure 5.1, as follows:

```
>> x = 0.75:0.0001:1.25;
>> y = test1D.f9(x);
>> plot(x, y)
```

Similarly, higher-dimensional functions are captured in the class `testFunctionsND.m`, and this is discussed in Section 5.6.

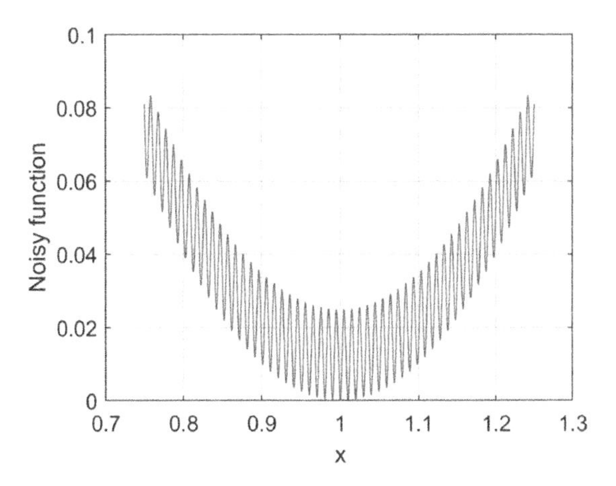

Figure 5.1 A function with numerous local minima.

5.2 One-Variable Minimization

As a special case of Equation (5.1), consider the *one-variable* unconstrained minimization problem

$$\underset{x}{\text{minimize}}\ f(x) \tag{5.6}$$

The choice of a numerical method for solving Equation (5.6) depends on the properties of $f(x)$. For example, observe that some of the functions in Equation (5.2) are differentiable, while others are not. This plays an important role in choosing the numerical method. Specifically, if $f(x)$ is *not* differentiable, for example $|x-1|$, then one must employ *zeroth-order methods*, i.e., methods that only require the function values, and not its derivatives. Methods such as golden section (see Section 5.5.2) fall into this category. If $f(x)$ is differentiable once, such as $f_5(x)$ in Equation (5.2), then *first-order methods*, which are usually more efficient, can be used. Methods such as quadratic bisection are first-order methods. Even more efficient are second-order methods such as Newton–Raphson, which require first- and second-order derivatives.

Once a numerical method has been chosen (see the detailed discussion that follows), the user must either provide an interval $[a, b]$ to search within, or an initial guess (depending on the method). Methods such as golden section and quadratic fitting require an interval, while methods such as Newton–Raphson require an initial guess. Obviously, the tighter the interval or the closer the initial guess is to the minimum, the better. Observe that some functions in Equation (5.2) exhibit only one global minimum (which ones?), while others exhibit numerous local minima (identify!). In general, numerical methods converge to the local minimum closest to the initial guess, or one of the local minima contained in the specified interval. *These concepts are important to remember when exploiting "black-box" optimization algorithms supported by MATLAB or other software packages.*

Finally, the desired *accuracy* must also be supplied by the user; numerical methods iteratively compute solutions x^k and terminate when *one* of the following conditions is satisfied:

$$
\begin{aligned}
&|x^k - x^{k-1}| \leq x_\varepsilon \\
&|f(x^k) - f(x^{k-1})| \leq f_\varepsilon \\
&|\nabla f(x^k)| \leq g_\varepsilon
\end{aligned}
\tag{5.7}
$$

Thus, three different measures of accuracy, namely $x_\varepsilon, f_\varepsilon$ and g_ε, must be specified by the user; otherwise default values are typically assumed. Depending on the method, one or more of these tolerances are used for termination. Tightening the tolerances typically increases the number of iterations. To ensure that the method terminates, one must additionally impose the maximum number of iterations permitted, or the maximum number of function calls, i.e., the number of times a method can evaluate $f(x)$.

To summarize, the user must:

1. pick an appropriate method depending on the differentiability of the function
2. understand the peculiarities and limitations of the method
3. provide an initial guess, or an interval to search within
4. provide tolerances for termination.

5.3 Unimodal Functions

Just as the concept of the differentiability of a function is important for stating and proving optimization theorems, the concept of a *unimodal* function is critical for understanding numerical methods. Informally, a function is unimodal if it is "U-shaped" in a given interval. More formally:

Definition: A 1D function $f(x)$ is *unimodal* in a given interval $[a, b]$ if:

1. the interval $[a, b]$ contains a unique minimum x^* of $f(x)$, and
2. the function $f(x)$ is *monotonically decreasing* in the sub-interval $[a, x^*]$ and *monotonically increasing* in the sub-interval $[x^*, b]$.

For example, $\cos(x)$ is unimodal in the interval $[0, 2\pi]$ since (1) it has a unique minimum $x^* = \pi$ within the interval, and (2) it is monotonically decreasing in the interval $[0, \pi]$ and monotonically increasing in the interval $[\pi, 2\pi]$. However, $\cos(x)$ is not unimodal in the interval $[0, 4\pi]$ or in the interval $[-\pi, \pi]$. Similarly, the function $f(x) = 1 - xe^{-x^2}$ (illustrated in Figure 5.2) is unimodal in the interval $[0, 2]$ but not in the interval $[-2, 2]$.

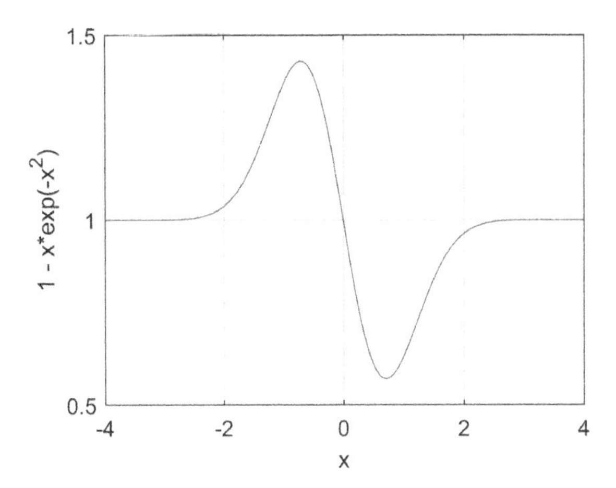

Figure 5.2 Plot of $1 - xe^{-x^2}$.

Observe that the concept of unimodality is not related to differentiability. For example, in the region $[-1, 1]$ the functions $f(x) = x^2$ and $f(x) = |x|$ are unimodal, but the latter is not differentiable.

5.4 Termination Criteria

Before we discuss specific methods, we briefly revisit the termination criteria presented in Equation (5.7), and their implementation in MATLAB. Recall that the user must set various criteria for the termination of optimization algorithms, including:

1. the tolerance x_ε on the design variables, i.e., the algorithm terminates when two successive solutions satisfy $|x^k - x^{k-1}| \leq x_\varepsilon$
2. the tolerance f_ε on the objective, i.e., the algorithm terminates when two successive objective evaluations satisfy $|f(x^k) - f(x^{k-1})| \leq f_\varepsilon$
3. the tolerance g_ε on the gradient, i.e., the algorithm terminates when the gradient satisfies $|\nabla f(x^k)| \leq g_\varepsilon$; note that this is relevant only for first- and second-order methods
4. the maximum number of function evaluations N_f, i.e., the algorithm terminates when the number of objective function evaluations has exceeded this value
5. the maximum number M of iterations, i.e., for certain algorithms such as the conjugate gradient, control of the number of permitted outer iterations.

In MATLAB, one can set and query these criteria using the `optimset` and `optimget` functions, as illustrated below.

```
>> opt = optimset('fminunc')
opt =
          Display: 'final'
      MaxFunEvals: '100*numberofvariables'
          MaxIter: 400
           TolFun: 1.0000e-06
             TolX: 1.0000e-06
      FunValCheck: 'off'
...
```

One can override the default values via

```
>> opt = optimset(opt,'TolX',1e-5);
```

Finally, one can query a specific value as follows:

```
>> opt.TolX
ans =
   1.0000e-05
>> opt.TolFun
ans =
   1.0000e-06
```

In the exercises and sample implementations, we will assume that the `opt` data structure is supplied by the user, and that the algorithm must respect the termination criteria.

5.5 1D Methods

We will now consider 1D methods for finding the minimum of unimodal functions. These methods often serve as the backbone for multi-dimensional unconstrained problems.

5.5.1 Trisection

A simple numerical method is the *trisection* method. It is a zeroth-order method and does not require the derivative of the function.

Let us assume that the user has supplied a function and a unimodal interval $[a, b]$ to search within (see Figure 5.3). In the trisection method, the interval $[a, b]$ is divided into three equal parts with intermediate points at x_1

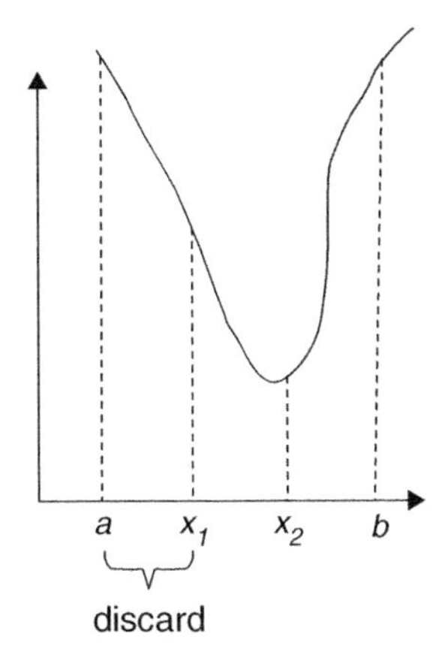

Figure 5.3 The trisection method.

and x_2; see Figure 5.3. The function is sampled at these points. Now suppose $f(x_1) > f(x_2)$ (as in Figure 5.3). Then the interval $[x_1, b]$ must contain a minimum (an exercise for the reader). Therefore the interval $[a, x_1]$ is discarded, and a new trisection search is performed over the interval $[x_1, b]$, i.e., the cycle repeats. Note that if $f(x_2) > f(x_1)$ the interval $[x_2, b]$ is discarded.

Observe that, in each cycle, the length of the interval reduces to 2/3 of the original size. Thus, the number N of iterations required to reduce the interval to a desired tolerance x_ε satisfies

$$\left(\frac{2}{3}\right)^N |b - a| = x_\varepsilon \tag{5.8}$$

i.e.,

$$N = \text{ceil}\left(\frac{\log \dfrac{x_\varepsilon}{|b - a|}}{\log(2/3)}\right) \tag{5.9}$$

where ceil(x) returns the least integer greater than or equal to x. For example, suppose $b - a = 1$ and $x_\varepsilon = 10^{-8}$; then $N = 46$. In other words, about 46 iterations may be needed to find the solution to the specified accuracy. Further, observe that the method uses two function evaluations per iteration, and the derivative of the function is not needed.

A MATLAB implementation of the trisection method is given below; the reader is encouraged to study the implementation, paying careful attention to the termination conditions. Most of these sample codes are included in the software package accompanying the text.

```
function [xMin,fMin,flag,output] = triSection(fun,a,b,opt)
% Minimize a function via trisection method
if (nargin < 4), opt = optimset('fminunc'); end % default parameters
iter = 0;
flag = 0;
fEvals = 0;
output.message = '#FunEvals exceeded'; % default
while(fEvals < opt.MaxFunEvals)
    x1 = a + (b-a)/3;
    x2 = a + (b-a)*2/3;
    if ((b-a) < opt.TolX)
        output.message = 'X tol met';
        flag = 1;
        break;
    end
    f1 = fun(x1);
    f2 = fun(x2);
    fEvals = fEvals + 2; % note 2 function calls every iteration
    if (f1 > f2)
        a = x1;
    else % f1 = f2, or f1 < f2
        b = x2;
    end
    iter = iter + 1;
end
xMin = (x1+x2)/2;
fMin = fun(xMin);
output.iterations = iter;
output.funcCount = fEvals;
```

An example to find the minimum of $\cos^2(x)$ in the range [0, 3] is given below. First, we will use default termination criteria:

```
>>test1D = testFunctions1D;
>>fun = @(x)test1D.f6(x);
>>[xMin,fMin,flag,output] = triSection(fun,0,3)
xMin =
   1.570585916189545
fMin =
     4.427262219097918e-08
flag =
     0
output =
  struct with fields:
       message: '#FunEvals exceeded'
    iterations: 21
     funcCount: 42
```

Observe that the computed solution differs from the exact solution is $\pi/2 \approx 1.570796326794897$ in the fourth decimal place. We will now change the default termination criteria, allowing for 10000 function evaluations:

```
>> opt = optimset('fminunc');
>> opt = optimset(opt,'MaxFunEvals',10000);
>> [xMin,fMin,flag,output] = triSection(fun,0,3,opt)
xMin =
   1.570796334683222
fMin =
     6.222567899026791e-17
flag =
     1
output =
  struct with fields:
       message: 'X tol met'
    iterations: 37
     funcCount: 74
```

Now the computed solution is more accurate.

5.5.2 Golden Section

The main drawback of the trisection method is that function evaluations are not reused. To elaborate, suppose the initial interval is $[0, L]$; then the function is sampled at $L/3$ and at $2L/3$. In the next iteration, if $[0, L/3]$ is discarded, the function is sampled at $5L/9$ and $7L/9$, and so on. None of the previous function evaluations are ever reused.

The *golden section* method overcomes this deficiency by carefully choosing the sampling points. Let the initial interval be $[a, b]$, where $b - a = L$. In this method, two samples are chosen, at αL and $(1 - \alpha)L$ from the starting point, where $0 < \alpha < 1$ is to be determined (in the trisection method, $\alpha = 1/3$). Thus the first two samples are $x_1 : a + \alpha L$ and $x_2 : a + (1 - \alpha)L$ (see Figure 5.4(a)).

Now suppose the interval $[a, x_1]$ is discarded (based on function comparison). Then, observe that the remaining length is $(1 - \alpha)L$; see Figure 5.4(b). Thus, the next two samples are $x_3 : x_1 + \alpha(1 - \alpha)L$ and $x_4 : x_1 + (1 - \alpha)(1 - \alpha)L$. To reuse function evaluations, we choose α such that points x_3 and x_2 coincide, i.e.,

$$x_1 + \alpha(1 - \alpha)L = a + (1 - \alpha)L \tag{5.10}$$

i.e.,

$$a + \alpha L + \alpha(1 - \alpha)L = a + (1 - \alpha)L \tag{5.11}$$

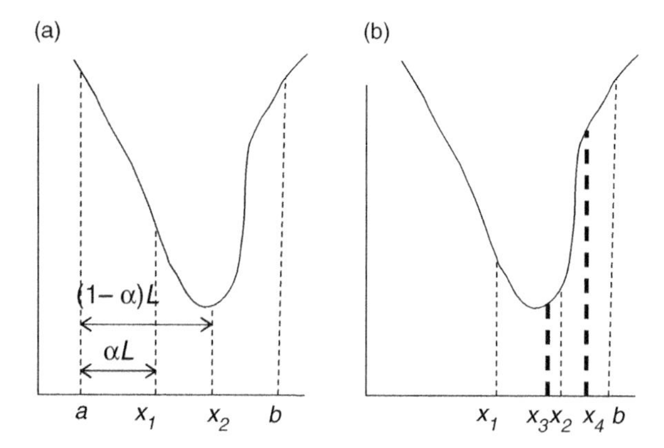

Figure 5.4 Golden section method: (a) the first set of samples; (b) the second set.

i.e.,

$$\alpha + \alpha(1-\alpha) = (1-\alpha) \tag{5.12}$$

i.e.,

$$\alpha^2 - 3\alpha + 1 = 0 \tag{5.13}$$

Thus, the optimal value of α is

$$\alpha = \frac{3-\sqrt{5}}{2} \approx 0.381966 \tag{5.14}$$

As a side-note, observe that α can be expressed as

$$\alpha = 1 - \frac{1}{g} \tag{5.15}$$

where

$$g = (1+\sqrt{5})/2 \tag{5.16}$$

is the famous golden ratio, hence the name "golden section" method.

In conclusion, given the initial interval $[a, b]$, we sample the function at $x_1 = a + \alpha(b-a)$ and $x_2 = a + (1-\alpha)(b-a)$. Then we discard one of the sub-intervals using the same logic as before. Since the length of the interval reduces by a factor of $(1-\alpha)$, the number N of iterations required is

$$N = \text{ceil}\left(\frac{\log \frac{x_\varepsilon}{|b-a|}}{\log(1-\alpha)}\right) \tag{5.17}$$

For example, if $b - a = 1$ and $x_\varepsilon = 10^{-8}$, then $N = 39$.

In other words, about 39 iterations may be needed to find the solution to the specified accuracy. *On average*, the number of golden section iterations is about 18 percent less than the number of iterations with the trisection method. Furthermore, only one function call is required during each golden section iteration (as opposed to two in the trisection method), making the golden section method more efficient. Once again, the golden section method is a zeroth-order method since the derivative of the function is not used.

The MATLAB implementation of the golden section method is an exercise for the reader.

5.5.3 Quadratic Interpolation

Another popular zeroth-order method is *quadratic interpolation* . In this method, the function is evaluated at the endpoints of $[a, b]$ and at its mid-point c.

Let the function values be f_a, f_b and f_c. Then a quadratic polynomial passing through these three points is determined, as illustrated in Figure 5.5a. The reader can easily prove that the quadratic function is given by

$$f(x) = A + B(x - a) + C(x - a)(x - b) \tag{5.18}$$

where

$$\begin{aligned} A &= f_a \\ B &= \frac{f_b - f_a}{b - a} \\ C &= \frac{(b - a)(f_c - f_a) - (f_b - f_a)(c - a)}{(c - a)(c - b)(b - a)} \end{aligned} \tag{5.19}$$

Next a new location d, where the derivative of the quadratic polynomial vanishes, is computed (Figure 5.5b). As the reader may verify, d is given by

$$d = \frac{1}{2}\left[\frac{f_a(b^2 - c^2) + f_b(c^2 - a^2) + f_c(a^2 - b^2)}{f_a(b - c) + f_b(c - a) + f_c(a - b)}\right] \tag{5.20}$$

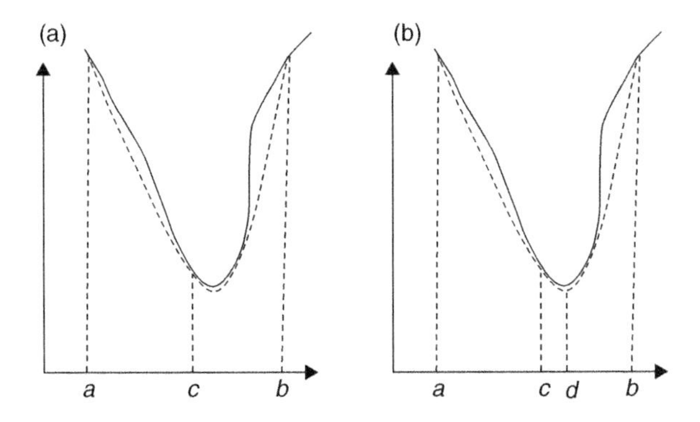

Figure 5.5 The quadratic interpolation illustrated.

The location d is used to reduce the interval size for the next iteration, as in the trisection method. Specifically, if $f(d) < f(c)$ then the interval $[a, c]$ is discarded; otherwise the interval $[d, b]$ is discarded. The process continues and terminates when the interval diminishes to within a given tolerance, as before.

This method is much more efficient than the trisection and golden section methods. However, numerical difficulties arise when the denominator in Equation (5.20) tends to zero. As we shall later see, MATLAB combines the robustness of the golden section method and the efficiency of quadratic interpolation through intelligent "switching."

The MATLAB implementation of quadratic interpolation is left as an exercise for the reader.

5.5.4 Bisection

In the *bisection* method, the gradient at the mid-point plays an important role in the selection of the appropriate sub-interval. Specifically, suppose $[a, b]$ is the unimodal interval, c is the mid-point and $\nabla f(c)$ is the gradient at the mid-point (see Figure 5.6). Then, (1) if $\nabla f(c) > 0$ then the local minimum must lie in the interval $[a, c]$; (2) if $\nabla f(c) < 0$ then the local minimum must lie in $[c, b]$; (3) if $\nabla f(c) = 0$ then the algorithm is terminated since c is the desired local minimum. Only one function call is made during each cycle, but the algorithm entails gradient computation. Thus, the bisection method is a first-order method.

The MATLAB implementation of bisection is left as an exercise for the reader.

5.5.5 Newton–Raphson

The *Newton–Raphson* method differs from the previous methods in the following ways:

1. It requires an initial guess x^0 rather than an interval.
2. The gradient and the second derivative of the function are required at each point.

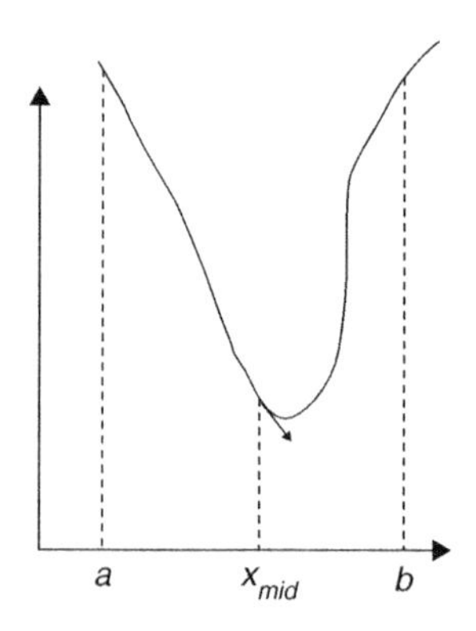

Figure 5.6 The bisection method illustrated.

3. Since the method seeks points where the gradient vanishes, it typically returns a stationary point *closest* to the initial guess x^0, i.e., not necessarily a minimum.
4. The method converges rapidly if the initial guess is close to a stationary point; if not, it may diverge.

To derive this method, recall that the gradient of the function must vanish at a stationary point. Let us suppose the gradient does not vanish at the initial guess x^0. The objective is to estimate the next point, where the gradient approaches zero. To find this point, consider the Taylor series of the gradient $\nabla f(x)$ about x^0:

$$\nabla f(x^0 + \Delta x) \approx \nabla f(x^0) + \Delta x^T H_f(x^0) \tag{5.21}$$

Since we want the gradient to vanish at $(x^0 + \Delta x)$, we have

$$\nabla f(x^0) + \Delta x^T H_f(x^0) = 0 \tag{5.22}$$

Therefore, the appropriate step size Δx is given by

$$\Delta x = -\left(H_f(x^0)\right)^{-1} \nabla f(x^0) \tag{5.23}$$

In other words, the next guess is

$$x^1 = x^0 - \left(H_f(x^0)\right)^{-1} \nabla f(x^0) \tag{5.24}$$

Because of the approximation in Equation (5.21), the gradient of $f(x)$ will not necessarily vanish at x^1. We must therefore repeat Equation (5.24) via

$$x^2 = x^1 - \left(H_f(x^1)\right)^{-1} \nabla f(x^1) \tag{5.25}$$

The process is repeated to generate points via the famous Newton–Raphson formula,

$$x^{i+1} = x^i - \left(H_f(x^i)\right)^{-1} \nabla f(x^i) \tag{5.26}$$

until the gradient is sufficiently small. Note that, for 1D functions, Equation (5.26) is essentially

$$x^{i+1} = x^i - \frac{\left.\frac{df}{dx}\right|_{x^i}}{\left.\frac{d^2f}{dx^2}\right|_{x^i}} \tag{5.27}$$

A MATLAB implementation of the Newton–Raphson method is provided below. It takes a function and an initial guess as inputs, and typically returns the "closest" stationary point (to within a fixed tolerance), the functional value at that point, etc. It is assumed that the function being passed as input not only returns the function value, but also its first and second derivatives. The algorithm terminates after a finite number of iterations or if the Hessian cannot be inverted; the termination criteria are controlled via the `opt` structure discussed earlier.

```
function [xStationary,fValue,flag,output] =
fStationaryNR(fun,x0,opt)
% Finds stationary values of a function via Newton Raphson
% Assumes fun(x) returns function value, gradient and
hessian
if (nargin == 2), opt = optimset('fminunc'); end % default
parameters
xStationary = x0(:);
iter = 1;
flag = 0;
output.message = '#FunEvals exceeded'; % default
[fValue,g,H] = fun(xStationary);
while(iter < opt.MaxFunEvals)
    if (cond(H) > 1e12)
        output.message = 'Hessian could not be inverted';
        flag = 0;
        break;
    end
    xNew = xStationary - H\g;
    err = norm(xStationary-xNew);
    if (err < opt.TolX)
        output.message = 'X Tol met';
        flag = 1;
        break;
    end
    xStationary = xNew;
    [fValueNew,g,H] = fun(xStationary);
     if (abs(fValueNew - fValue) < opt.TolFun)
        output.message = 'Func Tol met';
        flag = 1;
        break;
     end
    fValue = fValueNew;
    iter = iter + 1;
end
output.iterations = iter;
output.nFuncCalls = iter;
end
```

To illustrate, we consider the examples in Equation (5.2). As before, we have

```
>> test1D = testFunctions1D;
```

Then, to find the stationary points of $f_1(x) = (x - 1)^2$, we use `fStationaryNR.m` as follows:

```
>> [xStationary,fValue,exitflag,output] =
fStationaryNR(@(x)test1D.f1(x),0)
xStationary =
     1
fValue =
     0
exitflag =
     1
output =
    message: 'Newton-Raphson: X Tol met'
    iterations: 1
```

As a second example, to find the minimum of $f_7(x) = 1 - xe^{-x^2}$, with an initial guess of 0.2:

```
>> [xStationary,fValue,exitflag,output] = fStationaryNR
(@(x)test1D.f7(x),0.2)
xStationary =
    0.7071
fValue =
    0.5711
exitflag =
     1
output =
       message: 'Newton-Raphson: X Tol met
       iterations: 5
```

However, with an initial guess of 0.1, the method diverges:

```
>> [xStationary,fValue,exitflag,output] = fStationaryNR
(@(x)test1D.f7(x),.1)
xStationary =
  4.078710804125311
fValue =
  0.999999301497300
exitflag =
     1
output =
       message: 'Newton-Raphson: Func Tol met'
    iterations: 11
```

The Newton–Raphson method converges rapidly only when the initial guess is close to the minimum. On the other hand, it is not robust, i.e., it can diverge unpredictably. With an initial guess of −0.2, the method converges to a different stationary point (a local maximum):

```
>> [xStationary,fValue,exitflag,output] = fStationaryNR
(@(x)test1D.f7(x),-0.2)
xStationary =
  -0.707106780101211
fValue =
   1.428881941163530
exitflag =
     1
output =
         message: 'Newton-Raphson: Func Tol met'
      iterations: 4
```

Next, we modify the termination criteria via the `opt` structure. With a coarse tolerance, we have

```
>> opt = optimset('fminunc'); % default
>> opt = optimset(opt,'TolX',1e-3);% coarse
>> format long;
>> [xStationary,fValue,exitflag,output] = fStationaryNR
(@(x)test1D.f7(x),0.2,opt)
xStationary =
   0.707067600154114
fValue =
   0.571118058836470
exitflag =
     1
output =
         message: 'Newton-Raphson: X Tol met'
         iterations: 4
```

With a tight tolerance, we have

```
>> opt = optimset(opt,'TolX',1e-10);
>> [xStationary,fValue,exitflag,output] = fStationaryNR
(@(x)test1D.f7(x),0.2,opt)
xStationary =
   0.707106780101211
fMin =
   0.571118058836470
exitflag =
     1
output =
         message: 'Newton-Raphson: Func Tol met'
```

With a tight tolerance on the design variable, but default tolerance on the function value, the method exits when the function tolerance is met. With a tight tolerance on the design variable *and* function value, we have

```
>> opt = optimset(opt,'TolX',1e-10, 'TolFun',1e-10);
>> [xStationary,fValue,exitflag,output] = fStationaryNR
(@(x)test1D.f7(x),0.2,opt)
xStationary =
   0.707106781186548
fValue =
   0.571118057519647
exitflag =
     1
output =
       message: 'Newton-Raphson: Func Tol met'
    iterations: 5
```

It takes one more iteration, and, as the reader can verify, the solution is closer to the exact answer.

5.6 Multi-variable Minimization

Next, we focus on unconstrained *multi-variable* problems; typical examples include the following:

$$
\begin{aligned}
f_1(x,y) &= x^2 - xy + y^2 - x - 2y \\
f_2(x,y) &= (x-1)^4 + (y-1)^2 \\
f_3(x,y) &= x^2 + 4xy + y^2 \\
f_4(x,y) &= 100(y-x^2)^2 + (1-x)^2 \\
f_5(x,y) &= |x-1| + (y-1)^2 \\
f_6(\mathbf{x}) &= \sum_{i=1}^{N-1}\left[100(x_{i+1}-x_i^2)^2 + (1-x_i)^2\right] \\
&\dots
\end{aligned}
$$

$$
f_{10}(u,v) = \begin{bmatrix} \frac{1}{2}100\left(\sqrt{u^2+(v+1)^2}-1\right)^2 + \frac{1}{2}50\left(\sqrt{u^2+(v-1)^2}-1\right)^2 \\ -(10u+8v) \end{bmatrix} \tag{5.28}
$$

These functions are captured through the module `testFunctionsND.m`, a snippet of which is provided below. Observe that, for some of the functions, the gradient and Hessian are optionally returned.

```
classdef testFunctionsND
    % This class encapsulates numerous N-D test functions
    methods(Static)
        function [f,g,h] = f1(xVec)
            % returns  x^2-x*y+y^2-x-2*y and derivatives
            x = xVec(1);y = xVec(2);
            f = x.^2 - x.*y + y.^2 -x -2*y;
            if (nargout > 1), g = [2*x-y-1; -x+2*y-2];end
            if (nargout > 2), h = [2 -1; -1 2];end
        end
        ...
        function [f,g,h] = f6(x)
           % returns N-D Rosenbrock and its derivatives
            N = numel(x);f = 0;
            for i = 1:N-1, f = f + 100*(x(i+1)-x(i)^2)^2 +
(1-x(i))^2; end
            if (nargout > 1)
                g = zeros(N,1);
                for i = 1:N-1
                    g(i) = g(i) + 2*x(i)-400*x(i).*(-
x(i).^2+x(i+1))-2;
                    g(i+1) = g(i+1) + (-
200*x(i).^2+200*x(i+1));
                end
            end
            if (nargout > 2)
                h = zeros(N,N);
                for i = 1:N-1
                    h(i,i) = h(i,i) + 1200*x(i).^2-
400*x(i+1)+2;
                    h(i,i+1) = h(i,i+1) + (-400*x(i));
                    h(i+1,i) = h(i,i+1) + (-400*x(i));
                    h(i+1,i+1) = h(i+1,i+1) + 200;
                end
            end
        end
...
    end
end
```

Multi-variable optimization methods have the following characteristics:

1. The methods are classified as zeroth-order, first-order or second-order, depending on the maximum order of the derivative (of the objective function) required by the method. Thus, if a method only requires function values, but not its derivatives, it is zeroth-order.
2. Most multi-variable methods require an initial guess $\mathbf{x}^0$, rather than an interval.

3. All multi-variable methods attempt to compute increasingly better solutions $\mathbf{x}^k$ and terminate when *one* of the following conditions is satisfied:

$$
\begin{aligned}
&\|\mathbf{x}^k - \mathbf{x}^{k-1}\| \leq x_\varepsilon \\
&|f(\mathbf{x}^k) - f(\mathbf{x}^{k-1})| \leq f_\varepsilon \\
&\|\nabla f(\mathbf{x}^k)\| \leq g_\varepsilon
\end{aligned}
\tag{5.29}
$$

where x_ε, f_ε and g_ε are specified by the user. Tightening the tolerances typically increases the number of iterations.

Multi-variable methods are broadly classified as *trust-region* or *line-search* methods. In the remainder of this chapter, we will discuss line-search methods, which reduce a multi-variable problem to a series of single-variable problems. Trust-region methods are not covered in this text (for an introduction to these methods, see reference [8]).

5.7 Line-Search

The *line-search* concept is illustrated with a 2D example. Consider the function

$$f(x,y) = x^2 - xy + y^2 - x - 2y \tag{5.30}$$

The exact minimum for this function is $\mathbf{x}^* = (4/3,\ 5/3)$, as the reader can verify. Figure 5.7 illustrates the contour plot of the function. Here, the objective is to find the minimum *numerically*.

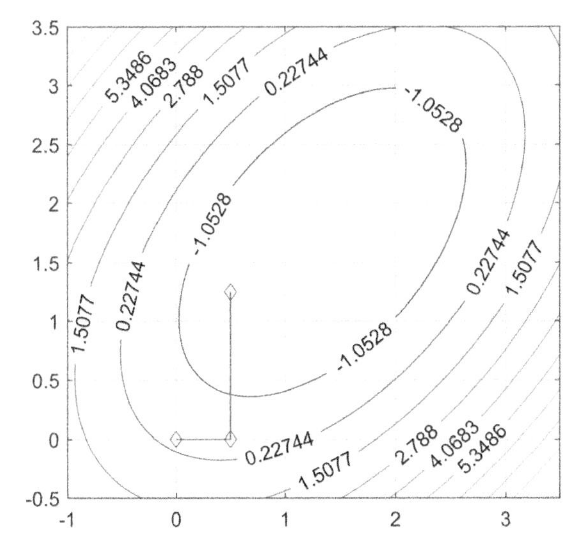

Figure 5.7 Line-search over contour plot of $f(x,y) = x^2 - xy + y^2 - x - 2y$.

Let us assume that the initial guess is $\mathbf{x}^0 = (0, 0)$; thus $f(\mathbf{x}^0) = 0$. In the line-search method, one must choose a direction of search. Later, we will discuss methods for computing optimal search directions. For now, let the first search direction be $\mathbf{d}^0 = (1, 0)$, i.e., we search for the next best estimate along the x-direction. All such points will be of the form

$$\mathbf{x}^1(\alpha) = (0, 0) + \alpha(1, 0) = (\alpha, 0) \tag{5.31}$$

Substituting $x = \alpha, y = 0$ in Equation (5.30) leads to the 1D function

$$f_{1D}(\alpha) = \alpha^2 - \alpha \tag{5.32}$$

Thus, by picking a search direction, we have reduced a 2D problem to a 1D problem. In general, an N-dimensional problem will also be reduced to a 1D problem, as demonstrated later on. The minimum of Equation (5.32), as the reader can verify, is $\alpha^* = 0.5$. In practice, the minimum for the 1D problem will be computed numerically.

Substituting this in Equation (5.31), we have $\mathbf{x}^1 = (0.5, 0)$. Further, observe that $f(\mathbf{x}^1) = -0.25 < f(\mathbf{x}^0)$, i.e., we have indeed reduced the function through 1D minimization. However, $\mathbf{x}^1$ is not the local minimum (check the gradient!). Next, we proceed to find the point $\mathbf{x}^2$ by searching along a different direction, say, $\mathbf{d}^1 = (0, 1)$, i.e.,

$$\mathbf{x}^2 = (0.5, 0) + \alpha(0, 1) = (0.5, \alpha) \tag{5.33}$$

Substituting $x = 1/2$, $y = \alpha$ in Equation (5.30) leads to the next 1D function,

$$f_{1D}(\alpha) = \alpha^2 - 5\alpha/2 - 1/4 \tag{5.34}$$

whose minimum is $\alpha^* = 5/4$, leading to $\mathbf{x}^2 = (0.5,\ 1.25)$, with $f(\mathbf{x}^2) = -1.8125 < f(\mathbf{x}^1)$. The function values continue to decrease.

The process is repeated, alternating between the directions $(1, 0)$ and $(0, 1)$ until the local minimum has been computed to within the desired tolerance. Figure 5.7 illustrates the path traced for the two steps. Table 5.1 summarizes the results for ten line-searches.

The line-search concept can be easily generalized to higher-dimensional problems. Suppose $\mathbf{x}^0$ is the initial guess for the local minimum; the objective is to find the next best estimate $\mathbf{x}^1$.

Suppose $\mathbf{d}^0$ is the search direction at $\mathbf{x}^0$, then, by definition, the next point $\mathbf{x}^1$ is of the form

$$\mathbf{x}^1(\alpha) = \mathbf{x}^0 + \alpha\mathbf{d}^0; \quad -\infty < \alpha < \infty \tag{5.35}$$

Substituting this in a multi-dimensional objective function leads to a 1D function:

$$f_{1D}(\alpha) = f(\mathbf{x}^0 + \alpha\mathbf{d}^0) \tag{5.36}$$

Table 5.1 Consecutive solutions in a line-search algorithm.

k	$\mathbf{x}^k$	$f(\mathbf{x}^k)$	$\mathbf{d}^k$
0	$(0, 0)$	0	$(1, 0)$
1	$(0.5, 0)$	-0.25	$(0, 1)$
2	$(0.5, 1.25)$	-1.8125	$(1, 0)$
3	$(1.125, 1.25)$	-2.2031	$(0, 1)$
4	$(1.125, 1.5625)$	-2.3008	$(1, 0)$
5	$(1.2812, 1.5625)$	-2.3252	$(0, 1)$
6	$(1.2812, 1.6406)$	-2.3313	$(1, 0)$
7	$(1.3203, 1.6406)$	-2.3328	$(0, 1)$
8	$(1.3203, 1.6602)$	-2.3332	$(1, 0)$
9	$(1.3301, 1.6602)$	-2.3333	...

where α is unknown. Once again, we have reduced an N-dimensional problem to a 1D problem. One can now rely on 1D numerical methods discussed in the previous chapter to find α. Often, it may not be computationally efficient to find the "exact" minimum of the 1D problem; an "approximate line-search" is typically sufficient ([8]). Once α has been determined, a new direction $\mathbf{d}^1$ is chosen at $\mathbf{x}^1$, and the next point $\mathbf{x}^2$ satisfies

$$\mathbf{x}^2(\alpha) = \mathbf{x}^1 + \alpha\mathbf{d}^1; \quad -\infty < \alpha < \infty \tag{5.37}$$

and so on. The process is repeated until the local minimum $\mathbf{x}^*$ has been computed to within the desired tolerance.

In the 2D example summarized in Table 5.1, the directions alternated between $(1, 0)$ and $(0, 1)$. In N dimensions, this may be generalized by selecting the N coordinate directions in sequence. In other words, the first direction is chosen to be $\mathbf{d}^0 = (1, 0, \ldots, 0)$, while the next direction is chosen to be $\mathbf{d}^1 = (0, 1, \ldots, 0)$, and so on until $\mathbf{d}^{N-1} = (0, 0, \ldots, 1)$. Once all N directions have been chosen, the sequence is restarted with $\mathbf{d}^N = (1, 0, \ldots, 0)$. This is referred to as the *coordinate cycling* method with line-search.

5.8 Search Directions

Unfortunately, the coordinate cycling method is inefficient, even in 2D. One can easily accelerate the search, provided good search directions are known, as the following example illustrates.

Example 5.1 Consider the function

$$f(x, y) = x^2 - xy + y^2 - x - 2y \tag{5.38}$$

Find the local minimum by searching along the two directions (1, 1) and (−1, 1).

Solution: As before, let the starting point be the origin, $\mathbf{x}^0 = (0, 0)$, leading to

$$\mathbf{x}^1(\alpha) = (0, 0) + \alpha(1, 1) = (\alpha, \alpha) \tag{5.39}$$

Substituting in Equation (5.38), we have

$$f_{1D}(\alpha) = \alpha^2 - 3\alpha \tag{5.40}$$

whose minimum is $\alpha^* = 3/2$, leading to $\mathbf{x}^1 = (3/2,\ 3/2)$. Next, we have

$$\mathbf{x}^2(\alpha) = (3/2,\ 3/2) + \alpha(-1,\ 1) = (3/2 - \alpha,\ 3/2 + \alpha) \tag{5.41}$$

From Equation (5.38) we have the 1D function

$$f_{1D}(\alpha) = 3\alpha^2 - \alpha - \frac{9}{4} \tag{5.42}$$

whose minimum is $\alpha^* = 1/6$, and thus $\mathbf{x}^2 = (4/3,\ 5/3)$. The reader can verify that $(4/3,\ 5/3)$ is the exact minimum! Figure 5.8 illustrates the contour plot and the path traced for the two steps.

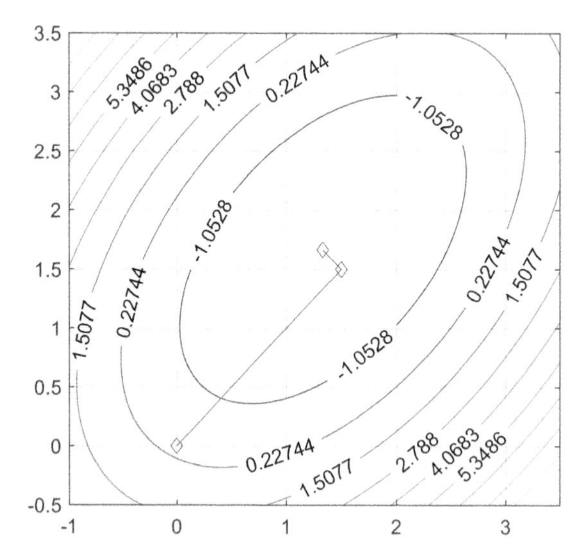

Figure 5.8 Optimal search directions over contour plot of $f(x, y) = x^2 - xy + y^2 - x - 2y$.

In this example, the two search directions $(1,\ 1)$ and $(-1,\ 1)$ were optimal. The challenge, of course, lies in generating such optimal search directions!

5.9 Eigendirections

Consider again the quadratic function

$$f(x,y) = x^2 - xy + y^2 - x - 2y \tag{5.43}$$

Observe that one can write this in the standard form as

$$f(x,y) = \frac{1}{2}\{x \quad y\}\begin{bmatrix} 2 & -1 \\ -1 & 2 \end{bmatrix}\begin{Bmatrix} x \\ y \end{Bmatrix} + \{x \quad y\}\begin{Bmatrix} -1 \\ -2 \end{Bmatrix} \tag{5.44}$$

Now consider the symmetric positive definite matrix $\mathbf{H}$:

$$\mathbf{H} = \begin{bmatrix} 2 & -1 \\ -1 & 2 \end{bmatrix} \tag{5.45}$$

Consider the eigenvectors of this matrix:

```
>> H = [2 -1; -1 2];
>> [V,D] = eig(H)
V =
   -0.7071   -0.7071
   -0.7071    0.7071
D =
     1     0
     0     3
```

Observe that the two eigenvectors of $\mathbf{H}$ are

$$\mathbf{v}_0 = \begin{Bmatrix} -1/\sqrt{2} \\ -1/\sqrt{2} \end{Bmatrix}; \ \mathbf{v}_1 = \begin{Bmatrix} -1/\sqrt{2} \\ 1/\sqrt{2} \end{Bmatrix} \tag{5.46}$$

If we set these vectors as the two search directions, the reader can confirm that the exact solutions can be found in exactly two line-searches. Indeed, the two eigenvectors are simple multiples of the directions $(1,\ 1)$ and $(-1,\ 1)$ discussed in Example 5.1. In other words, the eigenvectors of the $\mathbf{H}$ matrix are the ideal search directions, as captured via the following theorem.

Theorem 5.1 *Consider the quadratic function of order N*

$$f(\mathbf{x}) = \frac{1}{2}\mathbf{x}^T\mathbf{H}\mathbf{x} + \mathbf{x}^T\mathbf{g} + c \tag{5.47}$$

where $\mathbf{H}$ *is a symmetric positive definite matrix. Let* $\{\mathbf{v}_0, \mathbf{v}_1, \ldots, \mathbf{v}_{N-1}\}$ *be a set of eigenvectors of* $\mathbf{H}$*. Then an exact line-search along the directions* $\mathbf{d}^i = \mathbf{v}_i$ *will yield the exact minimum, starting from any point* $\mathbf{x}^0$.

Proof: See reference [8].

While the theorem is powerful, computing the eigenvectors of a matrix is more intensive than finding the minimum of the quadratic function! Thus, the theorem does not have a *direct* practical value, but it serves as a motivation for the next concept.

5.10 Conjugate Directions

Consider again the quadratic function of order N

$$f(\mathbf{x}) = \frac{1}{2}\mathbf{x}^T\mathbf{H}\mathbf{x} + \mathbf{x}^T\mathbf{g} + c \tag{5.48}$$

Let the eigenvectors of the symmetric positive definite matrix $\mathbf{H}$ be $\mathbf{v}_i;\ i = 0, 1, \ldots, N-1$, i.e.,

$$\mathbf{H}\mathbf{v}_i = \rho_i\mathbf{v}_i \tag{5.49}$$

Since the eigenvectors of a symmetric matrix are mutually orthonormal,

$$\mathbf{v}_j^T\mathbf{v}_i = 0;\ i \neq j \tag{5.50}$$

From Equations (5.49) and (5.50) we have

$$\mathbf{v}_j^T\mathbf{H}\mathbf{v}_i = 0;\ i \neq j \tag{5.51}$$

Equation (5.51) is called the *conjugate* property of an eigenvector with respect to the matrix $\mathbf{H}$. This turns out be a very important property in generating search directions, prompting the following definition.

Definition: Any two non-zero vectors d_0 and d_1 that satisfy

$$d_0^T\mathbf{H}d_1 = 0 \tag{5.52}$$

are referred to as $\mathbf{H}$-conjugate vectors.

Example 5.2 Let

$$\mathbf{H} = \begin{bmatrix} 2 & -1 \\ -1 & 2 \end{bmatrix} \tag{5.53}$$

Find a set of conjugate vectors of $\mathbf{H}$.

Solution: Let $d_0 = (1, 0)^T$. Observe that this is *not* an eigenvector of the $\mathbf{H}$ matrix. We can easily generate a vector d_1 that is $\mathbf{H}$-conjugate to d_0 as follows. Let $d_1 = \{a, b\}^T$. Since Equation (5.52) must be satisfied,

(cont.)

$$\{a \quad b\}\begin{bmatrix} 2 & -1 \\ -1 & 2 \end{bmatrix}\begin{Bmatrix} 1 \\ 0 \end{Bmatrix} = 0 \Rightarrow \{a \quad b\}\begin{Bmatrix} 2 \\ -1 \end{Bmatrix} = 0 \tag{5.54}$$

Thus, one may choose $d_1 = (1, 2)^T$ or any multiple of it. The reader can verify that, using these two vectors, one can reach the minimum in two steps, despite the fact that these search directions are *not* the eigenvectors of $\mathbf{H}$.

This result can be generalized via the following theorem.

Theorem 5.2 *Consider the quadratic function of order N*

$$f(\mathbf{x}) = \frac{1}{2}\mathbf{x}^T\mathbf{H}\mathbf{x} + \mathbf{x}^T\mathbf{g} + c$$

where $\mathbf{H}$ is a symmetric positive definite matrix. Let d_i; $i = 0, 1, 2, \ldots, N - 1$ be a set of mutually $\mathbf{H}$-conjugate vectors. Then an exact line-search along the directions d_i will yield the exact minimum, starting from any point $\mathbf{x}^0$.

Proof: See reference [8].

In order to apply Theorem 5.2, two questions remain to be answered:

1. For *quadratic* functions of order N, how does one find conjugate directions d_i; $i = 0, 1, 2, \ldots, N - 1$ in a computationally efficient manner?
2. How does one generalize Theorem 5.2 to *non-quadratic* functions?

We answer these two questions in the remainder of this chapter. Keep in mind that the primary objective is to generate an optimal set of directions to accelerate line-search.

5.11 Powell's Method

Powell's method is a popular zeroth-order algorithm that generates "optimal" search directions "on-the-fly," for arbitrary N-dimensional functions (see reference [16] for other zeroth-order – also referred to as *derivative-free* – methods). The method involves multiple *passes*, where each pass consists of N line-searches.

In the first pass, the search directions are chosen to be the N coordinate directions (as in coordinate cycling), i.e.,

$$\mathbf{D} = [\mathbf{e}^1, \mathbf{e}^2, \ldots, \mathbf{e}^N] \tag{5.55}$$

where e^i are unit vectors, i.e., $\mathbf{D}$ is the identity matrix. After the first pass, i.e., after N line-searches, let the converged point be $\mathbf{x}^1$. Powell suggested that $\mathbf{x}^1 - \mathbf{x}^0$ can serve as one of the search directions in the next pass. The reasoning is that, after one pass, the overall displacement gives a measure of the direction along which the true minimum lies. Specifically, the direction proposed by Powell is

$$\mathbf{d}' = \mathbf{x}^1 - \mathbf{x}^0 \tag{5.56}$$

He then recommended that one carry out a line-search along $\mathbf{d}'$, resulting in a point $\mathbf{x}'$. Then the direction set is updated as follows for the next pass:

$$\mathbf{D} = [\mathbf{e}^2, \mathbf{e}^3, \ldots, \mathbf{e}^N, \mathbf{d}'] \tag{5.57}$$

As one can observe in Equation (5.57), the first direction has been discarded and all other directions have been shifted forward to make room for a new direction. A second pass is carried out along the above N directions, starting at $\mathbf{x}'$, resulting in a new point $\mathbf{x}^2$. The process is repeated until convergence is achieved.

It can be shown that, in the limit, for a quadratic function, the N vectors in $\mathbf{D}$ form a set of optimal directions, also referred to as *conjugate directions.* Further, since non-quadratic functions are well approximated by quadratic functions as one approaches the stationary point, Powell's method can be used for arbitrary functions.

The algorithm is summarized below.

Powell's Algorithm

1. Let the initial guess be $\mathbf{x}^p$; $p = 0$ (here p denotes pass).
2. Initialize $\mathbf{D} = [\mathbf{d}^0, \mathbf{d}^1, \mathbf{d}^2, \ldots, \mathbf{d}^{N-1}] = [\mathbf{e}^1, \mathbf{e}^2, \mathbf{e}^3, \ldots, \mathbf{e}^N]$, where N is the dimension of the problem.
3. Carry out N line-searches starting at $\mathbf{x}^p$ along the directions in $\mathbf{D}$.
4. Let the converged point be $\mathbf{x}'$.
5. Compute $\mathbf{d}' = \mathbf{x}' - \mathbf{x}^p$.
6. Perform a line-search from $\mathbf{x}'$ along $\mathbf{d}'$; let the converged point be $\mathbf{x}^{p+1}$.
7. Update $\mathbf{D}$ as follows: $\mathbf{d}^k = \mathbf{d}^{k+1},\ k = 0, \ldots, N-2$ and $\mathbf{d}^{N-1} = \mathbf{d}'$.
8. Increment p; return to step 3 until convergence is achieved.

Example 5.3 Illustrate Powell's method for the quadratic function in Equation (5.38).

Solution: Since Powell's method in the first pass is exactly the coordinate cycling method, the two directions are

$$\mathbf{D} = \begin{bmatrix} 1 & 0 \\ 0 & 1 \end{bmatrix} \tag{5.58}$$

After searching along the two directions, we find the point to be

$$\mathbf{x}' = \begin{Bmatrix} 0.5 \\ 1.25 \end{Bmatrix} \tag{5.59}$$

Thus, the new direction is

$$\mathbf{d}' = \mathbf{x}' - \mathbf{x}^0 = \begin{Bmatrix} 0.5 \\ 1.25 \end{Bmatrix} \tag{5.60}$$

Performing a line-search, along $\mathbf{d}'$ starting at $\mathbf{x}'$, results in the point

(cont.)

$$\mathbf{x}^1 = \begin{Bmatrix} 0.6316 \\ 1.5789 \end{Bmatrix} \tag{5.61}$$

Thus, in the second pass, we start at $\mathbf{x}^1$, with the two directions

$$\mathbf{D} = \begin{bmatrix} 0 & 0.5 \\ 1 & 1.25 \end{bmatrix} \tag{5.62}$$

We will leave it as an exercise to the reader to determine the next two points.

Powell's method is a zeroth-order method in that it only requires function evaluations. A MATLAB implementation of Powell's method is provided below.

```
function [pMin,fMin,nFunEvals,flag] = fminPowell(f,p0,opt)
% Powell's method using line-search
if (nargin == 2), opt = optimset('fminunc'); end % default
parameters
opt = optimset(opt,'display','off'); % suppress annoying
messages
p0 = p0(:); % force into column vector
flag = 0;pass = 1;nFunEvals = 0;
N = length(p0); % dimension of the problem
pCurrent = p0;
D = eye(N,N); %% we need to keep track of the direction
maxFunEvals = 100*N;
while (1) % for each pass
    pStartOfPass = pCurrent;
    fStartOfPass = feval(f,pStartOfPass);
    nFunEvals = nFunEvals+1;
    for ls = 1:N % N line search for each pass
        d = D(:,ls); % the direction of the current pass
        [pNew,fNew,flag,funcCount] =
lineSearch(f,pCurrent,d,opt);
        pCurrent = pNew;
    end
    pEndOfPass = pCurrent;
    fEndOfPass = feval(f,pEndOfPass);
    nFunEvals = nFunEvals+funcCount;
    pass = pass+1;
    if (norm(pEndOfPass -pStartOfPass)<opt.TolX) || ...
       (abs(fEndOfPass-fStartOfPass) < opt.TolFun) || ...
       (nFunEvals > maxFunEvals)
        break; % terminate
    end
    % update directions
    D(:,1:end-1) = D(:,2:end);
    D(:,end) = (pEndOfPass-pStartOfPass);
```

```
    % One line search along new direction
    [pCurrent,fNew,flag,funcCount] =
lineSearch(f,pCurrent,D(:,end),opt);
    nFunEvals = nFunEvals+funcCount;
    if (rank(D) < N),D = eye(N,N); end%reset directions
end
pMin = pNew;
fMin = fNew;
%%%%%%%%%%%%%%%%%%%%%%%%%%%%%%%%%%%%%%%%%%%%%%%%%%%%%%%%%%%%
function [pMin,fMin,flag,funcCount] = lineSearch(f,p,d,opt)
% Line search of a N-dimensional function
d = d/norm(d); % ensure proper scaling
[alphaMin,fMin,flag,output] =
fminunc(@(alpha)f1D(f,p,d,alpha),0,opt);
pMin = p+alphaMin*d;
funcCount = output.funcCount;
%%%%%%%%%%%%%%%%%%%%%%%%%%%%%%%%%%%%%%%%%%%%%%%%%%%%%%%%%%%%
function [fValue] = f1D(f,p,d,alpha)
fValue = feval(f,p+alpha*d);% Evaluate f at p+alpha*d
```

A few observations are:

1. MATLAB's `fminunc` method is used for repeated 1D line-searches; this will be discussed in Chapter 6.
2. Given a point and a direction, 1D functions are constructed "on-the-fly."
3. After many passes, the **D** matrix in Powell's method may lose "rank," i.e., one or more directions may coincide; in this case, the matrix is reinitialized.

5.12 Gradient Methods

Powell's method is easy to implement, and much more efficient than coordinate cycling. It can be particularly useful when the function is very noisy. However, it can still be slow to converge. Better convergence can be expected with first-order methods where gradient information is used for acceleration. Such first-order methods rely on the following key observation: *the gradient of a function at any point captures the direction along which the function most rapidly increases or decreases at that point.*

5.12.1 Steepest-Descent Method

The *steepest-descent* method is based on the intuitive notion that the direction of maximum function change is along the gradient, i.e.,

$$\mathbf{d}^i = \nabla f(\mathbf{x}^i) \tag{5.63}$$

In other words, at $\mathbf{x}^0$, we search along

$$\mathbf{d}^0 = \nabla f(\mathbf{x}^0) \tag{5.64}$$

Once the line-search converges to $\mathbf{x}^1$, we search along

$$\mathbf{d}^1 = \nabla f(\mathbf{x}^1) \tag{5.65}$$

and so on. Note the following:

1. The sign of $\mathbf{d}^i$ is not critical since, during a line-search, we consider both positive and negative values of α in Equation (5.35).
2. In numerical implementations, the directions are usually normalized to unit length.

Example 5.4 Illustrate steepest-descent method for the quadratic function in Equation (5.38).

Solution: Note that the gradient of the function is given by

$$\nabla f = \begin{Bmatrix} 2x - y - 1 \\ -x + 2y - 2 \end{Bmatrix} \tag{5.66}$$

Let the starting point be the origin, i.e., $\mathbf{x}^0 = (0,0)$. Thus, the initial direction in the steepest-descent method is

$$\mathbf{d}^0 = \nabla f(0,0) = \begin{Bmatrix} -1 \\ -2 \end{Bmatrix} \tag{5.67}$$

Therefore

$$\mathbf{x}^1(\alpha) = (0,0) - \alpha(1,2) = -(\alpha, 2\alpha) \tag{5.68}$$

Substituting $x = -\alpha, y = -2\alpha$ in Equation (5.30), we have

$$f_{1D}(\alpha) = 3\alpha^2 + 5\alpha \tag{5.69}$$

As the reader can verify, $\alpha^* = -5/6$. Substituting this in Equation (5.31), we have $\mathbf{x}^1 = (5/6,\ 10/6)$. Observe that $f(\mathbf{x}^1) = -2.083 < f(\mathbf{x}^0)$, but $\mathbf{x}^1 \neq \mathbf{x}^*$. Next, we proceed to find the point $\mathbf{x}^2$ by searching along a different direction $\mathbf{d}^1$, where

$$\mathbf{d}^1 = \nabla f(\mathbf{x}^1) = \begin{Bmatrix} -1 \\ 1/2 \end{Bmatrix} \tag{5.70}$$

Thus,

$$\mathbf{x}^2 = (5/6,\ 10/6) + \alpha(-1,\ 1/2) = (5/6 - \alpha,\ 10/6 + \alpha/2) \tag{5.71}$$

(cont.)

Substituting in Equation (5.30), we have

$$f_{1D}(\alpha) = 7\alpha^2/4 + 5\alpha/4 - 25/12 \tag{5.72}$$

whose minimum is $\alpha^* = -5/14$, leading to $\mathbf{x}^2 = (1.19,\ 1.49)$, with $f(\mathbf{x}^2) = -2.30 < f(\mathbf{x}^1)$. Indeed, the function values continue to decrease (as the reader can verify for the next step by choosing $\mathbf{d}^2 = (1, 0)$). However, unlike the set of eigendirections or conjugate directions, we have not converged in two steps (see Figure 5.9).

Table 5.2 summarizes the results for six line-searches using steepest descent.

Table 5.2 Consecutive solutions in a steepest-descent algorithm.

k	$\mathbf{x}^k$	$f(\mathbf{x}^k)$	$\mathbf{d}^k$
0	$(0, 0)$	0	$(-0.45, -0.89)$
1	$(0.833, 1.67)$	-2.08	$(-0.89, 0.45)$
2	$(1.19, 1.49)$	-2.31	$(-0.45, -0.89)$
3	$(1.28, 1.67)$	-2.33	$(-0.89, 0.45)$
4	$(1.32, 1.65)$	-2.33	$(-0.45, -0.89)$
5	$(1.33, 1.67)$	-2.33	$(-0.89, 0.45)$

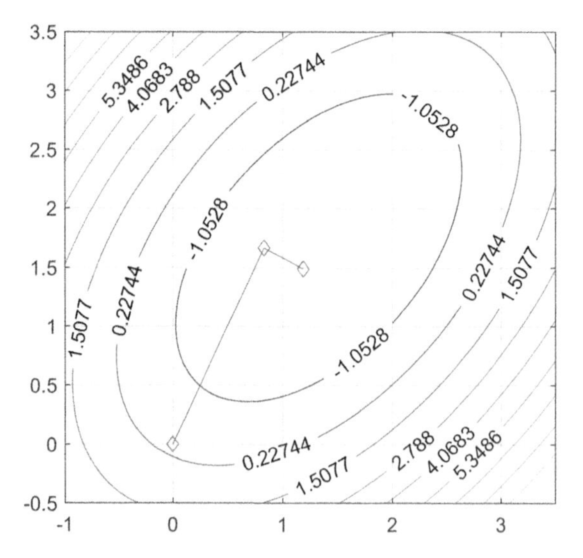

Figure 5.9 Steepest descent over contour plot of $f(x, y) = x^2 - xy + y^2 - x - 2y$.

Despite its intuitive appeal, steepest descent is *not* very efficient, since the gradients are not necessarily conjugate directions (if the directions were indeed conjugate, the search in the previous example should have terminated in two steps). The steepest-descent method is therefore rarely used in practice. Moreover, observe that the steepest-descent method is similar to coordinate cycling in that every direction is perpendicular to the previous one.

5.12.2 Linear Conjugate Gradient Method

The conjugate gradient method to be discussed next is perhaps the most popular first-order method today ([17]). In this method, the direction of search at the first step is along the steepest descent. However, after the first step, the search direction is modified to be explicitly conjugate with respect to all previous directions of search.

Specifically, consider the quadratic function

$$f(\mathbf{x}) = \frac{1}{2}\mathbf{x}^T\mathbf{H}\mathbf{x} + \mathbf{x}^T\mathbf{g} + c \tag{5.73}$$

At the first step, we start at $\mathbf{x}^0$ and search along

$$\mathbf{d}^0 = \nabla f(\mathbf{x}^0) = \mathbf{H}\mathbf{x}^0 + \mathbf{g} \tag{5.74}$$

After a line-search, let the next point be $\mathbf{x}^1$. Now let the search direction be a linear combination of the current gradient and the previous search direction, i.e.,

$$\mathbf{d}^1 = \nabla f(\mathbf{x}^1) + \beta\mathbf{d}^0 \tag{5.75}$$

The constant β is determined by enforcing the conjugacy condition with respect to $\mathbf{d}^0$, i.e.,

$$(\mathbf{d}^0)^T\mathbf{H}\mathbf{d}^1 = 0 \tag{5.76}$$

i.e.,

$$(\mathbf{d}^0)^T\mathbf{H}(\nabla f^1 + \beta\mathbf{d}^0) = 0 \tag{5.77}$$

leading to

$$\beta = -\frac{(\mathbf{d}^0)^T\mathbf{H}\nabla f(\mathbf{x}^1)}{(\mathbf{d}^0)^T\mathbf{H}\mathbf{d}^0} \tag{5.78}$$

Finally, the new direction is given by

$$\mathbf{d}^1 = \nabla f(\mathbf{x}^1) - \left(\frac{(\mathbf{d}^0)^T\mathbf{H}\nabla f(\mathbf{x}^1)}{(\mathbf{d}^0)^T\mathbf{H}\mathbf{d}^0}\right)\mathbf{d}^0 \tag{5.79}$$

The search direction is typically normalized for numerical reasons. Performing a line-search, we determine $\mathbf{x}^2$. For the next direction, we let

$$\mathbf{d}^2 = \nabla f(\mathbf{x}^2) + \beta\mathbf{d}^1 \tag{5.80}$$

The constant β is now determined by enforcing conjugacy with respect to $\mathbf{d}^1$:

$$(\mathbf{d}^1)^T\mathbf{H}\mathbf{d}^2 = 0 \tag{5.81}$$

leading to

$$\mathbf{d}^2 = \nabla f(\mathbf{x}^2) - \left(\frac{(\mathbf{d}^1)^T\mathbf{H}\nabla f(\mathbf{x}^2)}{(\mathbf{d}^1)^T\mathbf{H}\mathbf{d}^1}\right)\mathbf{d}^1 \tag{5.82}$$

One can show that $\mathbf{d}^2$ is also conjugate with respect to $\mathbf{d}^0$. Thus, it is sufficient that conjugacy with respect to the previous direction is established.

In conclusion, the conjugate directions are generated via a simple iterative process:

$$\begin{aligned}
&\mathbf{d}^0 = \nabla f(\mathbf{x}^0) \\
&\left.\begin{aligned}
&\mathbf{x}^i : \text{ linesearch}(f, \mathbf{x}^{i-1}, \mathbf{d}^{i-1}) \\
&\beta = -\left(\frac{(\mathbf{d}^{i-1})^T\mathbf{H}\nabla f(\mathbf{x}^i)}{(\mathbf{d}^{i-1})^T\mathbf{H}\mathbf{d}^{i-1}}\right) \\
&\mathbf{d}^i = \nabla f(\mathbf{x}^i) + \beta\mathbf{d}^{i-1}
\end{aligned}\right\} i \geq 1
\end{aligned} \tag{5.83}$$

Example 5.5 Illustrate the linear conjugate gradient method for the quadratic function in Equation (5.38).

Solution: As in the case of the steepest-descent method, the initial search direction is given by

$$\mathbf{d}^0 = \nabla f(\mathbf{x}^0) = \begin{Bmatrix} -1 \\ -2 \end{Bmatrix} \tag{5.84}$$

and $\mathbf{x}^1 = (5/6, 10/6)$. Next, we proceed to find the point $\mathbf{x}^2$ by searching along the direction $\mathbf{d}^1$, where

$$\mathbf{d}^1 = \nabla f(\mathbf{x}^1) - \left(\frac{(\mathbf{d}^0)^T\mathbf{H}\nabla f(\mathbf{x}^1)}{(\mathbf{d}^0)^T\mathbf{H}\mathbf{d}^0}\right)\mathbf{d}^0 \tag{5.85}$$

i.e.,

$$\begin{aligned}
\beta &= -\left(\frac{\{-1 \quad -2\}\begin{bmatrix} 2 & -1 \\ -1 & 2 \end{bmatrix}\begin{Bmatrix} -1 \\ 0.5 \end{Bmatrix}}{\{-1 \quad -2\}\begin{bmatrix} 2 & -1 \\ -1 & 2 \end{bmatrix}\begin{Bmatrix} -1 \\ -2 \end{Bmatrix}}\right) = 0.25 \\
\mathbf{d}^1 &= \begin{Bmatrix} -1 \\ 0.5 \end{Bmatrix} + 0.25\begin{Bmatrix} -1 \\ -2 \end{Bmatrix} = \begin{Bmatrix} -5/4 \\ 0 \end{Bmatrix}
\end{aligned} \tag{5.86}$$

Thus,

$$\mathbf{x}^2 = (5/6 - 5\alpha/4, 10/6) \tag{5.87}$$

Substituting we have

$$f_{1D}(\alpha) = 25\alpha^2/16 + 5\alpha/4 - 25/12 \tag{5.88}$$

whose minimum is $\alpha^* = -2/5$, leading to $\mathbf{x}^2 = (8/6, 10/6) = \mathbf{x}^*$. Indeed, we have converged to the minimum in two steps; this is illustrated in Figure 5.10.

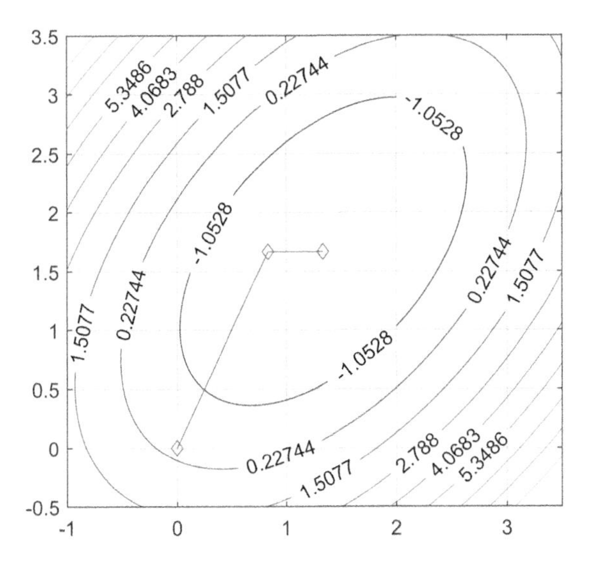

Figure 5.10 Conjugate gradient method over contour plot of $f(x,y) = x^2 - xy + y^2 - x - 2y$.

5.12.3 Non-linear Conjugate Gradient Method

For non-quadratic functions, observe that Equation (5.83) does not apply. However, it can be shown (see [2], [17]) that β in Equation (5.83) can also be expressed as

$$\beta = \frac{\left(\nabla f(\mathbf{x}^i)\right)^T \nabla f(\mathbf{x}^i)}{\left(\nabla f(\mathbf{x}^{i-1})\right)^T \nabla f(\mathbf{x}^{i-1})} \tag{5.89}$$

For Example 5.5,

$$\beta = \frac{\left(\nabla f(\mathbf{x}^1)\right)^T \nabla f(\mathbf{x}^1)}{\left(\nabla f(\mathbf{x}^0)\right)^T \nabla f(\mathbf{x}^0)} = \frac{\{-1 \quad 0.5\}\begin{Bmatrix}-1\\0.5\end{Bmatrix}}{\{-1 \quad -2\}\begin{Bmatrix}-1\\-2\end{Bmatrix}} = 0.25 \tag{5.90}$$

Thus, Equation (5.83) can be expressed in an equivalent form that makes no explicit reference to the Hessian matrix $\mathbf{H}$:

$$\begin{aligned}
&\mathbf{d}^0 = \nabla f(\mathbf{x}^0) \\
&\left.\begin{aligned}
&\mathbf{x}^i : \ \text{linesearch}(f, \mathbf{x}^{i-1}, \mathbf{d}^{i-1}) \\
&\beta = \frac{\left(\nabla f(\mathbf{x}^i)\right)^T \nabla f(\mathbf{x}^i)}{\left(\nabla f(\mathbf{x}^{i-1})\right)^T \nabla f(\mathbf{x}^{i-1})} \\
&\mathbf{d}^i = \nabla f(\mathbf{x}^i) + \beta \mathbf{d}^{i-1}
\end{aligned}\right\} \ i \geq 1
\end{aligned} \tag{5.91}$$

Equation (5.91) is now valid for non-linear functions as well. The Fletcher–Reeves conjugate gradient algorithm is based on Equation (5.91).

Fletcher–Reeves Conjugate Gradient Algorithm

1. At the initial guess $\mathbf{x}^i; i = 0$, find the gradient.
2. Then perform a line-search along the steepest-descent direction to find $\mathbf{x}^i; i = 1$.
3. Repeat the following steps until convergence is reached:
   a. Compute the new direction $\mathbf{d}^i$ using Equation (5.91).
   b. Perform a line-search to find $\mathbf{x}^{i+1}$.

Note that, unlike quadratic problems where exact termination is expected after N line-searches, the Fletcher–Reeves algorithm will not terminate after N line-searches (in general). Explicit termination criteria must be provided for non-linear conjugate gradient via user-specified tolerances.

5.13 Newton–Raphson Method

If the second derivative of the objective function is readily available, then one can appeal to highly efficient second-order methods such as the Newton–Raphson method, which is essentially a non-linear solver.

Recall that non-linear solvers find the stationary point (rather than the minimum) by numerically solving

$$\nabla f = 0 \tag{5.92}$$

The same method now applies to higher dimensions. Let

$$\mathbf{g} \equiv \nabla f \tag{5.93}$$

Observe that $\mathbf{g}$ is a vector. The objective is to find stationary points $\mathbf{x}^*$ such that

$$\mathbf{g}(\mathbf{x}^*) = \mathbf{0} \tag{5.94}$$

Suppose we are at a point $\mathbf{x}^i$ such that

$$\mathbf{g}(\mathbf{x}^i) \neq \mathbf{0} \tag{5.95}$$

One must pick the step size $\Delta\mathbf{x}$ such that

$$\mathbf{g}(\mathbf{x}^i + \Delta\mathbf{x}) \approx \mathbf{0} \tag{5.96}$$

Exploiting Taylor series,

$$\begin{aligned} &\mathbf{g}(\mathbf{x}^i + \Delta\mathbf{x}) \approx \mathbf{g}(\mathbf{x}^i) + \nabla\mathbf{g}(\mathbf{x}^i)\Delta\mathbf{x} \approx 0 \\ &\Rightarrow \quad \Delta\mathbf{x} = -[\nabla\mathbf{g}(\mathbf{x}^i)]^{-1}\mathbf{g}(\mathbf{x}^i) \end{aligned} \tag{5.97}$$

The matrix $[\nabla\mathbf{g}(\mathbf{x}^i)]$ is nothing but the Hessian, i.e.,

$$\Delta\mathbf{x} = -[\mathbf{H}(\mathbf{x}^i)]^{-1}\mathbf{g}(\mathbf{x}^i) \tag{5.98}$$

Thus, the next point in the iteration is

$$\mathbf{x}^{i+1} = \mathbf{x}^i - [\mathbf{H}(\mathbf{x}^i)]^{-1}\nabla f(\mathbf{x}^i) \tag{5.99}$$

The MATLAB implementation `fStationaryNR.m` discussed in Section 5.5.5 can handle multiple variables as well. The following example illustrates finding the local minimum of the 2D Rosenbrock function using [100, 100] as the starting point:

```
>> testND = testFunctionsND;
>> [xMin,fMin,exitflag,output] =
fstationaryNR(@(x)testND.f4(x),[100 100])
xMin =
    0.999999999999999
    0.999999997551214
fMin =
     2.448784047392388e-09
exitflag =
     1
output =
          message: 'Newton-Raphson: Func Tol met'
      iterations: 3
```

The reader is encouraged to try different starting points, termination criteria and other test functions.

5.14 Equilibrium of Spring Systems

Thus far, we have considered "analytically constructed" unconstrained problems. We will now study the equilibrium of elastic spring systems such as the one illustrated in Figure 5.11. We will show that this naturally leads to an unconstrained optimization problem. In Figure 5.11, nodes 3, 4, 5 and 6 are fixed (marked by a filled circle), while nodes 1 and 2 are free to move. An external force is applied to node 1, as shown. The task is to determine the displacements of nodes 1 and 2.

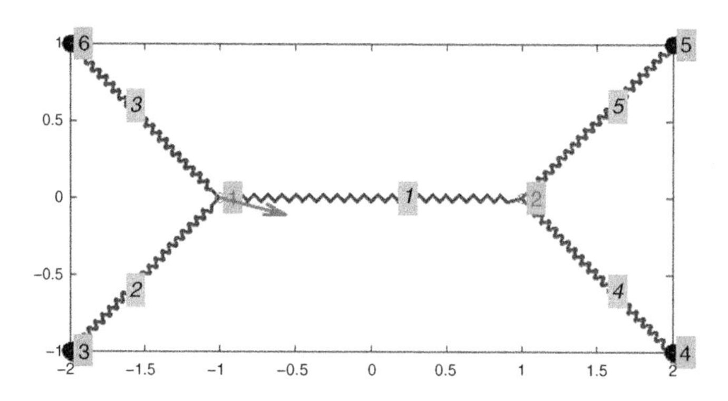

Figure 5.11 Undeformed five-spring system.

We explore two strategies for finding the spring displacement, namely the classic *force balance strategy* and the *minimization of potential energy strategy*.

We will first consider the *force balance principle*, which states that *when an external force is applied to an elastic body, the displacements are such that the internal and external forces are in balance in the deformed configuration*. We will illustrate this strategy using a simple two-spring system.

Example 5.6 Consider the spring system in Figure 5.12, where the stiffness of springs 1 and 2 are 100 and 50 units, respectively, and a force of (10, 8) is applied to node 1. Using force balance, find the equations governing the displacement of node 1.

Solution: Note that node 1 can move in the x- and y-directions; the displacement will be denoted by (u, v). Observe that the undeformed length of each spring is 1 unit. Further, given an arbitrary displacement (u, v) of node 1, the deformed lengths are (see Figure 5.13)

$$\begin{aligned} L_{12} &= \sqrt{(0-u)^2 + (-1-v)^2} = \sqrt{u^2 + (1+v)^2} \\ L_{13} &= \sqrt{(0-u)^2 + (1-v)^2} = \sqrt{u^2 + (1-v)^2} \end{aligned} \tag{5.100}$$

Therefore, the increase in length is given by

$$\begin{aligned} \Delta L_{12} &= \sqrt{u^2 + (1+v)^2} - 1 \\ \Delta L_{13} &= \sqrt{u^2 + (1-v)^2} - 1 \end{aligned} \tag{5.101}$$

Because of this change in length, node 1 experiences two spring forces (see Figure 5.13): $\mathbf{F}_{12}$ and $\mathbf{F}_{13}$. The magnitude of $\mathbf{F}_{12}$ is $k_{12}\Delta L_{12}$, where $k_{12} = 100$ is the stiffness of the spring connecting nodes 1 and 2. The direction of the force is along the unit vector:

(cont.)

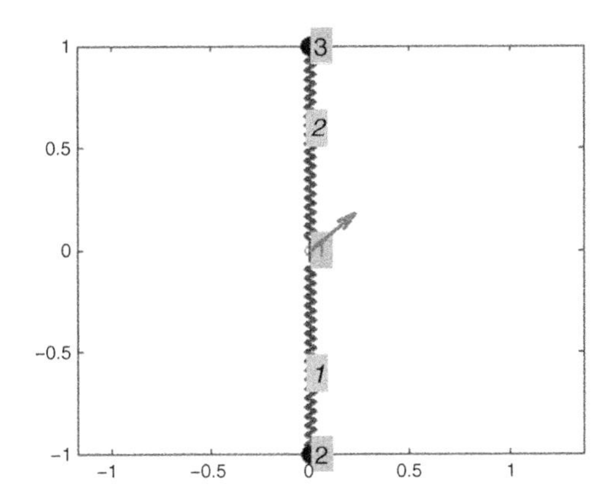

Figure 5.12 Undeformed spring system, where node 1 is free to move.

$$\hat{\mathbf{F}}_{12} = \frac{1}{L_{12}} \begin{Bmatrix} 0 - u \\ -1 - v \end{Bmatrix} \tag{5.102}$$

Thus, the force on node 1 due to the spring joining node 2 is

$$\mathbf{F}_{12} = k_{12} \Delta L_{12} \hat{\mathbf{F}}_{12} = 100 \frac{L_{12} - 1}{L_{12}} \begin{Bmatrix} -u \\ -1 - v \end{Bmatrix} \tag{5.103}$$

Similarly, the force on node 1 due to the spring joining node 3 is

$$\mathbf{F}_{13} = k_{13} \Delta L_{13} \hat{\mathbf{F}}_{13} \tag{5.104}$$

where $k_{13} = 50$ and the direction is given by

$$\hat{\mathbf{F}}_{13} = \frac{1}{L_{13}} \begin{Bmatrix} 0 - u \\ 1 - v \end{Bmatrix} \tag{5.105}$$

Therefore

$$\mathbf{F}_{13} = 50 \frac{L_{13} - 1}{L_{13}} \begin{Bmatrix} -u \\ 1 - v \end{Bmatrix} \tag{5.106}$$

Carrying out a force balance on node 1 (see Figure 5.13),

$$\mathbf{F}_{12} + \mathbf{F}_{13} + \begin{Bmatrix} 10 \\ 8 \end{Bmatrix} = \begin{Bmatrix} 0 \\ 0 \end{Bmatrix} \tag{5.107}$$

The reader can verify that this leads to

$$-100 \frac{L_{12} - 1}{L_{12}} \begin{Bmatrix} u \\ 1 + v \end{Bmatrix} - 50 \frac{L_{13} - 1}{L_{13}} \begin{Bmatrix} u \\ v - 1 \end{Bmatrix} + \begin{Bmatrix} 10 \\ 8 \end{Bmatrix} = \begin{Bmatrix} 0 \\ 0 \end{Bmatrix} \tag{5.108}$$

(cont.)

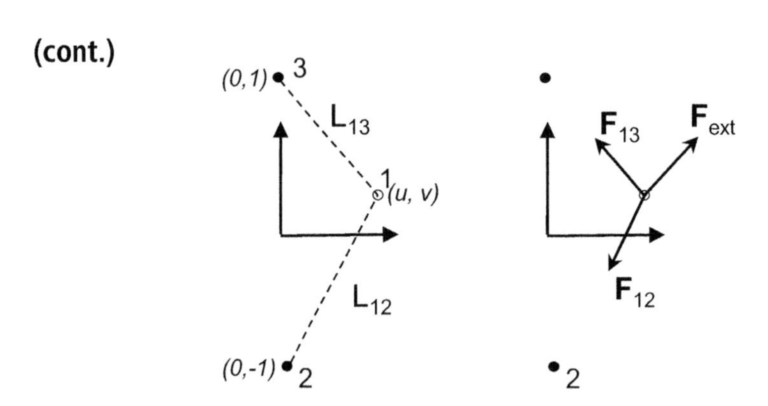

Figure 5.13 Deformed lengths and force balance on node 1.

i.e.,

$$
\begin{aligned}
&100\frac{L_{12}-1}{L_{12}}u + 50\frac{L_{13}-1}{L_{13}}u - 10 = 0 \\
&100\frac{L_{12}-1}{L_{12}}(1+v) + 50\frac{L_{13}-1}{L_{13}}(v-1) - 8 = 0
\end{aligned}
\tag{5.109}
$$

Note that L_{12} and L_{13} are both functions of u and v (see Equation (5.100)), i.e., Equations (5.109) are non-linear.

To solve for u and v, we create a `twoSpring.m` class containing a method called `forceBalance` which captures Equation (5.109):

```
function [residual] = forceBalance(~,uv)
    fx1 = 10;
    fy1 = 8;
    k1 = 100; % stiffness of spring 1
    k2 = 50;% stiffness of spring 2
    u = uv(1);v = uv(2);
    L12 = sqrt(u^2+(1+v)^2); % deformed lengths
    L13 = sqrt(u^2+(1-v)^2);
    residual(1) = k1*(L12-1)/L12*u + k2*(L13-1)/L13*u - fx1;
    residual(2) = k1*(L12-1)/L12*(1+v) + k2*(L13-1)/L13*(v-
1) - fy1;
end
```

We can then find the equilibrium location (u, v) by using MATLAB's `fsolve` (discussed in Chapter 3):

```
>> s = twoSpring();
>> [soln,PEMin,Flag,Output]= fsolve(@(x)s.forceBalance(x),
[0 0]);
>> disp(soln)
0.539505127726566   0.016644715574197
>> s.plotDeformed()
```

(cont.)

The resulting solution is displayed in Figure 5.14.

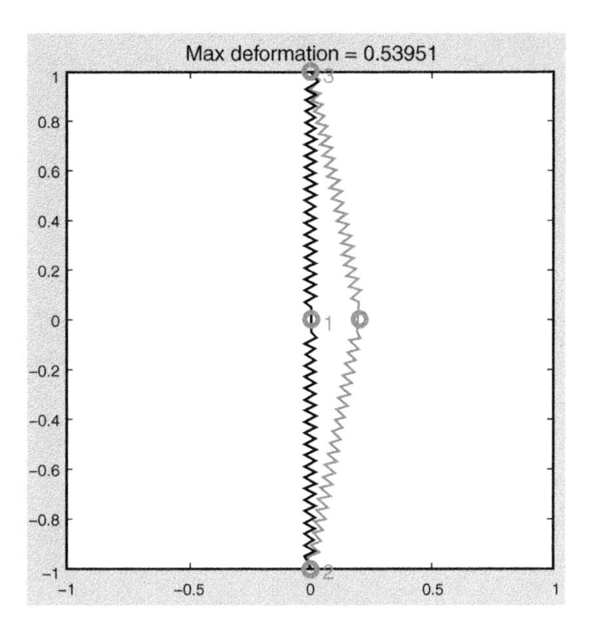

Figure 5.14 Deformed spring.

Next, we consider the potential energy strategy, which states that, at stable equilibrium, the displacements are such that the potential energy takes a minimum. For any elastic body the potential energy is defined as

$$\Pi(\mathbf{d}) = U(\mathbf{d}) - 2W(\mathbf{d}) \tag{5.110}$$

where U is the *internal elastic energy* and W is the *work done* by the external forces, both of which depend on the deformation $\mathbf{d} = \{u, v\}$. Thus, one can find the stable equilibrium points of elastic bodies by solving the optimization problem

$$\underset{\mathbf{d}}{\text{minimize}}\, \Pi(\mathbf{d}) \tag{5.111}$$

We will again consider the two-spring system to illustrate this principle.

Example 5.7 Consider again the spring system in Figure 5.15, where the stiffness values of springs 1 and 2 are 100 and 50 units, respectively, and a force of (10, 8) is applied to node 1. Find the displacement of node 1 by minimizing the potential energy.

(cont.)

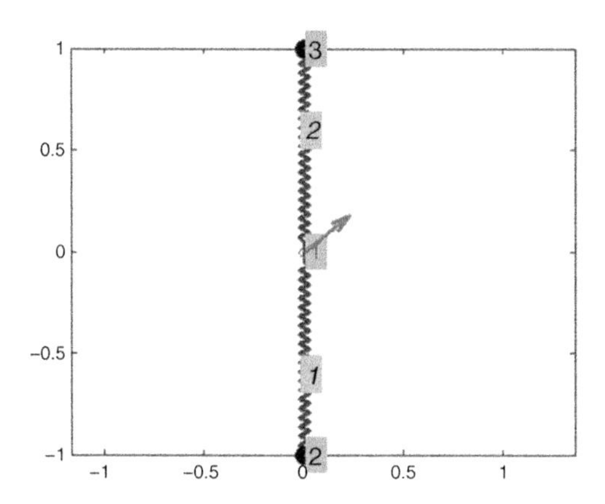

Figure 5.15 An undeformed spring system, where node 1 is free to move.

Solution: Observe that the values of elastic energy in the two springs are

$$\begin{aligned} U_1 &= 0.5k_1(\Delta L_{12})^2 = 0.5k_1(L_{12}-1)^2 \\ U_2 &= 0.5k_2(\Delta L_{13})^2 = 0.5k_2(L_{13}-1)^2 \end{aligned} \tag{5.112}$$

Given the spring stiffness, the total elastic energy is

$$U = 0.5[100(L_{12}-1)^2 + 50(L_{13}-1)^2] \tag{5.113}$$

The quasi-static external work done is

$$W = (1/2)(f_x u + f_y v) = (1/2)(10u + 8v) \tag{5.114}$$

Therefore, the potential energy is given by

$$\Pi(\mathbf{d}) = U(\mathbf{d}) - 2W(\mathbf{d}) \tag{5.115}$$

i.e.,

$$\Pi = 0.5[100(L_{12}-1)^2 + 50(L_{13}-1)^2] - (10u + 8v) \tag{5.116}$$

Substituting Equation (5.100), we arrive at

$$\Pi = 0.5\begin{bmatrix} 100\left(\sqrt{u^2+(1+v)^2}-1\right)^2 \\ +50\left(\sqrt{u^2+(1-v)^2}-1\right)^2 \end{bmatrix} - (10u + 8v) \tag{5.117}$$

Thus, we have the minimization problem

$$\underset{\{u,v\}}{\text{minimize}}\ \Pi \tag{5.118}$$

We now add a method to capture the potential energy within `twoSpring.m`.

```
function [PE] = potentialEnergy(obj,uv) %#ok<INUSL>
    % returns the potential energy of a 2-spring system
    fx1 = 10;
    fy1 = 8;
    k1 = 100; % stiffness of spring 1
    k2 = 50;% stiffness of spring 2
    u = uv(1);v = uv(2);
    L12 = sqrt(u^2+(1+v)^2); % deformed lengths
    L13 = sqrt(u^2+(1-v)^2);
      deltaL12 = (L12 -1);
      deltaL13 = (L13 -1);
      U = 0.5*k1*deltaL12.^2 + 0.5*k2*deltaL13.^2; %elastic
      W = 0.5*(fx1*u+fy1*v); % Work done
      PE =  U-2*W; % potential energy
end
```

Then we can find the equilibrium location (u, v) by using MATLAB's `fminunc` (to be discussed in Chapter 6):

```
>> s = twoSpring();
>> [soln,PEMin,Flag,Output]= fminunc(@(x)s.potentialEnergy(x), [0
0]);
>> disp(soln)
0.539505198774633    0.016644757067306
```

As one can observe, we have converged to the same solution.

Clearly, the two methods are equivalent, since they yield the same solution. Indeed, *the force balance equations can be obtained by setting the gradient of the potential energy function to zero*:

$$\nabla \Pi|_{\mathbf{d}} = 0 \tag{5.119}$$

Specifically, observe that

$$\frac{\partial \Pi}{\partial u} = 0.5 \begin{bmatrix} 2 * 100\left(\sqrt{u^2 + (1+v)^2} - 1\right) \dfrac{2u}{2\sqrt{u^2 + (1+v)^2}} \\ +2 * 50\left(\sqrt{u^2 + (1-v)^2} - 1\right) \dfrac{2u}{2\sqrt{u^2 + (1-v)^2}} \end{bmatrix} - (10) = 0 \tag{5.120}$$

and

$$\frac{\partial \Pi}{\partial v} = 0.5 \begin{bmatrix} 2 * 100\left(\sqrt{u^2 + (1+v)^2} - 1\right) \dfrac{2(1+v)}{2\sqrt{u^2 + (1+v)^2}} \\ +2 * 50\left(\sqrt{u^2 + (1-v)^2} - 1\right) \dfrac{2(1-v)(-1)}{2\sqrt{u^2 + (1-v)^2}} \end{bmatrix} - (8) = 0 \tag{5.121}$$

i.e.,

$$\begin{aligned}\frac{\partial \Pi}{\partial u} &= 100\frac{(L_{12}-1)u}{L_{12}} + 50\frac{(L_{13}-1)u}{L_{13}} - 10 = 0 \\ \frac{\partial \Pi}{\partial v} &= 100\frac{(L_{12}-1)(1+v)}{L_{12}} + 50\frac{(L_{13}-1)(v-1)}{L_{13}} - 8 = 0\end{aligned} \tag{5.122}$$

Equations (5.122) are identical to the force balance equations (5.109). We can capture this duality as follows: *The gradient of the potential energy of an elastic system yields the force balance equations*. Thus, the link between optimization and physical laws is deep-rooted!

5.15 Summary

Table 5.3 summarizes various algorithms covered in this chapter for solving the unconstrained minimization problem

$$\underset{\mathbf{x}}{\text{minimize}}\ f(\mathbf{x}) \tag{5.123}$$

Table 5.3 Typical algorithms to solve an unconstrained multi-variable minimization problem.

	Name	Description	Performance		Comments
			Quadratic	General	
0th order	Coordinate cycling	Perform line-search by cycling among unit vectors e_i	Converges in N line-searches if **H** is diagonal; otherwise can be very slow	Converges in N line-searches if Hessian is diagonal; otherwise can be very slow	
	Powell	Perform line-search by cycling among unit vectors e_i, then search along newly generated directions	Converges in N line-searches if H is diagonal; otherwise quadratic convergence	Converges in N line-searches if **H** is diagonal; otherwise quadratic convergence	Recommended among zeroth-order methods
1st order	Steepest descent	Perform line-search along gradient ∇f at each point	Converges in one line-search if **H** has identical eigenvalues; otherwise poor performance	Typically poor performance	
	Fletcher–Reeves Conjugate Gradient	Perform line-search along conjugate directions	Converges in N line-searches	Typically excellent performance	One of the best among first-order methods
2nd order	Newton–Raphson	Generate next point using Hessian and gradient	Converges in one step	Converges to stationary point	Expensive

and the unconstrained quadratic problem

$$\underset{\mathbf{x}}{\text{minimize}} \quad \frac{1}{2}\mathbf{x}^T\mathbf{H}\mathbf{x} + \mathbf{x}^T\mathbf{g} + c \tag{5.124}$$

In the next chapter, we will discuss the implementation of these ideas within MATLAB's optimization toolbox.

5.16 EXERCISES

Exercise 5.1 If $[a, b]$ is a unimodal interval for a continuous function $f(x)$, prove that if $a < c < d < b$ and $f(c) < f(d)$, then the minimum must lie in the interval $[a, d]$. Use the definition of a unimodal function.

Exercise 5.2 Suppose $[a, b]$ is a unimodal interval for a function $f(x)$, and c is the mid-point. Prove that if $\nabla f(c) > 0$ then the local minimum must lie in the interval $[a, c]$.

Exercise 5.3 Test the trisection algorithm on the following test cases, using the default termination criterion $x_\varepsilon = f_\varepsilon = 10^{-6}$.

(a) $1 - xe^{-x^2}$ in the range [0, 1]
(b) $1 - xe^{-x^2}$ in the range [−5, 1]
(c) $\sin x + \sin(10x/3)$ in the range [4.5, 6]
(d) $\sin x + \sin(10x/3)$ in the range [0, 6].

Use MATLAB's `uitable` to create a table with the following columns: `xMin`, `fMin`, gradient, Hessian, number of function calls and reason for termination. The rows should correspond to each of the cases. For each case explain the results, with figures if necessary.

Exercise 5.4 Write a MATLAB function for the golden section algorithm. Your code should respect the tolerances requested. The format of your function should be

```
function [xMin,fmin,flag,output] =
goldenSection(fun,a,b,opt)
```

Test your code, and create a table, following the instructions in Exercise 5.3. For each case explain the results.

Exercise 5.5 Test the golden section algorithm on the following test cases, with default termination criteria in the interval $[0, 5]$:

(a) $\cos(x)$
(b) $10^{-6}\cos(x)$
(c) $10^{6}\cos(x)$.

Create a table following the instructions in Exercise 5.3. For each case explain the results.

Exercise 5.6 Write a MATLAB function for quadratic interpolation. Your code should, for example, respect the tolerances requested. Further, you should detect "convergence failures" and exit gracefully with the flag set to −1. The format of your function should be

```
function [xMin,fmin,flag,output ] = quadraticInterpolation
(fun,a,b,opt)
```

Test your code, and create a table, following the instructions in Exercise 5.3. For each case explain the results.

Exercise 5.7 Write a MATLAB function for the bisection algorithm. You should detect "convergence failures" and exit gracefully with the flag set to −1. The format of your function should be

```
function [xMin,fmin,flag,output] = bisection(fun,a,b,opt)
```

Test your code, and create a table, following the instructions in Exercise 5.3. For each case explain the results.

Exercise 5.8 Use the Newton–Raphson method to find stationary points of the function

$$f_8(x) = 3\,\sin x - x + 0.1x^2 + 0.1\,\cos(2x) \tag{5.125}$$

with different initial starting points. Tabulate your results and observations.

Exercise 5.9 *Swann's algorithm* is a method for seeking a unimodal interval $[a,b]$ of a 1D function, starting from an initial guess x^0; it proceeds as follows:

1. The first step is to evaluate $f(x^0)$ and $f(x^0+\Delta x)$. If $f(x^0+\Delta x) < f(x^0)$, then the unimodal interval (if it exists) must lie to the right of x^0; otherwise it must lie to the left.
2. The next step is to advance the chosen direction. For example, suppose the interval lies to the right of x^0, then we continue to evaluate the function at $x^0+2\Delta x$, $x^0+4\Delta x, \ldots$ For any three consecutive points x^1, x^2, x^3: (1) if $f(x^2) < f(x^1)$ and $f(x^2) < f(x^3)$, then $[x^1, x^3]$ is a potential unimodal interval; (2) if $f(x^2) \geq f(x^1)$ and $f(x^2) \geq f(x^3)$, then a non-unimodal behavior has been detected.

Write a MATLAB implementation of Swann's algorithm. Your code should, for example, respect the tolerances requested by the user through the `opt` structure (created via `optimset`). The format of your function should be

```
function [a,b] = swanns(fun,x0,opt)
```

Test your code using test functions discussed in this chapter.

Exercise 5.10 For the function $f(x,y) = 3x^2 - xy + 4y^2 - x - 2y + 10$, find the local minimum analytically and the corresponding minimum value of the function. Then carry out coordinate cycling using three analytical line-searches starting from the origin, and tabulate the results. You can use the line-search provided to numerically verify your answers.

Exercise 5.11 For the function $f(x,y) = 3x^2 + 4y^2 - x - 2y + 10$, find the local minimum analytically and the corresponding value of the function. Then carry out coordinate cycling. Carry out the line-search analytically, and tabulate the results. How many line-searches do you need in order to find the local minimum? Contrast this with Exercise 5.10.

Exercise 5.12 For the function $f(x,y) = 3x^2 - xy + 4y^2 - x - 2y + 10$, find the two corresponding eigenvectors. Starting from the origin, confirm that two line-searches along the two eigenvectors are sufficient to find the minimum. Carry out the line-search analytically and tabulate the results.

Exercise 5.13 For the function $f(x,y) = 3x^2 - xy + 4y^2 - x - 2y + 10$, suppose the initial direction is $d = [1\ 0.5]$. Find the next conjugate direction. Starting at the origin, confirm that two line-searches using these two conjugate directions are sufficient to find the minimum. Carry out the line-search analytically and tabulate the results. You can use the line-search provided to verify your answers.

Exercise 5.14 For $f(x,y,z) = x^2 - xy + y^2 + 5z^2 + xz - 3yz - x - 2y + 4z$, suppose the initial direction is $\mathbf{d} = (1,\ 0,\ 0)$. Find the next two conjugate directions. Confirm that three line-searches using these three conjugate directions are sufficient. Carry out the line-search analytically and tabulate the results.

Exercise 5.15 Write a MATLAB function that implements the steepest-descent algorithm. Your code should respect the tolerances requested by the user through the `opt` structure (created via `optimset`). The format of your function should be

```
function [xMin,fmin,flag,output] = steepestDescent
(fun,x0,opt)
```

The output structure should contain: (a) a message on why the code terminated, (b) the number of iterations and (c) the total number of function calls (make sure to account for all line-search function calls). You can use the line-search provided as part of your code.

Use MATLAB's `uitable` to create a table with the following columns: `fMin`, norm of the gradient, number of function calls and reason for termination. The rows should correspond to each of the following cases:

(a) the simple function $f(x,y) = 3x^2 - xy + 4y^2 - x - 2y + 10$ with default tolerances
(b) the 2D Rosenbrock function with default tolerances, starting at the origin
(c) the 2D Rosenbrock function with default tolerances, but allow for 10 000 function calls, starting at the origin
(d) the two-spring potential energy function with default tolerances, starting at the origin.

For each case, explain the results. You can use the analytical results in Example 5.4 to verify your code.

Exercise 5.16 Write a MATLAB function that implements the Fletcher–Reeves version of the (non-linear) conjugate gradient. Your code should, for example, respect the tolerances requested by the user through the `opt` structure (created via `optimset`). You can use the line-search provided as part of your code. The format of your function should be

```
function [xMin,fmin,flag,output ] = conjugateGradient
(fun,x0,opt)
```

Follow the instructions in Exercise 5.15 to report the results. You can use the analytical results in Example 5.5 to verify your implementation.

Exercise 5.17 The *Broyden–Fletcher–Goldfarb–Shanno (BFGS) method* is another popular first-order optimization method. It is based on a secant approximation to the Hessian and is referred to as a quasi-Newton method. The method proceeds as follows:

(a) Initialize $k = 0$ and $\mathbf{x}^0$ (initial guess). Let $\mathbf{M}^0$ be the identity matrix; $\mathbf{M}^k$ is the inverse of the approximate Hessian.
(b) Obtain a direction of descent $\mathbf{d}^k$ via

$$\mathbf{d}^k = -\mathbf{M}^k \nabla f(\mathbf{x}^k) \tag{5.126}$$

(also see Equation (5.129)).
(c) Perform a line-search at $\mathbf{x}^k$ along $\mathbf{d}^k$ to find α^k, and let

$$\mathbf{x}^{k+1} = \mathbf{x}^k + \alpha^k \mathbf{d}^k \tag{5.127}$$

(d) If the norm $\|\mathbf{x}^{k+1} - \mathbf{x}^k\|$ is small or if $\|\nabla f(\mathbf{x}^{k+1})\|$ is small, exit.
(e) Otherwise, compute

$$\begin{aligned} \mathbf{s}^k &= \alpha^k \mathbf{d}^k \\ \mathbf{y}^k &= \nabla f(\mathbf{x}^{k+1}) - \nabla f(\mathbf{x}^k) \end{aligned} \tag{5.128}$$

(f) Update $\mathbf{M}^k$ via the *Sherman–Morrison–Woodbury formula*:

$$\mathbf{M}^{k+1} = \mathbf{M}^k + \left\{ \begin{array}{l} \dfrac{[(\mathbf{s}^k)^T\mathbf{y}^k + (\mathbf{y}^k)^T\mathbf{M}^k\mathbf{y}^k]\left(\mathbf{s}^k(\mathbf{s}^k)^T\right)}{\left((\mathbf{s}^k)^T\mathbf{y}^k\right)^2} \\ -\dfrac{\mathbf{M}^k\mathbf{y}^k(\mathbf{s}^k)^T + \mathbf{s}^k(\mathbf{y}^k)^T\mathbf{M}^k}{(\mathbf{s}^k)^T\mathbf{y}^k} \end{array} \right\} \tag{5.129}$$

(g) Increment k and return to step 2.

Write a MATLAB function that implements BFGS. The format of your function should be

```
function [xMin,fmin,flag,output ] = BFGS(fun,x0,opt);
```

Follow the instructions in Exercise 5.15 to report the results.

6 MATLAB Optimization Toolbox

Highlights

1. In Chapter 5, we studied a few fundamental ideas underlying numerical optimization. In this chapter, we will study implementations of these ideas in MATLAB's optimization toolbox.
2. In particular, we consider methods available to solve unconstrained problems. These methods include `fminbnd`, `fzero`, `fminsearch`, `fminunc` and `fsolve`. The merits and limitations of these methods are highlighted through examples.
3. Through this study, we will get a deeper understanding of numerical methods.
4. Finally, we will briefly explore global optimization methods that are important when an objective function exhibits multiple local minima, or is noisy. We will illustrate the use of MATLAB global optimization methods such as `MultiStart`, genetic algorithms and simulated annealing.

6.1 Overview

MATLAB offers a variety of methods for solving optimization problems. These methods are part of MATLAB's optimization toolbox; for an overview, type `help optim` in MATLAB's command window.

```
>> help optim
  Optimization Toolbox
  Version 7.4 (R2016a) 10-Feb-2016
  Nonlinear minimization of functions.
    fminbnd   - Scalar bounded nonlinear function
minimization.
    ...
```

In this chapter, we will focus on methods within this toolbox to solve unconstrained optimization problems of the form

$$\underset{\mathbf{x}}{\text{minimize}}\ f(\mathbf{x}) \tag{6.1}$$

These methods are summarized in Table 6.1; also see reference [2].

We will apply these methods on the test functions introduced in Chapter 5.

Table 6.1 MATLAB methods for solving unconstrained minimization problems.

Name	Description	Comments
fminbnd	Solve the scalar bounded minimization problem $\underset{x}{\text{minimize}}\ f(x)$ $a \le x \le b$ via a combination of golden section and quadratic fitting.	Since fminbnd is a zeroth-order method, it is relatively robust, and is particularly useful when the objective function is non-differentiable. Observe that a search interval is required; fminbnd can be used explicitly in 1D problems, or implicitly as a line-search technique in higher dimensions.
fzero	Solve the scalar non-linear problem $g(x) = 0$.	An initial guess is required for fzero. If $g(x) = df/dx$, the stationary points of $f(x)$ are computed.
fminsearch	Solve the multi-dimensional unconstrained problem $\underset{\mathbf{x}}{\text{minimize}}\ f(\mathbf{x})$ via zeroth-order Nelder–Mead.	An initial guess is required for fminsearch. This function can be slow but is relatively robust.
fminunc	Solve the multi-dimensional unconstrained problem. $\underset{\mathbf{x}}{\text{minimize}}\ f(\mathbf{x})$	An initial guess is required here. This is the preferred method of optimization for unconstrained problems. If analytical gradients are not provided, finite difference is used.
fsolve	Solve the multi-dimensional non-linear problem $g(\mathbf{x}) = 0$.	An initial guess is required for fsolve. If $g(\mathbf{x}) = \nabla f(\mathbf{x})$, the stationary points of $f(\mathbf{x})$ are computed.

6.2 The fminbnd Function

One of the simplest optimization functions in MATLAB is the zeroth-order method fminbnd. It solves the scalar bounded minimization problem

$$\underset{x}{\text{minimize}}\ f(x) \qquad a \le x \le b \tag{6.2}$$

via a combination of golden section and quadratic fitting. It requires the user to provide an interval to search within. We will start with a trivial example to illustrate fminbnd.

Example 6.1 Using MATLAB's fminbnd, solve

$$\underset{x}{\text{minimize}}\ (x-1)^2 \tag{6.3}$$

within the interval [0 2].

(cont.)

Solution: Note that this function is implemented as $f_1(x)$ in `testFunctions1D.m`. Therefore, we create an instance of the test function class:

```
>> test1D = testFunctions1D;
```

Then we use `fminbnd` as follows, with an interval of [0, 2] to search within:

```
>> [xMin,fMin,exitflag,output] =
fminbnd(@(x)test1D.f1(x),0,2)
xMin =
     1
fMin =
     0
exitflag =
     1
output =
    iterations: 5
     funcCount: 6
     algorithm: 'golden section search, parabolic
interpolation'
       message: [1x112 char]
```

Thus, the minimum is found in just six function calls. Further, increasing the interval to, say, [−1000, 1000] does not increase the number of function calls (why?).

We will now consider a continuous function, whose derivative is not defined at the minimum (since `fminbnd` is a zeroth-order method, the function need not be differentiable).

Example 6.2 Using MATLAB's `fminbnd`, solve

$$\underset{x}{\text{minimize}} \ |x - 1| \tag{6.4}$$

within (a) the interval [0 2], and (b) the interval (−100, 100).

Solution: Note that this function is implemented as $f_4(x)$ in `testFunctions1D.m`. We have

```
>> [xMin,fMin,exitflag,output] =
fminbnd(@(x)test1D.f4(x),0,2)
xMin =
     1
fMin =
     0
exitflag =
     1
output =
    iterations: 11
     funcCount: 12
```

(cont.)

```
      algorithm: 'golden section search, parabolic
interpolation'
        message: 'Optimization terminated:…'
```

Now observe the impact of increasing the interval to [−100, 100] on the number of function calls.

```
>> [xMin,fMin,exitflag,output= fminbnd(@(x)test1D.f4(x),-
100,100)
xMin =
   0.999985865063441
fMin =
     1.413493655888498e-05
exitflag =
     1
output =
    iterations: 19
     funcCount: 20
```

Thus, a tighter interval for search is preferable for non-differentiable functions.

Next, we will explore the role of tolerances.

Example 6.3 Using MATLAB's fminbnd, solve

$$\underset{x}{\text{minimize}} \; 1 - xe^{-x^2} \tag{6.5}$$

within the interval [0 2], with (a) default tolerances, and (b) tighter tolerances. This function is illustrated in Figure 6.1.

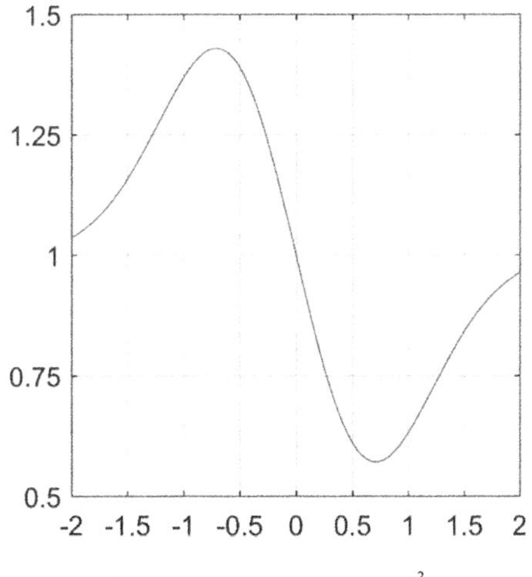

Figure 6.1 Plot of $1 - xe^{-x^2}$.

Solution: Note that this function is implemented as $f_7(x)$ in `testFunctions1D.m`, with an exact solution of $1/\sqrt{2}$. With default tolerances, and an interval [0, 2], we have

```
>> [xMin,fMin,exitflag,output] =
fminbnd(@(x)test1D.f7(x),0,2)
xMin =
   0.707113291040050
fMin =
   0.571118057555997
exitflag =
     1
output =
     iterations: 8
      funcCount: 9
      algorithm: 'golden section search, parabolic
interpolation'
        message: [1x112 char]
```

Let us now alter the tolerances using the `optimset` command. First, we create an `opt` variable as follows. The variable `tolfun` corresponds to f_ε and `tolx` corresponds to x_ε, discussed in Chapter 5.

```
>> opt = optimset('tolfun',1e-12,'tolx',1e-12);
```

The `opt` structure is then passed as an argument, as below. By tightening the tolerances, we observe that the accuracy of the solution improves, and the number of function calls increases.

```
>>>> [xMin,fMin,exitflag,output] =
fminbnd(@(x)test1D.f7(x),0,2,opt)
xMin =
   0.707106787581570
fMin =
   0.571118057519647
exitflag =
     1
output =
     iterations: 10
      funcCount: 11
      algorithm: 'golden section search, parabolic
interpolation'
        message: [1x112 char]
```

The solution error has dropped to 6.5×10^{-9} (and not 1×10^{-12}; why?). The reader is encouraged to experiment with other test functions and optimization parameters.

Finally, we will consider a function with numerous local minima, illustrated in Chapter 5.

Example 6.4 Using MATLAB's `fminbnd`, solve

$$\underset{x}{\text{minimize}} \;\; (x-1)^2 + 0.025 \sin^2(1000(x-1)/\pi) \tag{6.6}$$

within (a) the interval [0, 2], and (b) the interval (0.55, 2).

Solution: Note that the above function is implemented as $f_9(x)$ in `testFunctions1D.m`. With a symmetric interval around the exact minimum, the method finds the true minimum:

```
>> test1D = testFunctions1D;
>> [xMin,fMin,exitflag,output] = fminbnd(@(x)test1D.f9(x),0,2)
xMin =
   1.0000
fMin =
   1.9990e-27
exitflag =
     1
output =
    struct with fields:
    iterations: 8
     funcCount: 9
     algorithm: 'golden section search, parabolic interpolation'
       message: 'Optimization terminated:…
```

On the other hand, with an asymmetric interval, the method is unable to find the true minimum.

```
>> [xMin,fMin,exitflag,output] = fminbnd(@(x)test1D.f9(x),0.55,2)
xMin =
    0.8915
fMin =
    0.0118
exitflag =
     1
output =
   struct with fields:
    iterations: 14
     funcCount: 15
     algorithm: 'golden section search, parabolic interpolation'
       message: 'Optimization terminated:
```

In conclusion, `fminbnd` is a good choice if the function is unimodal in the given interval.

6.3 The `fzero` Function

Another popular method for scalar problems is `fzero`, which is used to solve the scalar non-linear problem

$$g(x) = 0$$

In optimization settings, the function $g(x)$ is typically the derivative of some objective function, i.e.,

$$g(x) = \frac{df}{dx}$$

Thus, `fzero` is often used to find the stationary points of a function.

Example 6.5 Using MATLAB's `fzero`, solve

$$\underset{x}{\text{minimize}} \ 1 - xe^{-x^2} \tag{6.7}$$

with (a) an initial guess of 1.0, and (b) an initial guess of -1.

Solution: The gradient of this function is $g(x) = -e^{-x^2} + 2x^2e^{-x^2}$. This is illustrated in Figure 6.2; the gradient is zero at $x = \pm 1/\sqrt{2}$.

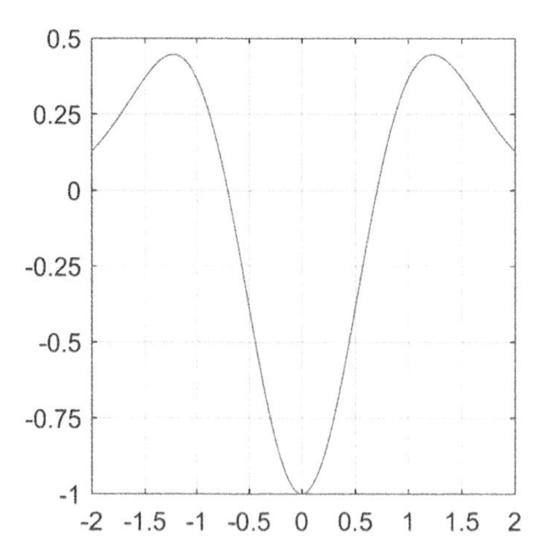

Figure 6.2 Plot of $g(x) = -e^{-x^2} + 2x^2e^{-x^2}$.

The gradient function has been coded as $g_7(x)$ in testFunctions1D.m. For example:

```
>> test1D.g7(1.0)
0.3679
```

We now use the method fzero, with an initial guess of 1.0, as follows:

```
>> [xStar,gVal,exitflag,output] =
fzero(@(x)test1D.g7(x),1.0)
xStar =
   0.707106781186548
gVal =
     1.110223024625157e-16
exitflag =
     1
output =
    intervaliterations: 8
            iterations: 6
             funcCount: 22
            algorithm: 'bisection, interpolation'
            message: 'Zero found in the interval [0.68,
1.22627]'
```

Observe that the stationary point (which also happens to be a local minimum) was found in 22 function calls. Now try providing −1 as the initial guess:

```
>> [xStar,gVal,exitflag,output] = fzero(@(x)test1D.g7(x),-
1.0)
xStar =
   -0.707106781186548
gVal =
     1.110223024625157e-16
exitflag =
     1
output =
    intervaliterations: 8
            iterations: 6
             funcCount: 22
             algorithm: 'bisection, interpolation'
             message: 'Zero found in the interval [-0.68, -
1.22627]'
```

The function converges to an alternative stationary point (which happens to be a local maximum); see Figure 6.1.

In conclusion:

1. `fzero` is a good choice if the scalar function is smooth and differentiable, and the gradient of the function is known in closed form.
2. However, `fzero` finds the closest stationary point, not necessarily the minimum.

6.4 The `fminsearch` Function

For higher-dimensional functions, a popular zeroth-order method in MATLAB is `fminsearch`, which is based on the Nelder–Mead simplex routine ([18]). Recall that zeroth-order methods are particularly useful when the function is non-differentiable or non-smooth. We will illustrate the use of `fminsearch` using a simple non-smooth function.

Example 6.6 Using MATLAB's `fminsearch`, solve

$$\underset{x}{\text{minimize}} \;\; (x-1)^2 + 0.025\sin^2(1000(x-1)/\pi) \tag{6.8}$$

starting at 1.1, 1.2 and 2.0.

Solution:

```
>> [xMin,fMin,exitflag,output] = fminsearch(@(x)test1D.f9(x),1.1)
xMin = 1.0987

>> [xMin,fMin,exitflag,output] = fminsearch(@(x)test1D.f9(x),1.2)
xMin = 0.9901

>> [xMin,fMin,exitflag,output] = fminsearch(@(x)test1D.f9(x),2.0)
xMin = 1.000
```

Thus, although the initial guess of 1.1 is closer to the global minimum, the method does not converge correctly with this initial guess.

Next, we consider a non-differentiable function in 2D.

Example 6.7 Using MATLAB's `fminsearch`, solve

$$\underset{\{x,y\}}{\text{minimize}} \; |x-2|+|y-3| \tag{6.9}$$

using an initial guess of (0, 0).

Solution: This function is implemented as function $f_{11}(x)$ in `testFunctionsND.m`:

```
>>[xMin,fMin,exitflag,output] =
fminsearch(@(x)testND.f11(x),[0 0])
xMin =
   2.000003111288824   3.000053397668920
fMin =
     5.650895774333975e-05
exitflag =
     1
output =
    iterations: 86
     funcCount: 161
     algorithm: 'Nelder-Mead simplex direct search'
       message: 'Optimization terminated:…'
```

The method can also be used for smooth functions as well, but it is not as efficient (compared to first-order methods).

Example 6.8 Using `fminsearch`, minimize the 2D Rosenbrock function, i.e., solve

$$\underset{x,y}{\text{minimize}}\{100(y-x^2)^2+(1-x)^2\} \tag{6.10}$$

using an initial guess of (0, 0).

Solution: With (0, 0) as the initial guess, we have

```
>> [xMin,fMin,flag,output] = fminsearch(@(x)testND.f4(x),[0
0])
xMin =
   1.000004385898617   1.000010640991648
fMin =
     3.686176915175908e-10
flag =
     1
output =
    iterations: 79
     funcCount: 146
     algorithm: 'Nelder-Mead simplex direct search'
       message: [1x196 char]
```

Later we will compare the performance of `fminsearch` against first-order methods.

We will now consider an N-dimensional version of the Rosenbrock function.

Example 6.9 Using `fminsearch`, solve

$$\underset{\mathbf{x}}{\text{minimize}} \sum_{i=1}^{N} [100(x_{i+1} - x_i^2)^2 + (1 - x_i)^2] \tag{6.11}$$

for $N = 10$, and the origin as the initial guess.

Solution: Again, by inspection, we can conclude that the minimum is located at $\mathbf{x} = [1, 1, \ldots, 1]$. The above function is implemented as $f_6(x, y)$ in `testFunctionsND.m`.

```
>> x = zeros(10,1);
>> [xMin,fMin,flag,output] = fminsearch(@(x)
testND.f6(x),x)
xMin =
   1.000009584005730
   1.000016150591684
   0.240493020465122
  -0.866424110174783
  -1.109628834024236
  -0.937202023119192
   0.789863534780492
  -0.044480972784228
   0.481726234023317
   1.028163322499352
fMin =
     1.002390808546588e-09
flag =
     1
output =
    iterations: 393
     funcCount: 599
     algorithm: 'Nelder-Mead simplex direct search'
       message: 'Optimization terminated:...'
```

`fminsearch` failed to converge.

In conclusion:

1. `fminsearch` is often the best choice for low-dimensional functions that are either noisy or non-differentiable.
2. It does not perform well for higher-dimensional problems.
3. It does not support constraints (see Section 7.6).

6.5 The fminunc Function

Next, we consider the first-order multi-dimensional method fminunc. It is one of the most popular methods for unconstrained problems. We will begin with a 1D example.

Example 6.10 Using MATLAB's fminunc, solve

$$\underset{x}{\text{minimize}} \ 1 - xe^{-x^2} \tag{6.12}$$

using an initial guess of 0, with (a) no gradient provided, and (b) explicit gradient provided.

Solution: Note that fminunc is a first-order method. If an explicit gradient is not provided by the user, finite-difference is used to estimate the gradient.

```
>> [xMin,fMin,flag,output] = fminunc(@(x) test1D.f7(x),0)
Warning: Gradient must be provided for trust-region
algorithm; using quasi-newton algorithm instead.
xMin =
   0.707107798027459
fMin =
   0.571118057520533
flag =
     1
output =
       iterations: 5
       funcCount: 12
       stepsize: 1.596933156960843e-04
       lssteplength: 1
       firstorderopt: 1.758337020874023e-06
       algorithm: 'quasi-newton'
       message: 'Local minimum found.…'
```

In this example, by default, it is assumed that the gradient is not provided. Since the gradient can be explicitly computed, the optimset method can be used to indicate this.

```
>> opt = optimset('GradObj','on');
```

Further, the MATLAB implementation of the function must return both the function value and its gradient, as illustrated below.

```
function [f,g,h] = f7(x)
% returns  1-x*exp(-x^2) and optionally its derivatives
  f = (1-x.*exp(-x.^2));
  if (nargout > 1), g = -exp(-x.^2) + 2*(x.^2).*exp(-
x.^2);end
  if (nargout > 2), h = 6*x.*exp(-x.*x)- 4*x.^3.*exp(-
x.*x);end
 end
```

(cont.)

Then the `opt` variable is passed to `fminunc` as follows:

```
>> [xMin,fMin,flag,output] = fminunc(@(x)
test1D.f7(x),0,opt)
xMin =
   0.707106781149014
fMin =
   0.571118057519647
flag =
     1
output =
         iterations: 7
         funcCount: 8
         stepsize: 7.292241589564029e-06
         cgiterations: 6
         firstorderopt: 6.438916066997535e-11
         algorithm: 'trust-region'
         message: 'Local minimum found....'
         constrviolation: []
```

As one can observe, the number of function calls has been reduced since an accurate and explicit gradient is provided. One can also compare its efficiency with zeroth-order methods.

Example 6.11 Using `fminunc`, solve

$$\underset{\mathbf{x}}{\text{minimize}} f(x,y) = 100(y-x^2)^2 + (1-x)^2 \tag{6.13}$$

with the origin as the initial guess. Note that the gradient of this function is given by

$$g = \nabla f = \begin{Bmatrix} -400(y-x^2)x - 2(1-x) \\ 200(y-x^2) \end{Bmatrix} \tag{6.14}$$

Solution: As noted in Example 6.8, this function is implemented as $f_4(x,y)$ in `testFunctionsND.m`. Further, the implementation returns both the function value and the gradient. First, we will use `fminunc` by relying on finite-difference for the gradient (default).

(cont.)

```
>> [xMin,fMin,flag,output] = fminunc(@(x)testND.f4(x),[ 0
0])
Warning: Gradient must be provided for trust-region
algorithm; using quasi-newton algorithm instead.
xMin =

   0.999995588079578   0.999991166415393

fMin =

     1.947457387335612e-11

flag =
     1

output =

       iterations: 20

        funcCount: 81
...
```

Since the gradient is explicitly available, the `optimset` method can be used to indicate this.

```
>> opt = optimset('GradObj','on');
>> [xMin,fMin,flag,output] = fminunc(@(x) testND.f4(x),[0
0],opt)
xMin =

   0.999999885761474   0.999999761974053

fMin =

     2.216860603381429e-14

flag =

     3

output =

         iterations: 16

          funcCount: 17

           ...
```

As one can observe, providing an explicit gradient reduces the number of function calls significantly (additional computation is needed to evaluate the gradient).

As a third example, we will use the `fminunc` method to solve the multidimensional Rosenbrock function.

Example 6.12 Using `fminunc`, solve

$$\underset{\mathbf{x}}{\text{minimize}} f(\mathbf{x}) = \sum_{i=1}^{N}[100(x_{i+1} - x_i{}^2)^2 + (1 - x_i)^2] \tag{6.15}$$

(cont.)

with $N = 10$, and the origin as the initial guess. Note that the gradient of this function is given by

$$g = \nabla f = \begin{Bmatrix} -400(x_2 - x_1{}^2)x_1 - 2(1 - x_1) \\ 200(x_2 - x_1{}^2) - 400(x_3 - x_2{}^2)x_2 - 2(1 - x_2) \\ \vdots \\ 200(x_i - x_{i-1}{}^2) - 400(x_{i+1} - x_i{}^2)x_i - 2(1 - x_i) \\ \vdots \\ 200(x_{N+1} - x_N{}^2) \end{Bmatrix} \tag{6.16}$$

Solution: As noted earlier, this function is implemented as $f_6(x, y)$ in `testFunctionsND.m`. Further, the implementation returns both the function value and the gradient. First, we will use `fminunc` by relying on finite difference for the gradient (default).

```
>> x = zeros(10,1);
>> [xMin,fMin,flag,output] = fminunc(@(x) testND.f6(x),x)
Warning: Gradient must be provided for trust-region
algorithm; using quasi-newton algorithm instead.
xMin =
    1.000000018388358
    0.999999994628223
    0.999999931945232
    0.999999831410932
    0.999999633469772
    0.999999147006020
    0.999998422421269
    0.999996618256450
    0.999993129520646
    0.999986276242075
fMin =
      7.211747269670000e-11
flag =
      2
output =
        iterations: 52
         funcCount: 748
        ...
```

Since the gradient is explicitly available, the `optimset` method can be used to indicate this.

(cont.)

```
>> opt = optimset('GradObj','on');
>> [xMin,fMin,flag,output] = fminunc(@(x)
testND.f6(x),x,opt)
xMin =
    0.999988984682601
    0.999981402386161
    0.999968333666822
    0.999939346268861
    0.999880503239110
    0.999764509247733
    0.999530498383390
    0.999061704163463
    0.998127852745470
    0.996254305366014
fMin =
      4.691048783744318e-06
flag =
      3
output =
            iterations: 81
             funcCount: 82
             ...
```

One can observe the significant improvement in efficiency (i.e., reduction in function calls) when first-order methods are used in conjunction with analytical gradient.

When the objective function exhibits numerous local minima, first-order methods will often converge to the closest local minimum.

Example 6.13 Minimize the function

$$\underset{x}{\text{minimize}} \;\; (x-1)^2 + 0.025\,\sin^2(1000(x-1)/\pi) \tag{6.17}$$

using `fminunc`, starting at 1.1, 1.2 and 2.0.

Solution: We have

(cont.)

```
>> [xMin] = fminunc(@(x)test1D.f9(x),1.1)
xMin = 1.0789

>> [xMin] = fminunc(@(x)test1D.f9(x),1.2)
xMin = 1.0789

>> [xMin] = fminunc(@(x)test1D.f9(x),2.0)
xMin = 1.000
```

Thus, although 1.1 and 1.2 are closer to the global minimum, the method finds a different local minimum.

In conclusion:

1. `fminunc` is the best choice for minimizing smooth, unconstrained, multi-dimensional functions.
2. Its performance improves significantly if analytical gradients are provided.

6.6 The `fsolve` Function

The `fsolve` function is the equivalent of `fzero` in higher dimensions, and it solves a system of non-linear equations

$$\mathbf{g}(\mathbf{x}) = 0$$

Example 6.14 Solve the pair of equations

$$\begin{aligned} 2x_1 - x_2 &= e^{-x_1} \\ -x_1 + 2x_2 &= e^{-x_2} \end{aligned} \tag{6.18}$$

using an initial guess of $(-5, -5)$.

Solution: The first task is to rewrite the two equations in the form $\mathbf{g}(\mathbf{x}) = 0$, i.e.,

$$\begin{aligned} 2x_1 - x_2 - e^{-x_1} &= 0 \\ -x_1 + 2x_2 - e^{-x_2} &= 0 \end{aligned} \tag{6.19}$$

(cont.)

Next, we will capture these non-linear equations in anonymous format:

```
g = @(x) [2*x(1) - x(2) - exp(-x(1));
         -x(1) + 2*x(2) - exp(-x(2))];
```

Observe that the variable **x** is interpreted as a 2D vector. Next, we can pass this function to `fsolve` with [−5 −5] as an initial guess.

```
>> [xStar,gVal,flag,output] = fsolve(@(x)g(x),[-5 -5])
xStar =
    0.5671    0.5671
gVal =
   1.0e-06 *
   -0.4059
   -0.4059
flag =
  ...
```

In optimization settings, the function $\mathbf{g}(\mathbf{x})$ is typically the gradient of some objective function, i.e.,

$$\mathbf{g}(\mathbf{x}) = \nabla f \tag{6.20}$$

Example 6.15 Using `fsolve`,

$$\underset{x,y}{\text{minimize}}\ 100(y - x^2)^2 + (1 - x)^2 \tag{6.21}$$

using an initial guess of (0, 0).

Solution: Note that the gradient of the function is given by

$$g = \nabla f = \begin{Bmatrix} -400(y - x^2)x - 2(1 - x) \\ 200(y - x^2) \end{Bmatrix} \tag{6.22}$$

It is implemented as $g_4(x, y)$ in `testFunctionsND.m`. We provide this to `fsolve`:

```
>> [xStar,gVal,flag,output] = fsolve(@(x) testND.g4(x),[0
0]);
xStar =
   0.999999999997126   0.999999999994188
>> output.funcCount
ans =
  86
```

Example 6.16 Using `fsolve`,

$$\underset{\mathbf{x}}{\text{minimize}} \sum_{i=1}^{N} [100(x_{i+1} - x_i^2)^2 + (1 - x_i)^2] \tag{6.23}$$

with $N = 10$, and the origin as the initial guess.

Solution:

```
>> [xStar,gVal,flag,output] = fsolve(@(x)
testND.g6(x),zeros(1,10));
>> xStar =
  -0.555376076092360
   0.322445494402679
   0.115178205622253
   0.023506197759278
   0.010661379252702
   0.010217972669098
   0.010208534022082
   0.010204214636605
   0.010004085144901
   0.000100081719586
>> gVal =
   1.0e-12 *
  -0.428101998295460
   ...
```

Observe that, at the converged point, the gradient vanishes, i.e., it corresponds to a stationary point.

In conclusion:

1. `fsolve` is typically used for solving non-linear sets of equations.
2. It can also be used for finding stationary points of an optimization problem, provided the gradient of the objective is known in closed form.
3. However, it is not a recommended method for solving minimization problems.

6.7 Global Optimization Toolbox

All optimization methods discussed thus far seek a local minimum. However, in certain scenarios, an objective function may exhibit multiple local minima. As an example, consider the function

$$f(x) = 3 \sin x - x + 0.1x^2 + 0.1 \cos(2x) \tag{6.24}$$

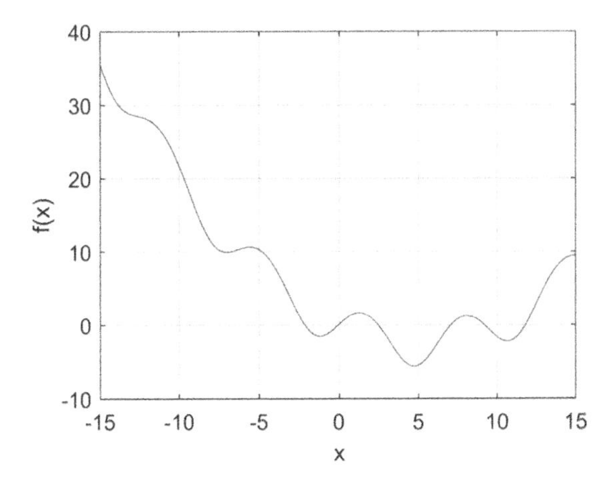

Figure 6.3 Plot of $f(x) = 3\ \sin x - x + 0.1x^2 + 0.1\ \cos(2x)$.

which we encountered in Chapter 4. Observe in Figure 6.3 that the function has numerous local minima and a global minimum at $x \approx 4.8$. Methods such as `fminunc` will not necessarily converge to the global minimum (depending on the initial guess). Global optimization methods ([19], [20]), briefly discussed below, have a better chance of finding the global minimum.

We will consider global optimization methods such as `MultiStart`, genetic algorithms and simulated annealing, supported by MATLAB. These methods are part of MATLAB's global optimization toolbox, which is distinct from the standard optimization toolbox. To determine whether your installation of MATLAB includes the global optimization toolbox, try a `help` on one of the methods such as `MultiStart`. If you get a message along the following lines, then you can proceed with the remainder of this section:

```
>> help MultiStart
 MultiStart A multi-start global optimization solver.
```

6.7.1 MultiStart

As the name suggests, the `MultiStart` algorithm relies on classic gradient-based methods, but with multiple starting points, to converge to possibly different local minima (with a better chance that one of them may be a global minimum). To illustrate, consider the function illustrated in Figure 6.3. In MATLAB, we capture the function as follows:

```
>> f = @(x) 3*sin(x)-x+0.1*x^2+0.1*cos(2*x);
```

To use `MultiStart`, we must first create a problem structure by declaring that `fmincon` should be used as the local optimizer. When creating the problem structure, we will also pass this function and the origin as a seed point. The seed point is used to generate several starting points.

```
>> problem =
createOptimProblem('fmincon','objective',@(x)f(x),'x0',0);
```

Next the problem structure is passed to `MultiStart`, with desired number of starting points set to 50:

```
>> [x,fval,eflag,output,manymins] =
run(MultiStart, problem,50)
MultiStart completed the runs from all start points.
All 50 local solver runs converged with a positive local
solver exit flag.
x =
   4.7284
fval =
  -5.5922
...
output =
  struct with fields:
                funcCount: 1266
```

As one can observe, `MultiStart` found the global minimum, using 1266 function calls. Of course, the number of function calls can be reduced if the number of starting points is reduced, but this would also reduce the chances of finding the global minimum.

6.7.2 GlobalSearch

There is an alternative method called `GlobalSearch`, whose behavior is almost identical to `MultiStart`, except that (1) the underlying algorithm is different and (2) we do not provide the number of starting points. This is internally controlled by default parameters:

```
>> [x,fval,eflag,output,manymins] =
run(GlobalSearch,problem)
GlobalSearch stopped because it analyzed all the trial
points.
All 6 local solver runs converged with a positive local
solver exit flag.
x =
   4.7284
fval =
  -5.5922
output =
  struct with fields:
                funcCount: 2092
...
```

`GlobalSearch` found the global minimum, using 2092 function calls.

6.7.3 Genetic Algorithms

While `MultiStart` and `GlobalSearch` rely on classic gradient-based local optimizers, genetic algorithms work on the principle of natural selection. The idea is to start with a population of points, and then evaluate the objective function at each of these points. Then points with high objective values are discarded. The remaining samples are retained, and additional samples are inserted, based on genetic principles of mixing and mutation. The hypothesis is that this process mimics the principle of evolution and that the fittest (lowest objective) point will survive.

The syntax for the genetic algorithm method is as follows, where the second argument is the number of variables that the function expects:

```
>> [x,Fval,exitFlag,Output] = ga(f,1)
x =
    4.7306
Fval =
   -5.5922
Output =
        funccount: 4753
...
```

The genetic algorithm (`ga`) method found the global minimum, using 4753 function calls. Once again, the user can change the default parameters to reduce the number of function calls, and so on.

6.7.4 Simulated Annealing

Annealing is a metallurgical process of repeated heating and cooling of a material to reduce defects. Simulated annealing ([21]) mimics this process in optimization, with cooling analogous to descending to a local minimum, and heating analogous to escaping from it. Through simulated annealing (i.e., repeated descending and ascending), the hypothesis is that one can avoid getting trapped in a local minimum.

In MATLAB, the syntax for the simulated annealing method is as follows (where the second argument is the initial guess):

```
>> [x,fval,exitflag,output] = simulannealbnd(@(x)f(x),0)
x =
    4.7279
fval =
   -5.5922
output =
      funccount: 711
...
```

The global minimum was found using 711 function calls. A limitation of simulated annealing (as implemented in MATLAB) is that it cannot handle nonlinear constraints.

6.7.5 Multiple Local Minima

We will compare the performance of the global optimization methods of Sections 6.7.1 through 6.7.4 for minimizing the function (see Figure 6.4)

$$f(x) = (x - 1)^2 + a \sin^2(1000(x - 1)/\pi) \tag{6.25}$$

The first component is the true objective, while the second component contributes to numerous local minima, where a controls the depth of the minima. The global minimum is at $x = 1$.

We will use each one of the global optimization methods, for different levels of the a value. However, to be consistent between methods, a time limit of 0.025 sec is imposed on all methods (unfortunately, this is not strictly adhered to by the methods). The MATLAB script `globalOptimizationExample1.m` used in generating the results below is included in the code accompanying the text.

Table 6.2 summarizes the error in the final solution for various algorithms and values of a. Table 6.3 summarizes the number of function calls, while Table 6.4 summarizes the actual time taken.

A few points are worth noting:

1. We have relied on the default settings for each of the algorithms; the results can be different if the settings are changed.
2. Some of the methods, such as genetic algorithms, are stochastic in nature, i.e., the results can change from one run to the next.

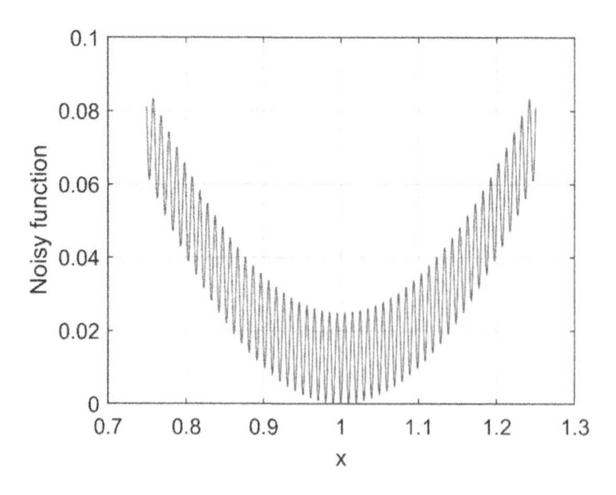

Figure 6.4 A function with numerous local minima.

3. While we have imposed a time limit of 0.025 sec, some of the methods exceed this time limit, while others terminate before this time limit is reached (probably because they got stuck at a local minimum).
4. Multi-search and global search have a larger overhead than simulated annealing and genetic algorithms; they use fewer function calls, but take longer.

Table 6.2 Error in solution, i.e., $|x - 1|$.

	0.001	0.01	0.025	0.05	0.1
`MultiStart`	0.0391	0.0197	0.0592	0	0
`GlobalSearch`	0.0489	0.0197	0.1480	0	0
`GA`	0.0352	0.0196	0.0190	0.0903	0.1384
`SimAnnealing`	9.51×10^{-5}	0.0098	0.0301	0.0491	0.0099

Table 6.3 Number of function calls.

	0.001	0.01	0.025	0.05	0.1
`MultiStart`	73	91	92	65	96
`GlobalSearch`	1 038	1 030	1 031	1 031	1 031
`GA`	2 450	4 236	2 450	3 481	3 061
`SimAnnealing`	278	294	300	277	299

Table 6.4 Actual time taken.

	0.001	0.01	0.025	0.05	0.1
`MultiStart`	0.0378	0.0389	0.0385	0.0306	0.0387
`GlobalSearch`	0.0476	0.0447	0.0443	0.05461	0.0458
`GA`	0.0178	0.0281	0.0186	0.0298	0.0271
`SimAnnealing`	0.0268	0.0266	0.0270	0.0272	0.0277

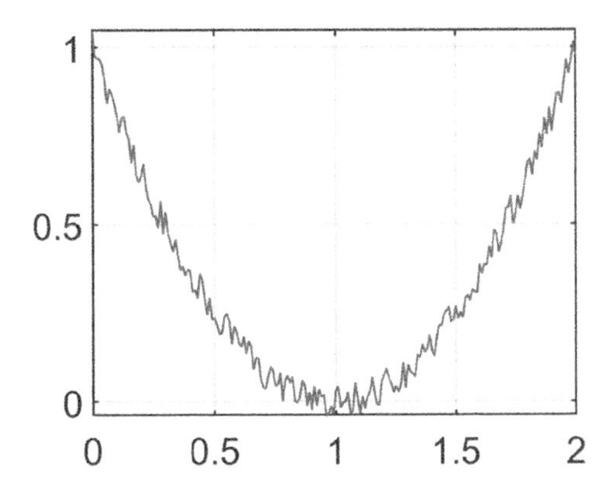

Figure 6.5 A function with noise.

6.7.6 Noisy Objective

Next, we will compare the performance of the above global optimization methods for minimizing an objective function that has inherent noise:

$$f(x) = (x - 1)^2 + a(0.5 - \text{rand}) \tag{6.26}$$

illustrated in Figure 6.5. Unlike the previous function, observe that this function is not differentiable. Here the value of the parameter a controls the level of "noise." The global minimum is approximately at x = 1.

Table 6.5 summarizes the error, Table 6.6 summarizes the number of function calls and Table 6.7 summarizes the actual time taken.

In addition to the observations made earlier, we also note that, for non-differentiable functions, the first three methods fail in a few scenarios. For this particular example, simulated annealing was able to successfully converge to solutions close to the global minimum.

Table 6.5 Error in solution, i.e., $|x - 1|$.

	0.001	0.01	0.025	0.05	0.1
`MultiStart`	0.0020	0.9916	0.0383	0.0581	0.8977
`GlobalSearch`	0.0041	1.0000	0.0240	0.0480	0.0690
`GA`	0.0637	0.4019	0.0543	0.0659	3.284×10^{-4}
`SimAnnealing`	0.0428	0.0562	0.0261	0.0214	0.0924

Table 6.6 Number of function calls.

	0.001	0.01	0.025	0.05	0.1
`MultiStart`	59	64	66	57	48
`GlobalSearch`	1 052	1 049	64	67	61
`GA`	350	300	300	350	350
`SimAnnealing`	137	143	177	138	141

Table 6.7 Actual time taken.

	0.001	0.01	0.025	0.05	0.1
`MultiStart`	0.0362	0.0356	0.0403	0.0383	0.0364
`GlobalSearch`	0.1355	0.1312	0.0392	0.0344	0.0380
`GA`	0.0310	0.0315	0.0324	0.0294	0.0329
`SimAnnealing`	0.0273	0.0270	0.0273	0.0268	0.0279

6.8 EXERCISES

Exercise 6.1 Using MATLAB's `fminbnd` and default tolerances, find the local minimum of $\sin x + \sin(10x/3)$ in the range (a) [4.5, 6], (b) [2.7, 7.5] and (c) [0, 10]. Create a table with the following columns: `xSolution`, `fSolution`, gradient at `xSolution`, Hessian at `xSolution`, and number of function calls (the "reason for termination" is often too long to include in a table; you can use that in your explanation). The rows should correspond to each of the cases. For each case explain the results.

Exercise 6.2 Using MATLAB's `fminsearch` and default tolerances, find the local minimum of $\sin x + \sin(10x/3)$, starting from (a) 5, (b) 4 and (c) 3.5. Follow the instructions in Exercise 6.1 to explain your results.

Exercise 6.3 Using MATLAB's `fzero` and default tolerances, find the stationary points of $\sin x + \sin(10x/3)$, starting from (a) 5, (b) 4 and (c) 0. Note that you need to pass the gradient of the function to `fzero`. Follow the instructions in Exercise 6.1 to explain your results.

Exercise 6.4 Using MATLAB's `fminbnd` and default tolerances, find the minimum of $-(1.4 - 3x)\sin(18x)$ in the range (a) [0, 0.2], (b) [0, 0.5] and (c) [−1, 0.5]. Follow the instructions in Exercise 6.1 to explain your results.

Exercise 6.5 Using MATLAB's `fminsearch` and default tolerances, find the minimum of $-(1.4 - 3x)\sin(18x)$, starting from (a) 0.1, (b) 0.25 and (c) −0.25. Follow the instructions in Exercise 6.1 to explain your results.

Exercise 6.6 Using MATLAB's `fzero` and default tolerances, find the stationary points of $-(1.4 - 3x)\sin(18x)$, starting from (a) 0.1, (b) 0.25 and (c) −0.25. Note that you need to pass the gradient of the function to `fzero`. Follow the instructions in Exercise 6.1 to explain your results.

Exercise 6.7 Use MATLAB's `fminbnd` to find the minimum for the following functions, with default tolerances, within the interval [0, 5]: (a) $\cos(x)$, (b) $10^6 \cos(x)$, (c) $10^{-6} \cos(x)$. Follow the instructions in Exercise 6.1 to explain your results.

Exercise 6.8 Use MATLAB's `fminsearch` to find the minimum for the following functions, with default tolerances, starting from 2.5: (a) $\cos(x)$, (b) $10^6 \cos(x)$, (c) $10^{-6} \cos(x)$. Follow the instructions in Exercise 6.1 to explain your results.

Exercise 6.9 Use MATLAB's `fzero` to find the stationary points for the following functions, with default tolerances, starting from 2.5: (a) $\cos(x)$, (b) $10^6 \cos(x)$, (c) $10^{-6} \cos(x)$. Note that you need to pass the gradient of the function to `fzero`. Follow the instructions in Exercise 6.1 to explain your results.

Exercise 6.10 Use MATLAB's `fminbnd` and `fminsearch` to find the minimum of

$$f(x) = (1 - xe^{-x^2}) + a \sin(1000x)$$

with default tolerances, for various values of a: 0, 10^{-4}, 10^{-3}, 10^{-2}, 10^{-1}, and different intervals: [0, 1], [−1, 1] and [0, 5]. Tabulate your results, showing the *error* in solution and the number of function calls. Assume that the exact answer is $1/\sqrt{2}$.

Exercise 6.11 Use `fminunc` to find the local minimum of the following functions, with default tolerances starting from the origin:

(a) the simple function $f(x, y) = 3x^2 - xy + 4y^2 - x - 2y + 10$ with default tolerances
(b) the 2D Rosenbrock function with default tolerances, starting at the origin
(c) the two-spring potential energy function with default tolerances, starting at the origin.

Tabulate your results with `fMin`, norm of the gradient and number of function calls for columns, and each of the cases as rows. For each case, explain the results.

Exercise 6.12 Use `fminsearch` to find the local minimum for each of the examples in Exercise 6.11 with default tolerances, starting from the origin. Follow the instructions in that exercise for reporting results.

Exercise 6.13 Use `fsolve` to find the stationary point for each of the examples in Exercise 6.11 with default tolerances, starting from the origin. Note that you need to pass the gradient of the function to `fsolve`. Follow the instructions in that exercise for reporting results.

Exercise 6.14 Use `fminunc` to find the local minimum of the N-dimensional Rosenbrock function, with default tolerances, starting from the origin, for the following cases: (a) $N = 4$, (b) $N = 10$ and (c) $N = 20$. Tabulate your results with `fMin`, norm of the gradient and number of function calls for columns, and each of the cases as rows. Was the global minimum found in all cases?

Exercise 6.15 Repeat Exercise 6.14 using `fminsearch`.

Exercise 6.16 Repeat Exercise 6.14 using `fsolve`, remembering to pass the gradient of the function.

Exercise 6.17 Consider the three-spring system in Figure 6.6. The three springs are each of length 1 and have a spring constant of 100 units. A force of (12, 18) acts on node 1.

Formulate the minimization problem governing the displacements (u, v) of node 1 using the principle of minimum potential energy (see Section 2.5).

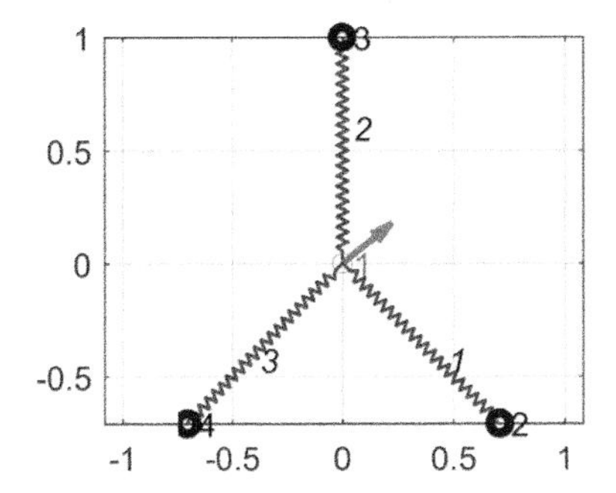

Figure 6.6 A three-spring system.

Create the class `threeSpring.m` as a sub-class of `springPlotter` with one method:

- `function [PE, PEgradient] = potentialEnergy(obj, uv)`, which returns the potential energy and its gradient for any displacement (u, v).

Then use `fminunc` to find the equilibrium point for node 1 by passing a pointer to `potentialEnergy`.

Exercise 6.18 Repeat the problem in Exercise 6.17, except now add a method:

- `function [residual] = forceBalance(obj, uv)`, which returns the force balance equations for any displacement (u, v).

Then use `fsolve` to find the equilibrium point by passing a pointer to the method `forceBalance`.

Exercise 6.19 Use MATLAB's global optimization methods to find the minimum of

$$f(x) = (1 - xe^{-x^2}) + a(0.5 - \text{rand})$$

for various values of a: 0, 10^{-4}, 10^{-3}, 10^{-2}, 10^{-1}, limiting the time for each method to around 0.025 sec. Tabulate your results showing the *error* in solution, the number of function calls and the actual time taken.

Exercise 6.20 Using the global optimization methods discussed in this chapter, minimize the N-dimensional Rosenbrock function, i.e., solve

$$\underset{\mathbf{x}}{\text{minimize}} \sum_{i=1}^{N} [100(x_{i+1} - x_i^2)^2 + (1 - x_i)^2] \tag{6.27}$$

with $N = 10$ and the origin as the initial guess, limiting the time for each method to around 0.5 sec. Tabulate your results with two columns: (norm of the error in solution and number of function calls), and four rows (for each of the algorithms).

7 Constrained Optimization

Highlights

1. In this chapter, we first consider equality-constrained problems of the form

$$\begin{aligned} &\underset{\mathbf{x}}{\text{minimize}} \; f(\mathbf{x}) \\ &\text{s.t.} \quad h_i(\mathbf{x}) = 0; \; i = 1, 2, ..., M \end{aligned}$$

2. As with the theorems for unconstrained optimization problems, we will state and illustrate the optimality theorem for equality-constrained optimization problems.
3. In the process, we will introduce the concept of Lagrange multipliers, which can be interpreted as the sensitivity of the objective function with respect to a constraint. This concept will be useful later in the text.
4. We will then consider a more generic class of equality- and inequality-constrained problems, of the form

$$\begin{aligned} &\underset{\mathbf{x}}{\text{minimize}} \; f(\mathbf{x}) \\ &\text{s.t.} \quad h_i(\mathbf{x}) = 0; \; i = 1, 2, ..., M \\ &\qquad g_j(\mathbf{x}) \leq 0; \; j = 1, 2, ..., L \end{aligned}$$

5. The optimality theorem for this class of problems is stated and illustrated.
6. Finally, we will explore methods within MATLAB's optimization toolbox for solving constrained problems.

7.1 Equality Constraints

First, we consider *equality-constrained* problems:

$$\begin{aligned} &\underset{\mathbf{x}}{\text{minimize}} \; f(\mathbf{x}) \\ &\text{s.t.} \quad h_i(\mathbf{x}) = 0; i = 1, 2, ..., M \end{aligned} \tag{7.1}$$

If the constraints h_i are linear, it may be possible to eliminate them, thereby reducing the problem to an unconstrained problem, as Example 7.1 illustrates. On the other hand, if the constraints are non-linear, elimination (if possible) can lead to erroneous results, as Example 7.2 illustrates.

Example 7.1 Solve the constrained problem

$$\begin{aligned} \underset{x,y}{\text{minimize}} \quad & x^2 + y^2 \\ \text{s.t.} \quad & x + y - 1 = 0 \end{aligned} \tag{7.2}$$

Solution: Observe that the equality constraint is linear. We will therefore eliminate the constraint by substituting

$$y = 1 - x \tag{7.3}$$

in the objective function. This leads to

$$\underset{x}{\text{minimize}} \; x^2 + (1 - x)^2 \tag{7.4}$$

i.e.,

$$\underset{x}{\text{minimize}} \; 2x^2 - 2x + 1 \tag{7.5}$$

Observe that Equation (7.5) is a single-variable, unconstrained minimization problem, whose minimum is $x = 1/2$. Consequently, from Equation (7.3), we have $y = 1/2$.

Thus the computed solution to Equation (7.2) is (1/2, 1/2). To confirm that this is indeed the correct solution to the original problem, one can interpret Equation (7.2) geometrically. First, observe that the constraint $x + y - 1 = 0$ defines a *feasible region*, illustrated in Figure 7.1. In other words, all solutions must lie on the straight line $x + y - 1 = 0$. Further, since the objective function is $x^2 + y^2$, Equation (7.2) is equivalent to finding the point (x, y) on the straight line that is closest to the origin. Thus, the geometric solution to the problem is the point (1/2, 1/2).

Note that the gradient of the objective function, $\nabla f = (2x, 2y)^T$, is not zero at (1/2, 1/2), i.e., for constrained minimization problems, the gradient of the objective function does not necessarily vanish at the optimal solution.

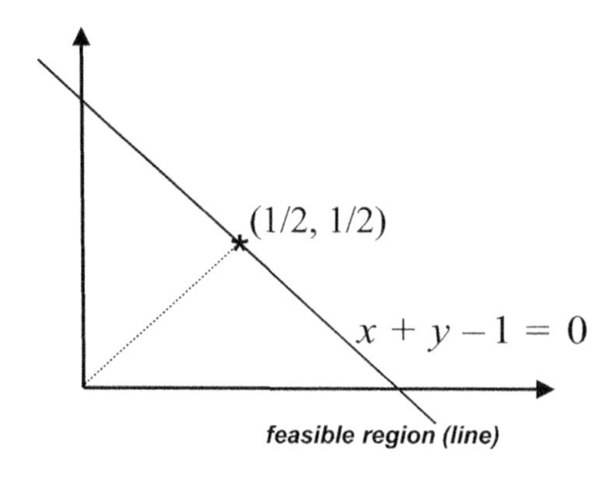

Figure 7.1 Geometric interpretation of Equation (7.2).

Example 7.2 Solve the constrained problem

$$\begin{aligned} &\underset{x,y}{\text{minimize}} \quad x^2 + y^2 \\ &\text{s.t.} \qquad (x-2)^2 + y^2 - 1 = 0 \end{aligned} \tag{7.6}$$

Solution: Observe that the equality constraint is *not* linear. Therefore eliminating the constraint is *not* recommended. To understand the implications, substitute

$$y^2 = 1 - (x-2)^2 \tag{7.7}$$

in the objective function. This leads to

$$\underset{x}{\text{minimize}} \; x^2 + 1 - (x-2)^2 \tag{7.8}$$

i.e., upon simplification,

$$\underset{x}{\text{minimize}} \quad 4x - 3 \tag{7.9}$$

Equation (7.9) is unbounded as $x \to -\infty$, implying that the problem in Equation (7.6) has no solution.

On the other hand, observe that the constraint $(x-2)^2 + y^2 - 1 = 0$ defines a circle on which the solution must lie; see Figure 7.2. Thus, one must find the point (x, y) that lies on the circle that is closest to the origin (since the objective is $x^2 + y^2$). The correct solution to this geometric problem is the point $(1, 0)$. The reader can ponder over the "flaw" in the elimination method! Once again, observe that the gradient of the objective does not vanish at $(1, 0)$.

Finally, elimination may not always be possible; for example, consider the problem

$$\begin{aligned} &\underset{x,y}{\text{minimize}} \quad x^2 + y^2 \\ &\text{s.t.} \qquad x + e^x \sin y - x^2 \cos xy = 0 \end{aligned} \tag{7.10}$$

Here it is not easy to express y as a function of x (or vice versa). Therefore, retaining them will often be essential. Finally, constraints reveal certain important characteristics of the problem, to be discussed in Section 7.5.

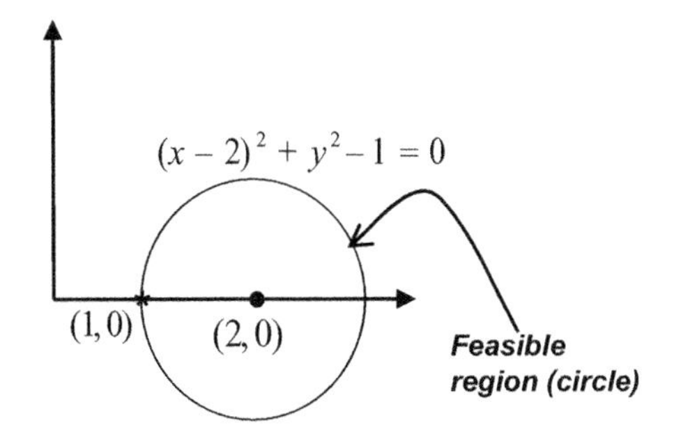

Figure 7.2 Geometric interpretation of Equation (7.6).

7.2 Optimality Criteria: Equality Constraints

We will now explore formal optimality theorems for equality-constrained problems.

7.2.1 Single Equality Constraint

Recall that, for unconstrained problems, we introduced the concept of *stationary points* (necessary conditions), using first-order derivatives, followed by *sufficient conditions*, using second-order derivatives. A similar process is adopted here.

Consider the equality-constrained optimization problem

$$\begin{aligned} \underset{\mathbf{x}}{\text{minimize}} \quad & f(\mathbf{x}) \\ \text{s.t.} \quad & h(\mathbf{x}) = 0 \end{aligned} \tag{7.11}$$

Similar to the optimality conditions for unconstrained problem, we can state the optimality conditions for this constrained problem.

Theorem 7.1 *Consider the equality-constrained optimization problem*

$$\begin{aligned} \underset{\mathbf{x}}{\text{minimize}} \quad & f(\mathbf{x}) \\ \text{s.t.} \quad & h(\mathbf{x}) = 0 \end{aligned} \tag{7.12}$$

A stationary point $\mathbf{x}^*$ *satisfies*

$$\begin{aligned} \nabla f(\mathbf{x}^*) + \mu^* \nabla h(\mathbf{x}^*) = 0 \\ h(\mathbf{x}^*) = 0 \end{aligned} \tag{7.13}$$

for some scalar μ^* *(referred to as the* Lagrange multiplier*). Further, the stationary point is a solution to the minimization problem if the matrix*

$$\mathbf{H}_f(\mathbf{x}^*) + \mu^* \mathbf{H}_h(\mathbf{x}^*) \tag{7.14}$$

is positive definite, where $\mathbf{H}_f$ *is the Hessian of the objective function* f *and* $\mathbf{H}_h$ *is the Hessian of the constraint function* h.

Proof: See the Appendix, Section A.2.

In Section 7.2.4 we will discuss the mathematical and physical importance of the Lagrange multiplier, but, first, we illustrate the theorem with a couple of examples.

Example 7.3 Using the optimality condition, find the stationary point for the problem

$$\begin{aligned} \underset{x,y}{\text{minimize}} \quad & x^2 + y^2 \\ \text{s.t.} \quad & x + y - 1 = 0 \end{aligned} \tag{7.15}$$

and determine whether it is a local minimum.

(cont.)

Solution: We have

$$\nabla f = \begin{Bmatrix} 2x \\ 2y \end{Bmatrix}; \ \nabla h = \begin{Bmatrix} 1 \\ 1 \end{Bmatrix} \tag{7.16}$$

Thus, from Theorem 7.1 we must find x, y and μ such that

$$\begin{aligned} &\begin{Bmatrix} 2x \\ 2y \end{Bmatrix} + \mu \begin{Bmatrix} 1 \\ 1 \end{Bmatrix} = 0 \\ &x + y - 1 = 0 \end{aligned} \tag{7.17}$$

Observe that there are exactly three equations and three unknowns. The linear system is

$$\begin{bmatrix} 2 & 0 & 1 \\ 0 & 2 & 1 \\ 1 & 1 & 0 \end{bmatrix} \begin{Bmatrix} x \\ y \\ \mu \end{Bmatrix} = \begin{Bmatrix} 0 \\ 0 \\ 1 \end{Bmatrix} \tag{7.18}$$

and its solution is $x^* = 1/2$, $y^* = 1/2$, $\mu^* = -1$. Observe that (1/2, 1/2) is precisely the solution we obtained via the elimination method. To verify whether the stationary point is a solution, we consider

$$\mathbf{H}_f(\mathbf{x}^*) + \mu^* \mathbf{H}_h(\mathbf{x}^*) = \begin{bmatrix} 2 & 0 \\ 0 & 2 \end{bmatrix} \tag{7.19}$$

Since the matrix is positive definite, the stationary point is a local minimum.

Example 7.4 Using the optimality condition, find the stationary point for the problem

$$\begin{aligned} &\underset{x,y}{\text{minimize}} \quad x^2 + y^2 \\ &\text{s.t.} \qquad (x-2)^2 + y^2 - 1 = 0 \end{aligned} \tag{7.20}$$

and determine whether it is a local minimum.

Solution: We have

$$\nabla f = \begin{Bmatrix} 2x \\ 2y \end{Bmatrix}; \ \nabla h = \begin{Bmatrix} 2x - 4 \\ 2y \end{Bmatrix} \tag{7.21}$$

(cont.)

Thus, from Theorem 7.1 we must find x, y and μ such that

$$\begin{Bmatrix} 2x \\ 2y \end{Bmatrix} + \mu \begin{Bmatrix} 2x - 4 \\ 2y \end{Bmatrix} = 0$$
$$x^2 - 4x + y^2 + 3 = 0 \tag{7.22}$$

i.e.,

$$\begin{aligned} &2x + 2\mu x - 4\mu = 0 \\ &2y + 2\mu y = 0 \\ &x^2 - 4x + y^2 + 3 = 0 \end{aligned} \tag{7.23}$$

Unlike Example 7.3, we obtain here a set of non-linear equations. We can use MATLAB's `fsolve` to arrive at two sets of solutions: (a) $x^* = 1, y^* = 0, \mu^* = 1$ and (b) $x^* = 3, y^* = 0, \mu^* = -3$ (exercise for the reader).

Geometrically, one can interpret the first solution as a local minimum, and the second solution as a local maximum (see Figure 7.2). Indeed, consider the second-order condition

$$\mathbf{H}_f(\mathbf{x}^*) + \mu^* \mathbf{H}_h(\mathbf{x}^*) = \begin{bmatrix} 2 & 0 \\ 0 & 2 \end{bmatrix} + \mu^* \begin{bmatrix} 2 & 0 \\ 0 & 2 \end{bmatrix} \tag{7.24}$$

At $x^* = 1, y^* = 0, \mu^* = 1$, we have

$$\mathbf{H}_f(\mathbf{x}^*) + \mu^* \mathbf{H}_h(\mathbf{x}^*) = \begin{bmatrix} 4 & 0 \\ 0 & 4 \end{bmatrix}$$

which is positive definite, i.e., the point is a minimum. On the other hand, at $x^* = 3, y^* = 0, \mu^* = -1.5$,

$$\mathbf{H}_f(\mathbf{x}^*) + \mu^* \mathbf{H}_h(\mathbf{x}^*) = \begin{bmatrix} -1 & 0 \\ 0 & -1 \end{bmatrix}$$

which is negative definite, i.e., the point is a local maximum.

7.2.2 Quadratic Problems

We will now consider the special case of *quadratic* problems. Recall that, for *unconstrained* quadratic problems of the form

$$\underset{\mathbf{x}}{\text{minimize}} \; \frac{1}{2}\mathbf{x}^T\mathbf{H}\mathbf{x} + \mathbf{x}^T\mathbf{g} + c \tag{7.25}$$

the stationary point satisfies

$$\mathbf{H}\mathbf{x}^* = -\mathbf{g} \tag{7.26}$$

Thus, a linear solver can be used to find the local minimum. Further, the stationary point is a minimum if the matrix $\mathbf{H}$ is positive definite.

We define an *equality-constrained* quadratic problem as one where (1) the objective is quadratic, and (2) each of the constraints is linear. Since the constraints are linear, the optimization problem may be posed as

$$\begin{aligned} &\underset{\mathbf{x}}{\text{minimize}} && \frac{1}{2}\mathbf{x}^T\mathbf{H}\mathbf{x} + \mathbf{x}^T\mathbf{g} + c \\ &\text{s.t.} && \mathbf{A}\mathbf{x} + \mathbf{b} = 0 \end{aligned} \tag{7.27}$$

The first-order conditions for this quadratic problem are

$$\begin{bmatrix} \mathbf{H} & \mathbf{A}^T \\ \mathbf{A} & 0 \end{bmatrix} \begin{Bmatrix} \mathbf{x} \\ \boldsymbol{\mu} \end{Bmatrix} = \begin{Bmatrix} -\mathbf{g} \\ -\mathbf{b} \end{Bmatrix} \tag{7.28}$$

The matrix in Equation (7.28) is referred to as the Karush–Kuhn–Tucker (KKT) matrix for a quadratic problem.

7.2.3 Multiple Constraints

We now generalize Theorem 7.1 for multiple equality constraints.

***Theorem** 7.2 Consider the equality-constrained optimization problem*

$$\begin{aligned} &\underset{\mathbf{x}}{\text{minimize}} && f(\mathbf{x}) \\ &\text{s.t.} && h_i(\mathbf{x}) = 0;\ i = 1, 2, ..., M \end{aligned} \tag{7.29}$$

A stationary point $\mathbf{x}^*$ satisfies

$$\begin{aligned} \nabla f(\mathbf{x}^*) + \sum_{i=1}^{M} \mu_i^* \nabla h_i(\mathbf{x}^*) = 0 \\ h_i(\mathbf{x}^*) = 0;\ i = 1, 2, \ldots, M \end{aligned} \tag{7.30}$$

for some set of scalars $\{\mu_1^*\ \mu_2^*\ \ldots\ \mu_M^*\}$. Further, the stationary point is a local minimum if the matrix

$$\mathbf{H}_f(\mathbf{x}^*) + \sum_{i=1}^{M} \mu_i^* \mathbf{H}_{h_i}(\mathbf{x}^*) \tag{7.31}$$

is positive definite.

Proof: See reference [8].

Example 7.5 Using the optimality condition, find the stationary points for the problem

$$\begin{aligned} \underset{x,y}{\text{minimize}} \quad & x^2 + y^2 + z^2 \\ \text{s.t.} \quad & x + y + z - 1 = 0 \\ & x + 2y + 3z - 4 = 0 \end{aligned} \tag{7.32}$$

and determine whether they are local minima.

Solution: We have

$$\nabla f = \begin{Bmatrix} 2x \\ 2y \\ 2z \end{Bmatrix}; \ \nabla h_1 = \begin{Bmatrix} 1 \\ 1 \\ 1 \end{Bmatrix}; \ \nabla h_2 = \begin{Bmatrix} 1 \\ 2 \\ 3 \end{Bmatrix} \tag{7.33}$$

Thus, the first-order conditions are

$$\begin{aligned} &\begin{Bmatrix} 2x \\ 2y \\ 2z \end{Bmatrix} + \mu_1 \begin{Bmatrix} 1 \\ 1 \\ 1 \end{Bmatrix} + \mu_2 \begin{Bmatrix} 1 \\ 2 \\ 3 \end{Bmatrix} = \begin{Bmatrix} 0 \\ 0 \\ 0 \end{Bmatrix} \\ &x + y + z - 1 = 0 \\ &x + 2y + 3z - 4 = 0 \end{aligned} \tag{7.34}$$

We can write this in a matrix form as

$$\begin{bmatrix} 2 & 0 & 0 & 1 & 1 \\ 0 & 2 & 0 & 1 & 2 \\ 0 & 0 & 2 & 1 & 3 \\ 1 & 1 & 1 & 0 & 0 \\ 1 & 2 & 3 & 0 & 0 \end{bmatrix} \begin{Bmatrix} x \\ y \\ z \\ \mu_1 \\ \mu_2 \end{Bmatrix} = \begin{Bmatrix} 0 \\ 0 \\ 0 \\ 1 \\ 4 \end{Bmatrix} \tag{7.35}$$

The solution is $(-2/3, 1/3, 4/3)$, with $\mu_1 = 10/3$, $\mu_2 = -2$. It is easy to show that the stationary point is a minimum.

7.2.4 Significance of Lagrange Multipliers

By now, the reader must have observed that the variables, denoted by μ and referred to as Lagrange multipliers, are essential in solving constrained optimization problems. In this section, we discuss an important mathematical significance of these Lagrange multipliers.

Consider the single-constraint problem

$$\begin{aligned} \underset{\mathbf{x}}{\text{minimize}} \quad & f(\mathbf{x}) \\ \text{s.t.} \quad & h(\mathbf{x}) = 0 \end{aligned} \tag{7.36}$$

The necessary condition is

$$\begin{aligned} &\nabla f(\mathbf{x}^*) + \mu^* \nabla h(\mathbf{x}^*) = 0 \\ &h(\mathbf{x}^*) = 0 \end{aligned} \tag{7.37}$$

for some scalar μ^*. Now consider a "perturbed" problem

$$\begin{aligned} &\underset{\mathbf{x}}{\text{minimize}} \quad f(\mathbf{x}) \\ &\text{s.t.} \qquad h(\mathbf{x}) + \varepsilon = 0 \end{aligned} \tag{7.38}$$

where we have added a scalar variable ε to the constraint equation. When $\varepsilon = 0$, we recover the original problem. Clearly, the optimal solution depends on ε, i.e., we have $\mathbf{x}^*(\varepsilon)$. Similarly, the minimum objective value will depend on ε, i.e., we will have a dependency via $f(\mathbf{x}^*(\varepsilon))$.

Consider the derivative

$$\frac{df\left(\mathbf{x}^*(\varepsilon)\right)}{d\varepsilon} = \frac{\partial f}{\partial x_1}\frac{\partial x_1^*}{\partial \varepsilon} + \frac{\partial f}{\partial x_2}\frac{\partial x_2^*}{\partial \varepsilon} + \cdots \tag{7.39}$$

At $\varepsilon = 0$,

$$\left.\frac{df\left(\mathbf{x}^*(\varepsilon)\right)}{d\varepsilon}\right|_{\varepsilon=0} = (\nabla f)^T \nabla_\varepsilon \mathbf{x}^*|_{\varepsilon=0} \tag{7.40}$$

Also consider the derivative of the constraint in Equation (7.38) with respect to ε at $\varepsilon = 0$:

$$(\nabla h)^T \nabla_\varepsilon \mathbf{x}^*|_{\varepsilon=0} + 1 = 0 \tag{7.41}$$

Multiplying Equation (7.41) by the Lagrange multiplier at $\varepsilon = 0$, i.e., μ^*, and adding this to Equation (7.40), we have

$$\left.\frac{df\left(\mathbf{x}^*(\varepsilon)\right)}{d\varepsilon}\right|_{\varepsilon=0} = (\nabla f)^T \nabla_\varepsilon \mathbf{x}^*|_{\varepsilon=0} + \mu^* \left[(\nabla h)^T \nabla_\varepsilon \mathbf{x}^* \Big|_{\varepsilon=0} + 1 \right] \tag{7.42}$$

i.e.,

$$\left.\frac{df\left(\mathbf{x}^*(\varepsilon)\right)}{d\varepsilon}\right|_{\varepsilon=0} = [\nabla f(\mathbf{x}^*) + \mu^* \nabla h(\mathbf{x}^*)]^T \nabla_\varepsilon \mathbf{x}^* + \mu^* \tag{7.43}$$

However, from Equation (7.37), the term within the square brackets vanishes. Thus,

$$\left.\frac{df\left(\mathbf{x}^*(\varepsilon)\right)}{d\varepsilon}\right|_{\varepsilon=0} = \mu^* \tag{7.44}$$

That is, *the change in the objective when a constraint is perturbed slightly is captured precisely by the Lagrange multiplier*. Thus, we don't need to solve the perturbed problem if the main goal is to estimate the impact on the objective due to the perturbation.

Example 7.6 Recall the problem

$$\begin{aligned} &\underset{x,y}{\text{minimize}} \quad x^2 + y^2 \\ &\text{s.t.} \qquad x + y - 1 = 0 \end{aligned} \tag{7.45}$$

whose solution is given by $(0.5, 0.5)$, $\mu^* = -1$, and the objective at the local minimum is 0.5. Now consider a perturbed problem:

$$\begin{aligned} &\underset{x,y}{\text{minimize}} \quad x^2 + y^2 \\ &\text{s.t.} \qquad (x + y - 1) + 0.01 = 0 \end{aligned} \tag{7.46}$$

As before, the stationary point satisfies

$$\begin{bmatrix} 2 & 0 & 1 \\ 0 & 2 & 1 \\ 1 & 1 & 0 \end{bmatrix} \begin{Bmatrix} x \\ y \\ \mu \end{Bmatrix} = \begin{Bmatrix} 0 \\ 0 \\ 0.99 \end{Bmatrix} \tag{7.47}$$

whose solution is given by $(0.495, 0.495)$, and the new objective is 0.49, i.e.,

$$\frac{df^*(\alpha)}{d\alpha} \approx \frac{\Delta f}{\Delta \alpha} = \frac{0.49 - 0.50}{0.01} = -1$$

which is equal to the Lagrange multiplier of the unperturbed problem. Thus, we do not need to solve the perturbed problem.

The sensitivity relationship is particularly useful in many engineering applications. For example, suppose we are solving an optimization problem in which the objective is to minimize the weight of a structure subject to a stress constraint:

$$\begin{aligned} &\underset{\mathbf{x}}{\text{minimize}} \quad \text{weight} \\ &\text{s.t.} \qquad \sigma = \sigma_{\text{allowable}} \end{aligned} \tag{7.48}$$

Suppose the optimal weight has been determined. Now, the designer may be interested in determining the optimal weight if a different material (with a different allowable stress) is chosen. Equation (7.44) can be used to answer such questions without having to solve the modified optimization problem.

7.3 Optimality Criteria: Inequality Constraints

Next, we consider problems with equality and inequality constraints of the form

$$\begin{aligned} &\underset{\mathbf{x}}{\text{minimize}} \quad f(\mathbf{x}) \\ &\text{s.t.} \qquad h_i(\mathbf{x}) = 0;\ i = 1, 2, \ldots, M \\ &\qquad\quad\ \ g_j(\mathbf{x}) \le 0;\ j = 1, 2, \ldots, L \end{aligned} \tag{7.49}$$

Recall that the *feasible region* is the set of points that satisfy

$$\begin{aligned} h_i(\mathbf{x}) &= 0;\ i = 1, 2, \ldots, M \\ g_j(\mathbf{x}) &\leq 0;\ j = 1, 2, \ldots, L \end{aligned} \tag{7.50}$$

Therefore, the objective is to find $\mathbf{x}^*$ in the feasible region which minimizes $f(\mathbf{x})$. We will start with a simple problem to illustrate the role of such constraints.

Example 7.7 Using geometric reasoning, find the optimal solution to the problem

$$\begin{aligned} &\underset{x,y}{\text{minimize}} \quad x^2 + y^2 \\ &\text{s.t.} \qquad x + y - 1 \leq 0 \end{aligned} \tag{7.51}$$

Solution: The line $x + y - 1 = 0$ divides the space into two regions (see Figure 7.3), namely (a) a *feasible* region that contains all points satisfying the constraint $x + y - 1 \leq 0$, and (b) an *infeasible* region. Since the origin $(0, 0)$ yields the least possible value for the objective and it lies in the feasible region, it is the optimal solution. No further analysis is needed; the constraint is said to be *inactive*.

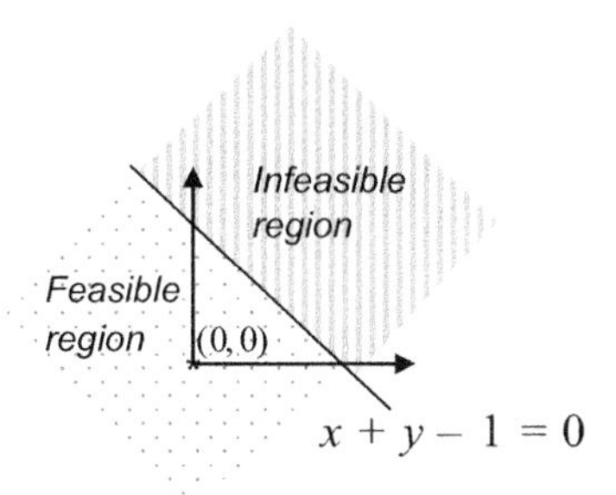

Figure 7.3 Feasible and infeasible regions for $x + y - 1 \leq 0$.

Example 7.8 Using geometric reasoning, find the optimal solution to the problem

$$\begin{aligned} &\underset{x,y}{\text{minimize}} \quad x^2 + y^2 \\ &\text{s.t.} \qquad 1 - x - y \leq 0 \end{aligned} \tag{7.52}$$

Solution: The feasible and infeasible regions are now reversed (see Figure 7.4). The minimum value of the unconstrained objective occurs at (0, 0); this point lies in the infeasible zone, and is therefore unacceptable. In other words, the actual solution must lie on the boundary of the feasible zone. It is easy to see from Figure 7.4 that the point $(0.5, 0.5)$ is closest to the origin within the feasible region, and is therefore the local minimum. In this case, the constraint is said to be *active*.

(cont.)

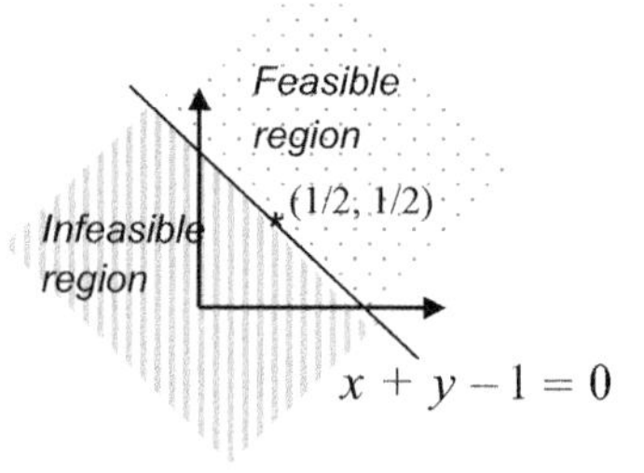

Figure 7.4 Feasible and infeasible regions for $1 - x - y \leq 0$.

Example 7.9 Consider next a problem that contains both equality and inequality constraints:

$$\begin{aligned} \underset{x,y}{\text{minimize}} \quad & x^2 + y^2 \\ \text{s.t.} \quad & x - 1 = 0 \\ & (x-2)^2 + y^2 - 1 \leq 0 \end{aligned} \tag{7.53}$$

Solution: Geometrically, the feasible region is the *intersection* of the line $x - 1 = 0$ and the region within (and including) the circle $(x-2)^2 + y^2 - 1 = 0$, i.e., the feasible region is the single point $(1, 0)$; see Figure 7.5. Therefore, the optimal solution is the point $(1, 0)$.

In the above examples, we used simple geometric reasoning to find the optimal solution. For more complex problems, we will need a formal theory and numerical algorithms.

To understand the underlying theory, consider a problem with a single inequality constraint (and no equality constraints):

$$\begin{aligned} \underset{\mathbf{x}}{\text{minimize}} \quad & f(\mathbf{x}) \\ \text{s.t.} \quad & g(\mathbf{x}) \leq 0 \end{aligned} \tag{7.54}$$

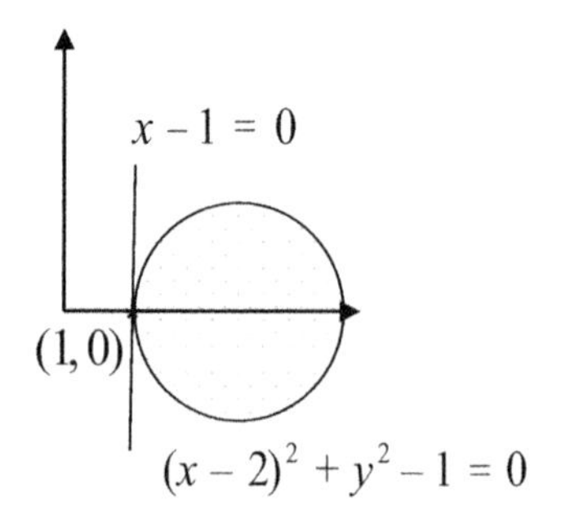

Figure 7.5 Feasible region for Equation (7.53) consists of a single point $(1, 0)$.

Let $\mathbf{x}^*$ be a local minimum to this problem. There are two possible scenarios: either (a) $g(\mathbf{x}^*) < 0$, or (b) $g(\mathbf{x}^*) = 0$. In other words, the point $\mathbf{x}^*$ lies either in the interior of the feasible region, or on the boundary of the constraint. Let us consider the two cases.

Case (a): If $\mathbf{x}^*$ lies in the interior, i.e., if $g(\mathbf{x}^*) < 0$, the constraint is said to be *inactive*. Since the point lies inside the feasible region, deleting the constraint does not affect the solution (as in Example 7.7), i.e., $\mathbf{x}^*$ can be determined by solving the unconstrained problem

$$\underset{\mathbf{x}}{\text{minimize}} \quad f(\mathbf{x}) \tag{7.55}$$

Therefore, $\mathbf{x}^*$ must satisfy the following first-order condition:

$$\begin{aligned} &\nabla f(\mathbf{x}^*) = 0 \\ &g(\mathbf{x}^*) < 0 \end{aligned} \tag{7.56}$$

Case (b): If $\mathbf{x}^*$ lies on the boundary, i.e., if $g(\mathbf{x}^*) = 0$, the constraint is said to be *active*. By definition, Equation (7.54) reduces to the equality-constrained problem

$$\begin{aligned} &\underset{\mathbf{x}}{\text{minimize}} \; f(\mathbf{x}) \\ &\text{s.t.} \qquad g(\mathbf{x}) = 0 \end{aligned} \tag{7.57}$$

Therefore, $\mathbf{x}^*$ must satisfy the following first-order condition:

$$\begin{aligned} &\nabla f(\mathbf{x}^*) + \lambda^* \nabla g(\mathbf{x}^*) = 0 \\ &g(\mathbf{x}^*) = 0 \end{aligned} \tag{7.58}$$

for some λ^*.

Thus, in conclusion, for the problem in Equation (7.54), either Equation (7.56) or Equation (7.58) must be satisfied at a local minimum $\mathbf{x}^*$. Further, observe that Equations (7.56) and (7.58) can be combined into a single equation:

$$\begin{aligned} &\nabla f(\mathbf{x}^*) + \lambda^* \nabla g(\mathbf{x}^*) = 0 \\ &\lambda^* g(\mathbf{x}^*) = 0 \\ &g(\mathbf{x}^*) \leq 0 \end{aligned} \tag{7.59}$$

where $\lambda^* g(\mathbf{x}^*) = 0$, referred to as the *complementarity condition*, states that either $\lambda^* = 0$ or $g(\mathbf{x}^*) = 0$. The reader can verify that both Equations (7.56) and (7.58) are captured in Equation (7.59). An additional constraint must, however, be imposed on λ^*, as the following argument reveals.

Consider again the case where $\mathbf{x}^*$ lies on the boundary, i.e., $g(\mathbf{x}^*) = 0$. Since the gradient $\nabla g(\mathbf{x}^*)$ always points in the direction of increasing constraint value $g(\mathbf{x})$, it must be directed from the feasible region (where $g(\mathbf{x}) < 0$) to the infeasible region (where $g(\mathbf{x}) > 0$). Further, from Equation (7.59),

$$\nabla f(\mathbf{x}^*) = -\lambda^* \nabla g(\mathbf{x}^*) \tag{7.60}$$

Now suppose $\lambda^* < 0$. Since $\nabla g(\mathbf{x}^*)$ points from feasible to infeasible region, $\nabla f(\mathbf{x}^*)$ must point in the same direction. In other words, the function f is increasing toward the infeasible region and decreasing toward the feasible region. Thus, we can always take an infinitesimal step into the feasible region and reduce f, i.e., $\mathbf{x}^*$ cannot possibly be the minimum. Thus, we conclude that $\lambda^* \geq 0$. Adding this additional constraint to Equation (7.59), we arrive at the following theorem.

Theorem 7.3 *For the problem*

$$\begin{aligned} &\underset{\mathbf{x}}{\text{minimize}} \; f(\mathbf{x}) \\ &\text{s.t.} \qquad g(\mathbf{x}) \leq 0 \end{aligned} \tag{7.61}$$

the local minimum x^* must satisfy the following first-order necessary condition:

$$\begin{aligned} &\nabla f(\mathbf{x}^*) + \lambda^* \nabla g(\mathbf{x}^*) = 0 \\ &g(\mathbf{x}^*) \leq 0 \\ &\lambda^* g(\mathbf{x}^*) = 0 \\ &\lambda^* \geq 0 \end{aligned} \tag{7.62}$$

Further, a point $\mathbf{x}^*$ that satisfies Equation (7.62) is a local minimum if the matrix

$$\mathbf{H}_f(\mathbf{x}^*) + \lambda^* \mathbf{H}_g(\mathbf{x}^*) \tag{7.63}$$

is positive definite.

Proof: See reference [8].

Let us now revisit the examples considered in the previous section, and verify that Equations (7.62) and (7.63) are indeed satisfied.

Example 7.10 Consider the problem of Equation (7.51), where we established that $\mathbf{x}^* = (0, 0)$ is the solution. Show that the solution satisfies Theorem 7.3.

Solution: Since $g(\mathbf{x}^*) < 0$, the constraint is inactive, i.e., from the complementarity condition, $\lambda^* = 0$. Further,

$$\begin{aligned} \nabla f(\mathbf{x}^*) &= \begin{Bmatrix} 2x \\ 2y \end{Bmatrix}_{\mathbf{x}^*} = \begin{Bmatrix} 0 \\ 0 \end{Bmatrix} \\ \nabla g(\mathbf{x}^*) &= \begin{Bmatrix} 1 \\ 1 \end{Bmatrix}_{\mathbf{x}^*} = \begin{Bmatrix} 1 \\ 1 \end{Bmatrix} \end{aligned} \tag{7.64}$$

Thus it is now easy to verify that Equation (7.62) is satisfied. Further, observe that

$$\begin{aligned} \mathbf{H}_f(\mathbf{x}^*) &= \begin{bmatrix} 2 & 0 \\ 0 & 2 \end{bmatrix} \\ \mathbf{H}_g(\mathbf{x}^*) &= \begin{bmatrix} 0 & 0 \\ 0 & 0 \end{bmatrix} \end{aligned} \tag{7.65}$$

Thus $H_f(\mathbf{x}^*) + \lambda^* H_g(\mathbf{x}^*)$ is positive definite.

Example 7.11 Consider the problem in Example 7.8, with the solution $\mathbf{x}^* = (0.5, 0.5)$. Show that the solution satisfies Theorem 7.3.

Solution: Since $g(\mathbf{x}^*) = 0$, the constraint is active, i.e., we must find a suitable $\lambda^* > 0$. Observe that

$$\begin{aligned} \nabla f(\mathbf{x}^*) &= \begin{Bmatrix} 2x \\ 2y \end{Bmatrix}_{\mathbf{x}^*} = \begin{Bmatrix} 1 \\ 1 \end{Bmatrix} \\ \nabla g(\mathbf{x}^*) &= \begin{Bmatrix} -1 \\ -1 \end{Bmatrix}_{\mathbf{x}^*} = \begin{Bmatrix} -1 \\ -1 \end{Bmatrix} \end{aligned} \tag{7.66}$$

From Equation (7.62), we must find $\lambda^* > 0$ satisfying

$$\nabla f(\mathbf{x}^*) + \lambda^* \nabla g(\mathbf{x}^*) = 0 \tag{7.67}$$

It is easy to verify that $\lambda^* = 1$ satisfies Equation (7.67). With this choice, one can verify that the remaining parts of Equation (7.62) are satisfied. Moreover, $\mathbf{H}_f(\mathbf{x}^*) + \lambda^* \mathbf{H}_g(\mathbf{x}^*)$ is positive definite.

Theorem 7.3 may be generalized to multiple inequality constraints.

Theorem 7.4 *For the problem*

$$\begin{aligned} &\underset{\mathbf{x}}{\text{minimize}} \; f(\mathbf{x}) \\ &\text{s.t.} \qquad g_j(\mathbf{x}) \leq 0; \; j = 1, 2, \ldots, L \end{aligned} \tag{7.68}$$

the minimum $\mathbf{x}^*$ *must satisfy the first-order condition*

$$\begin{aligned} &\nabla f(\mathbf{x}^*) + \sum_{j=1}^{L} \lambda_j^* \nabla g_j(\mathbf{x}^*) = 0 \\ &\left\{ \begin{matrix} g_j(\mathbf{x}^*) \leq 0 \\ \lambda_j^* g_j(\mathbf{x}^*) = 0 \\ \lambda_j^* \geq 0 \end{matrix} \right\}; j = 1, 2, \ldots, L \end{aligned} \tag{7.69}$$

for some set of scalars λ_j^*. *Further, a point* $\mathbf{x}^*$ *that satisfies Equation (7.69) is a minimum if the matrix*

$$\mathbf{H}_f(\mathbf{x}^*) + \sum_{j=1}^{L} \lambda_j^* \mathbf{H}_{g_j}(\mathbf{x}^*) \tag{7.70}$$

is positive definite.

Proof: See reference [8].

Finally, we combine equality constraints and inequality constraints into a single Karush–Kuhn–Tucker (KKT) theorem.

Theorem 7.5 *For the problem*

$$\begin{aligned} &\underset{\mathbf{x}}{\text{minimize}} \; f(\mathbf{x}) \\ &\text{s.t.} \quad h_i(\mathbf{x}) = 0;\; i = 1, 2, ..., M \\ &\qquad\; g_j(\mathbf{x}) \le 0;\; j = 1, 2, ..., L \end{aligned} \tag{7.71}$$

the minimum $\mathbf{x}^*$ *must satisfy the first-order condition*

$$\begin{aligned} &\nabla f(\mathbf{x}^*) + \sum_{i=1}^{M} \mu_i^* \nabla h_i(\mathbf{x}^*) + \sum_{j=1}^{L} \lambda_j^* \nabla g_j(\mathbf{x}^*) = 0 \\ &h_i(\mathbf{x}^*) = 0;\; i = 1, 2, \ldots, M \\ &\left\{ \begin{array}{l} g_j(\mathbf{x}^*) \le 0 \\ \lambda_j^* g_j(\mathbf{x}^*) = 0 \\ \lambda_j^* \ge 0 \end{array} \right\};\; j = 1, 2, \ldots, L \end{aligned} \tag{7.72}$$

for some set of scalars μ_i^* *and* λ_j^*. *Further, a point* $\mathbf{x}^*$ *that satisfies Equation (7.72) is a minimum if the matrix*

$$\mathbf{H}_f(\mathbf{x}^*) + \sum_{i=1}^{M} \mu_i^* \mathbf{H}_{h_i}(\mathbf{x}^*) + \sum_{j=1}^{L} \lambda_j^* \mathbf{H}_{g_j}(\mathbf{x}^*) \tag{7.73}$$

is positive definite.

Proof: See reference [8].

Example 7.12 Consider the problem in Example 7.9 with the solution $(1, 0)$. Show that the solution satisfies Theorem 7.5.

Solution: Observe that

$$\begin{aligned} \nabla f(\mathbf{x}) &= \left\{ \begin{array}{c} 2x \\ 2y \end{array} \right\} \\ \nabla h(\mathbf{x}) &= \left\{ \begin{array}{c} 1 \\ 0 \end{array} \right\} \\ \nabla g(\mathbf{x}) &= \left\{ \begin{array}{c} 2(x-2) \\ 2y \end{array} \right\} \end{aligned} \tag{7.74}$$

At the local minimum, $\mathbf{x}^* = (1, 0)$,

$$\begin{aligned} \nabla f(\mathbf{x}^*) &= \left\{ \begin{array}{c} 2 \\ 0 \end{array} \right\} \\ \nabla h(\mathbf{x}^*) &= \left\{ \begin{array}{c} 1 \\ 0 \end{array} \right\} \\ \nabla g(\mathbf{x}^*) &= \left\{ \begin{array}{c} -2 \\ 0 \end{array} \right\} \end{aligned} \tag{7.75}$$

(cont.)

Thus, we must find μ^* and $\lambda^* \geq 0$ satisfying

$$\nabla f(\mathbf{x}^*) + \mu^* \nabla h(\mathbf{x}^*) + \lambda^* \nabla g(\mathbf{x}^*) = 0 \tag{7.76}$$

i.e.,

$$\begin{Bmatrix} 4 \\ 0 \end{Bmatrix} + \mu^* \begin{Bmatrix} 1 \\ 0 \end{Bmatrix} + \lambda^* \begin{Bmatrix} -2 \\ 0 \end{Bmatrix} = \begin{Bmatrix} 0 \\ 0 \end{Bmatrix} \tag{7.77}$$

There are an infinite number of solutions; for example, $\mu^* = -4$ and $\lambda^* = 0$ satisfies Equation (7.77). With this choice, one can verify that the remaining parts of the theorem are also satisfied. Moreover, one can show that $\mathbf{H}_f(\mathbf{x}^*) + \mu^* \mathbf{H}_f(\mathbf{x}^*) + \lambda^* \mathbf{H}_g(\mathbf{x}^*)$ is positive definite.

7.4 The `fmincon` Method

Thus far we have considered the theory behind constrained optimization problems. Next, we consider numerical methods for solving such problems. There are several popular methods, including the penalty method, the augmented Lagrangian method, the active set method and so on. These are built upon the unconstrained methods discussed earlier, and the reader will be asked to implement these as exercises. Here, we will demonstrate the use of `fmincon`, a popular constrained optimization method in the MATLAB toolbox. The `fmincon` method is designed to solve the following constrained problem:

$$\begin{aligned}
&\underset{\mathbf{x}}{\text{minimize}} && f(\mathbf{x}) \\
&\text{s.t.} && \mathbf{A}\mathbf{x} \leq \mathbf{b} \\
& && \mathbf{A}_{eq}\mathbf{x} = \mathbf{b}_{eq} \\
& && \mathbf{x}_{\min} \leq \mathbf{x} \leq \mathbf{x}_{\max} \\
& && h_i(\mathbf{x}) = 0; \ i = 1, 2, ..., M \\
& && g_j(\mathbf{x}) \leq 0; \ j = 1, 2, ..., L
\end{aligned} \tag{7.78}$$

From a theoretical perspective, the linear and bound constraints are just "regular" constraints. However, if some of the constraints are linear, it is advantageous to separate them out prior to numerical optimization.

The syntax for `fmincon` is

```
>> X=fmincon(FUN,X0,A,b,Aeq,beq,xMin,xMax,NONLCON,OPTIONS)
```

We will study the use of `fmincon` through a few examples; the reader is encouraged to study its full description on the MathWorks homepage (www.mathworks.com).

7.4.1 Basics

We will start with equality-constrained problems.

Example 7.13 Using MATLAB's `fmincon`, solve the problem

$$\begin{aligned} \underset{x,y}{\text{minimize}} \quad & x^2 + y^2 \\ \text{s.t.} \quad & x + y - 1 = 0 \end{aligned} \tag{7.79}$$

Solution: This problem is equivalent to

$$\begin{aligned} \underset{x_1,x_2}{\text{minimize}} \quad & x_1^2 + x_2^2 \\ \text{s.t.} \quad & x_1 + x_2 - 1 = 0 \end{aligned} \tag{7.80}$$

The objective is easily captured via MATLAB's *anonymous function*:

```
>> f = @(x) x(1)^2 + x(2)^2;
```

There are multiple ways of capturing the constraint. Since we have linear constraints of the form

$$[1 \quad 1]\begin{Bmatrix} x_1 \\ x_2 \end{Bmatrix} = 1 \tag{7.81}$$

we define

```
>> Aeq = [1 1];
>> Beq = [1];
```

and

```
>> x0 = [0 0];
```

Finally, the `fmincon` method may be called as below. Observe that the third and fourth arguments are empty since we do not have an inequality constraint. The local minimum is as expected.

```
>> [xmin,fmin,flag,output,lambda]=
fmincon(f,x0,[],[],Aeq,Beq);

>> Local minimum found that satisfies the constraints.
xmin =

  0.500000000000000   0.500000000000000

fmin =
  0.500000000000000
```

(cont.)

```
flag=
      1
output=
          iterations:2
           funcCount:9

lambda
          eqlin:   -1.000000014901161
       eqnonlin:   [0x1 double]
        ineqlin:   [0x1 double]
          lower:   [2x1 double]
          upper:   [2x1 double]
     ineqnonlin:   [0x1 double]
```

The Lagrange multiplier value of -1, corresponding to the constraint, is part of the lambda structure above.

Example 7.14 Using MATLAB's `fmincon`, solve the problem

$$\begin{aligned} &\underset{x,y}{\text{minimize}} && x^2 + y^2 \\ &\text{s.t.} && x + y - 1 \geq 0 \end{aligned} \tag{7.82}$$

Solution: In the standard form

$$\begin{aligned} &\underset{x_1,x_2}{\text{minimize}} && x_1^2 + x_2^2 \\ &\text{s.t.} && -x_1 - x_2 \leq -1 \end{aligned} \tag{7.83}$$

```
>> f = @(x) x(1)^2 + x(2)^2;
```

we define

```
>> A = [-1 -1];
>> B = [-1];
```

and

```
>> x0 = [0 0];
```

(cont.)

Finally,

```
>> [xmin,fmin,flag,output,lambda]= fmincon(f,x0,A,B);
>> Local minimum found that satisfies the constraints.
xmin =
   0.499999989702791   0.500000030301298
fmin =
   0.500000020004090
flag =
     1
output =
         iterations: 8
          funcCount: 28
...
lambda =
         eqlin: [0x1 double]
      eqnonlin: [0x1 double]
       ineqlin: 1.000000029617534
         lower: [2x1 double]
         upper: [2x1 double]
    ineqnonlin: [0x1 double]
```

Observe that the number of function calls increases when the constraint is changed from an equality to an inequality. The Lagrange multiplier is once again part of the lambda structure above.

Example 7.15 Using MATLAB's `fmincon`, solve the problem

$$\begin{aligned} \underset{x,y}{\text{minimize}} \quad & x^2 + y^2 + z^2 \\ \text{s.t.} \quad & x + y + z - 1 = 0 \\ & x + 2y + 3z - 4 = 0 \end{aligned} \tag{7.84}$$

Solution: Once again, the constraints are linear; therefore, the problem may be captured via

```
>> f = @(x) x(1)^2 + x(2)^2 + x(3)^2;
>> Aeq = [1 1 1; 1 2 3];
>> Beq = [1 4];
>> x0 = [0 0 0];
```

(cont.)

Then the minimum and the Lagrange multipliers may be computed as before:

```
>> [xmin,fmin,flag,output,lambda]=
fmincon(f,x0,[],[],Aeq,Beq);
>> xmin
xmin =
   -0.6667    0.3333    1.3333
```

The Lagrange multipliers corresponding to the constraints are part of the lambda structure.

```
>> lambda.eqlin
ans =
   1.6667
   -1.0000
```

These are half of what was obtained earlier, since the multipliers are defined differently within MATLAB.

Example 7.16 Using MATLAB's `fmincon`, solve the problem

$$\begin{aligned} \underset{x,y}{\text{minimize}} \quad & x^2 + y^2 \\ \text{s.t.} \quad & (x-2)^2 + y^2 - 1 = 0 \end{aligned} \tag{7.85}$$

Solution: The objective and initial guess are as before:

```
>> f = @(x) x(1)^2 + x(2)^2;
>> x = [0 0];
```

However, the constraint is non-linear; we therefore create a script file `constraint1.m`, with the contents shown below. Note that since there are no inequality constraints, the first argument is empty.

```
function [cineq,ceq] = constraint1(x)
cineq = [];
ceq = (x(1)-2)^2 + x(2)^2 - 1;
```

Then,

```
>> [xmin,fmin,flag,output,lambda] =
fmincon(f,x0,[],[],[],[],[],[],@constraint1);
>> xmin
xmin =
    1.0000    0.0000
```

The Lagrange multiplier corresponding to the non-linear equality constraint can be extracted:

```
>> lambda.eqnonlin
1.0000
```

Recall that, for solving unconstrained problems, `fminunc` was not very effective when the objective was noisy. We can draw a similar conclusion about `fmincon` through the following example.

Example 7.17 Using MATLAB's `fmincon`, solve

$$\begin{aligned} \underset{x}{\text{minimize}} \quad & (x-1)^2 + 0.01(0.5 - \text{rand}(0,1)) \\ \text{s.t.} \quad & x \leq 2 \end{aligned} \tag{7.86}$$

with $x = 2$ as an initial guess; the exact answer (without the noise) is $x = 1$.

Solution: We have

```
>> f = @(x) (x-1)^2 + 0.01*(0.5-rand(1));
>> x0 = 2;
>> A = 1;
>> B = 2;

>> [xmin,fmin,flag,output,lambda] =fmincon(f,x0,A,B);
xmin =
   1.325409461543814
fmin =
   0.100929970495966
flag =
     2
output =
  struct with fields:
         iterations: 11
          funcCount: 90
```

The method failed to converge.

7.4.2 Exploiting Gradients

In the previous examples, we did not explicitly provide the gradient of the objective or the gradient of the constraint to `fmincon`, i.e., the method implicitly uses finite difference (to be discussed in Chapter 11) to compute the gradients. We will now redo some of these examples with explicit gradients, thereby improving the efficiency of the algorithm.

Example 7.18 Consider the problem

$$\begin{aligned} \underset{x,y}{\text{minimize}} \quad & (x-3)^4 + (y-1)^2 + z^2 \\ \text{s.t.} \quad & x + y + z - 1 = 0 \\ & x + 2y + 3z - 4 = 0 \end{aligned} \tag{7.87}$$

(cont.)

Use fmincon to solve this with and without an explicit objective gradient.

Solution: Note that the gradient of the objective is given by

$$\nabla f = \begin{Bmatrix} 4(x-3)^3 \\ 2(y-1) \\ 2z \end{Bmatrix} \tag{7.88}$$

The script function below optionally returns the gradient.

```
function [f,g] = fminconTest(xVec)
x = xVec(1);y = xVec(2);z = xVec(3);
f = (x-3).^4 + (y-1).^2 + z.^2;
if (nargout > 1), g = [4*(x-3).^3; 2*(y-1); 2*z];end
end
```

Since the constraints are linear, gradients are captured directly by the coefficients! There is no need to provide gradient information.

```
>> Aeq = [1 1 1; 1 2 3];
>> Beq = [1 4];
```

We will use the origin as the initial guess.

```
>> x0 = [0 0 0];
```

First, we will find the minimum, without the gradient (default).

```
[xmin,fmin,flag,output,lambda]= fmincon(@(x)fminconTest(x),
x0,[],[],Aeq,Beq,[],[],[]);
>> xmin =
   1.186334775358108  -3.372669550716216   3.186334775358108
>> output.funcCount
ans =
    36
```

We will now use the optimoptions method to indicate that an explicit gradient is available. Note that, for fmincon, the correct method of setting optimization options is through optimoptions, rather than through optimset.

```
>> opt=optimoptions('fmincon',
'SpecifyObjectiveGradient',true);
```

Then the minimum and the Lagrange multiplier may be determined as before, but with an additional argument at the end.

```
>> [xmin,fmin,flag,output,lambda]=
fmincon(@(x)fminconTest(x), x0,[],[],Aeq,Beq,[],[],[],opt);
>> xmin =
   1.186334787882923  -3.372669575765846
3.186334787882924
>> output.funcCount
ans =
    12
```

(cont.)

Observe that the number of function calls has reduced by a factor of 3.

One can also exploit constraint gradients, if available (see Section 7.5).

7.5 Constrained Spring Problem

To illustrate the use of `fmincon` for solving non-analytic problems, we will revisit the spring potential energy minimization problem, discussed in Chapter 5, but we will add a non-linear constraint.

Example 7.19 Solve the equality-constrained two-spring problem

$$\begin{aligned} \underset{u,v}{\text{minimize}} \quad & \Pi = \left\{ \begin{aligned} &\frac{1}{2}100\left(\sqrt{u^2+(v+1)^2}-1\right)^2 \\ &+\frac{1}{2}50\left(\sqrt{u^2+(v-1)^2}-1\right)^2 - (10u+8v) \end{aligned} \right\} \\ \text{s.t.} \quad & u^3 - v = 0 \end{aligned} \tag{7.89}$$

using `fmincon`. Essentially, the constraint forces the free node to follow a track $y = x^3$.

Solution: Observe that the gradient of the objective is given by

$$\nabla\Pi = \left\{ \begin{aligned} &\frac{100\,(L_{12}-1)u}{L_{12}} + \frac{50\,(L_{13}-1)u}{L_{13}} - 10 \\ &\frac{100\,(L_{12}-1)(1+v)}{L_{12}} - \frac{50\,(L_{13}-1)(1-v)}{L_{13}} - 8; \end{aligned} \right\}$$

where

$$\begin{aligned} L_{12} &= \sqrt{(u^2+(1+v)^2)} \\ L_{13} &= \sqrt{(u^2+(1-v)^2)} \end{aligned} \tag{7.90}$$

The gradient of the constraint is given by

$$\nabla h = \left\{ \begin{matrix} 3u^2 \\ -1 \end{matrix} \right\} \tag{7.91}$$

For convenience, the objective, constraint and gradients are captured in the class below. Observe the following:

- The class is built on top of a utility class that helps plot the spring system.
- The potential energy method returns the objective and its gradient for an arbitrary displacement (u, v).

(cont.)

- The `constraintsEquality` method returns the equality constraint of Equation (7.89) and an empty inequality constraint, and their respective gradients.
- A `constraintsInequality` method is also implemented where the equality constraint is replaced by an inequality constraint, and will be discussed in Example 7.20.
- An additional method, namely force balance, is included, and will be used in a later example.

```
classdef twoSpringWithConstraints < spring2dPlotter
    methods
        function obj = twoSpringWithConstraints()
            % two-spring nodes and connectivity
            nodeXY = [0 0; 0 -1; 0 1]'; % 2 x nNodes
            connectivity = [1 2; 1 3]'; % 2 x nSprings
            obj = obj@spring2dPlotter(nodeXY,connectivity);
        end
        function [PE,gradient_f] = potentialEnergy(obj,uv)
            % returns the potential energy of a 2-spring
            fx1 = 10;
            fy1 = 8;
            k1 = 100; % stiffness of spring 1
            k2 = 50;% stiffness of spring 2
            u = uv(1);v = uv(2);
            L12 = sqrt(u^2+(1+v)^2); % deformed lengths
            L13 = sqrt(u^2+(1-v)^2);
            deltaL12 = (L12 -1);
            deltaL13 = (L13 -1);
            U = 0.5*k1*deltaL12.^2 + 0.5*k2*deltaL13.^2;%
            W = 0.5*(fx1*u+fy1*v); % Work done
            PE = U-2*W; % potential energy
            if (nargout == 2)
                gradient_f(1) = k1*(L12-1)/L12*u + k2*(L13-
1)/L13*u - fx1;
                gradient_f(2) = k1*(L12-1)/L12*(1+v) +
k2*(L13-1)/L13*(v-1) - fy1;
            end
        end
        function [residual] = forceBalance(~,uv)
            fx1 = 10;
```

(cont.)

```
                fyl = 8;
                k1 = 100; % stiffness of spring 1
                k2 = 50;% stiffness of spring 2
                u = uv(1);v = uv(2);
                L12 = sqrt(u^2+(1+v)^2); % deformed lengths
                L13 = sqrt(u^2+(1-v)^2);
                residual(1) = k1*(L12-1)/L12*u + k2*(L13-
1)/L13*u -fx1;
                residual(2) = k1*(L12-1)/L12*(1+v) + k2*(L13-
1)/L13*(v-1) - fyl;
            end
            function [cineq,ceq,grad_cineq, grad_ceq] =
constraintsEquality(~,uv)
                u = uv(1);v = uv(2);
                cineq = [];
                ceq(1) = u.^3 - v;
                grad_cineq = [];
                grad_ceq = [3*u.^2; -1];
            end
           function [cineq,ceq,grad_cineq, grad_ceq] =
constraintsInequality(~,uv)
                u = uv(1);v = uv(2);
                ceq = [];
                cineq(1) =  u.^3 - v;
                grad_ceq = [];
                grad_cineq = [3*u.^2; -1];
            end
        end
end
```

We first solve the constrained spring problem without explicit use of the gradient, as follows:

```
>> s = twoSpringWithConstraints();
>> s.plot(); % plot spring without deformation
>> [soln,PEMin,Flag,Output]=
fmincon(@(x)s.potentialEnergy(x), [0 0],
[],[],[],[],[],[],@(x)s.constraintsEquality(x));
>> disp(soln)
0.431522334089445    0.080354431823819
```

The solution, as one can verify, satisfies the constraint $v - u^3 = 0$; see Figure 7.6. The number of function calls can be determined as follows:

```
>> Output.funcCount
ans =
    35
```

Finally, one can plot the deformed spring as follows:

(cont.)

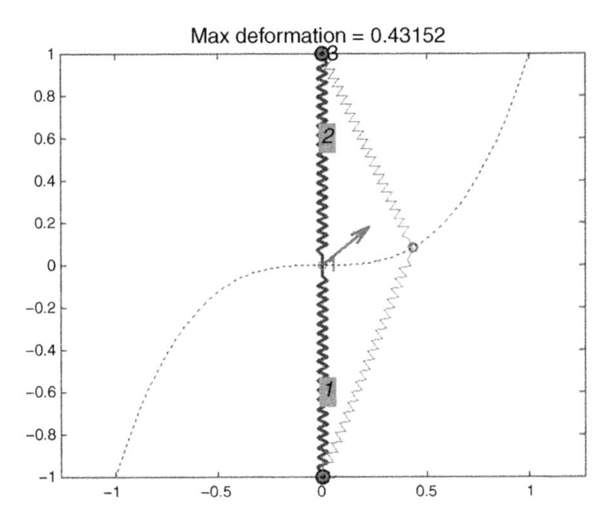

Figure 7.6 Solution to the equality-constrained spring system.

```
>> s = s.setUVAtNode(1,soln);
>> s.plotDeformed();
>> u = -1:0.01:1;
>> v = u.^3;
>> plot(u,v,':b'); % plot the constraint
```

We will now use the `optimoptions` method to indicate that explicit gradients for the objective and constraint are available:

```
opt=optimoptions ('fmincon',
'SpecifyObjectiveGradient',true,'SpecifyConstraintGradient'
, true);
[soln,PEMin,Flag,Output]= fmincon(@(x)s.potentialEnergy(x),
[0 0], [],[],[],[],[],[],@(x)s.constraintsEquality(x),opt);
disp(soln)
0.431522342592882    0.080354436574133
```

We converge approximately to the same solution, but the number of function calls has decreased:

```
>> Output.funcCount
ans = 15
```

In the next example, we change the equality constraint into an inequality constraint.

Example 7.20 Solve the inequality-constrained two-spring problem

$$\begin{aligned} \underset{u,v}{\text{minimize}} \quad & \Pi = \left\{ \begin{aligned} & \frac{1}{2}100\left(\sqrt{u^2+(v+1)^2}-1\right)^2 \\ & +\frac{1}{2}50\left(\sqrt{u^2+(v-1)^2}-1\right)^2 - (10u+8v) \end{aligned} \right\} \\ \text{s.t.} \quad & u^3 - v \le 0 \end{aligned} \tag{7.92}$$

using fmincon, with gradients. The inequality constraint now forces the free node to be on or above the track $y = x^3$.

Solution: As before, we exploit the gradient:

```
opt=optimoptions ('fmincon',
'SpecifyObjectiveGradient',true,'SpecifyConstraintGradient',
true);
```

However, for the constraint argument, we pass the inequality function:

```
[soln,PEMin,Flag,Output]= fmincon(@(x)s.potentialEnergy(x),
[0 0], [],[],[],[],[],[],@(x)s.constraintsInequality
(x),opt);
disp(soln)
0.431522174035336   0.080354652400860
```

We converge approximately to the same solution. However, the number of function calls has increased:

```
>> Output.funcCount
ans = 21
```

In other words, *if the constraint is known to be active, it is more efficient to use the equality constraint than the inequality constraint.*

7.6 The `fminsearchcon` Function

In Section 6.4, we observed that fminsearch can be very effective in handling noisy and non-differentiable unconstrained problems. As an extension to this method, a user-created function, fminsearchcon (created by Dr. John D'Errico), extends fminsearch to handle inequality constraints, and is available through MATLAB central. We illustrate its usage below; in later chapters, we will use this function to solve practical design problems (also see Exercise 7.21).

$$\begin{aligned} \underset{\mathbf{x}}{\text{minimize}} \quad & f(\mathbf{x}) \\ \text{s.t.} \quad & \mathbf{x}_{\min} \le \mathbf{x} \le \mathbf{x}_{\max} \\ & A\mathbf{x} \le b \\ & g_j(\mathbf{x}) \le 0; \ j = 1, 2, ..., L \end{aligned} \tag{7.93}$$

The syntax for fminsearchcon is

```
>>X=fminsearchcon(fun,x0,LB,UB,A,b,nonlcon,options)
```

We will study the use of fminsearchcon through a couple of examples.

Example 7.21 Solve Example 7.17 using fminsearchcon.

Solution: We have

```
>> [soln,fMin,Flag,Output]=fminsearchcon(f,2,[],[],A,B);
soln =
   0.994140624999999
fMin =
  -0.004752474221021
Flag =
     1
Output =
  struct with fields:
    iterations: 22
     funcCount: 53
```

We observe that fminsearchcon does a much better job than fmincon.

Example 7.22 Solve the inequality-constrained problem in Example 7.20 using fminsearchcon.

Solution: We cannot exploit gradient information in fminsearchcon (the method addresses problems where gradient information is not available). Therefore we have

```
[soln,PEMin,Flag,Output]=fminsearchcon(@(x)s.potentialEnerg
y(x), [0 0],[],[],[],[],@(x)s.constraintsInequality(x));
disp(soln)
0.431492131579173   0.080337612778862
```

Once again, we converge approximately to the same solution. However, the number of function calls has now increased significantly.

```
>> Output.funcCount
ans = 166
```

7.6.1 Numerical Challenges

In the numerical solution of constrained problems, several pathological conditions might arise. We will consider a few of them in this section.

The most obvious pathological case is where the constraints are linearly dependent. Consider the following example.

Example 7.23 For the problem

$$\begin{aligned}
&\underset{x,y}{\text{minimize}} && x^2 + y^2 + z^2 \\
&\text{s.t.} && x + y + z - 1 = 0 \\
& && 2x + 2y + 2z - 2 = 0
\end{aligned} \tag{7.94}$$

apply the KKT theorem to find the stationary points.

Solution:

$$\nabla f = \begin{Bmatrix} 2x \\ 2y \\ 2z \end{Bmatrix}; \; \nabla h_1 = \begin{Bmatrix} 1 \\ 1 \\ 1 \end{Bmatrix}; \; \nabla h_2 = \begin{Bmatrix} 2 \\ 2 \\ 2 \end{Bmatrix} \tag{7.95}$$

Thus, the first-order conditions are

$$\begin{aligned}
&\begin{Bmatrix} 2x \\ 2y \\ 2z \end{Bmatrix} + \mu_1 \begin{Bmatrix} 1 \\ 1 \\ 1 \end{Bmatrix} + \mu_2 \begin{Bmatrix} 2 \\ 2 \\ 2 \end{Bmatrix} = \begin{Bmatrix} 0 \\ 0 \\ 0 \end{Bmatrix} \\
&x + y + z - 1 = 0 \\
&2x + 2y + 2z - 2 = 0
\end{aligned} \tag{7.96}$$

We can write this in a matrix form as

$$\begin{bmatrix} 2 & 0 & 0 & 1 & 2 \\ 0 & 2 & 0 & 1 & 2 \\ 0 & 0 & 2 & 1 & 2 \\ 1 & 1 & 1 & 0 & 0 \\ 2 & 2 & 2 & 0 & 0 \end{bmatrix} \begin{Bmatrix} x \\ y \\ z \\ \mu_1 \\ \mu_2 \end{Bmatrix} = \begin{Bmatrix} 0 \\ 0 \\ 0 \\ 1 \\ 2 \end{Bmatrix} \tag{7.97}$$

Observe that the 5th row is a multiple of the 4th row, i.e., the matrix is not invertible. In conclusion, linearly dependent constraints must be eliminated before numerical solution of constrained problems.

Similar problems arise when the gradients of the constraints are linearly dependent at a stationary point (even though the constraints themselves may be linearly independent).

Example 7.24 For the problem

$$\begin{aligned}
&\underset{x,y}{\text{minimize}} && x^2 + y^2 \\
&\text{s.t.} && x - 1 = 0 \\
& && (x - 2)^2 + y^2 - 1 = 0
\end{aligned} \tag{7.98}$$

apply the KKT theorem to find the stationary points.

(cont.)

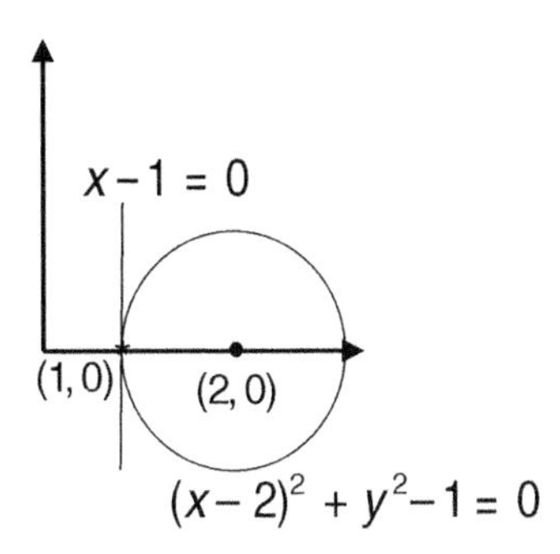

Figure 7.7 A geometric interpretation of Equation (7.98).

Solution: The constraints are linearly independent. Further, it is easy to establish, from a geometric argument, that the local minimum occurs at (1, 0); see Figure 7.7.

Now consider the gradients of the two constraints:

$$\nabla h_1 = \begin{Bmatrix} 1 \\ 0 \end{Bmatrix}; \ \nabla h_2 = \begin{Bmatrix} 2(x-2) \\ 2y \end{Bmatrix} \tag{7.99}$$

At the stationary point at (1, 0), we have

$$\nabla h_1 = \begin{Bmatrix} 1 \\ 0 \end{Bmatrix}; \ \nabla h_2 = \begin{Bmatrix} -2 \\ 0 \end{Bmatrix} \tag{7.100}$$

i.e., the two gradients are linearly dependent. Thus, the problem becomes ill-conditioned as we approach the stationary point. This is yet another caveat that readers must be aware of.

Finally, a similar problem arises when the gradient of a constraint vanishes at a stationary point.

Example 7.25 For the problem

$$\begin{aligned} \underset{x,y}{\text{minimize}} \quad & x^2 + y^2 \\ \text{s.t.} \quad & (x-2)^2 = 0 \end{aligned} \tag{7.101}$$

apply the KKT theorem to find the stationary points.

Solution: Although one can geometrically determine that the local minimum is $(2, 0)$, most algorithms will fail since the gradient vanishes at $(2, 0)$.

7.7 Conclusions

- The `fmincon` method in MATLAB is the work-horse routine for solving constrained minimization problems.
- Its performance improves if the gradient of the objective is known.
- However, `fmincon` is not robust when the objective is noisy.
- On the other hand, `fminsearchcon` is better at handling noisy objectives, but is not recommended when the objective is smooth.
- If optimization methods fail to find a solution to a constrained optimization problem, the user must check whether it is due to any pathological conditions.

7.8 EXERCISES

Exercise 7.1 Consider the constrained problem

$$\begin{aligned} \underset{x,y}{\text{minimize}} \quad & x^2 + (y+1)^2 \\ \text{s.t.} \quad & x - y - 2 = 0 \end{aligned} \tag{7.102}$$

Solve this problem by three methods:

(a) using geometric arguments
(b) by eliminating the constraint
(c) by setting up and solving the KKT equations.

What is the minimum objective value?

Exercise 7.2 Consider the following problem (a modified version of Exercise 7.1):

$$\begin{aligned} \underset{x,y}{\text{minimize}} \quad & x^2 + (y+1)^2 \\ \text{s.t.} \quad & x - y - 2.01 = 0 \end{aligned} \tag{7.103}$$

Solve by eliminating the constraint. What is the minimum objective value? What is the predicted objective based on the Lagrange multiplier of Exercise 7.1?

Exercise 7.3 Consider the constrained problem

$$\begin{aligned} \underset{x,y}{\text{minimize}} \quad & x^2 + (y-2)^2 \\ \text{s.t.} \quad & y - x^2 = 0 \end{aligned} \tag{7.104}$$

Find all local minima by eliminating the constraint. Then set up the KKT equations for the constrained problem; use the computed solution to find the value(s) of the Lagrange multiplier. What is the minimum objective value?

Exercise 7.4 Consider the constrained problem

$$\begin{aligned} \underset{x,y}{\text{minimize}} \quad & x^2 + (y-2)^2 \\ \text{s.t.} \quad & x^2 + y^2 - 1 = 0 \end{aligned} \tag{7.105}$$

Solve this problem using geometric arguments. Does eliminating the constraint lead to the same solution? Set up the KKT equations; use the computed solution to find the value(s) of the Lagrange multiplier. What is the minimum objective value?

Exercise 7.5 Consider the constrained problem

$$\begin{aligned} \underset{x,y,z}{\text{minimize}} \quad & x^2 + (y-2)^2 + (z-1)^2 \\ \text{s.t.} \quad & x + y - z = 1 \\ & x - 2y + z = 6 \end{aligned} \tag{7.106}$$

Solve this problem by:

(a) eliminating the constraints
(b) solving the KKT equations.

What is the minimum objective value?

Exercise 7.6 Consider the problem

$$\begin{aligned} \underset{x,y,z}{\text{minimize}} \quad & x^2 + (y-2)^2 + (z-1)^2 \\ \text{s.t.} \quad & x + y - z = 1.01 \\ & x - 2y + z = 5.99 \end{aligned} \tag{7.107}$$

Solve this problem by eliminating the constraint. What is the minimum objective value? What is the predicted objective based on the Lagrange multipliers of Exercise 7.5?

Exercise 7.7 Consider the problem

$$\begin{aligned} \underset{x,y}{\text{minimize}} \quad & x^2 + (y-2)^2 \\ \text{s.t.} \quad & x + y - 1 \leq 0 \end{aligned} \tag{7.108}$$

Geometrically argue whether the constraint is active. Then solve for the local minimum and Lagrange multiplier. What is the corresponding objective value?

Exercise 7.8 Consider the problem

$$\begin{aligned} \underset{x,y}{\text{minimize}} \quad & x^2 + (y-2)^2 \\ \text{s.t.} \quad & x + y - 1.03 \leq 0 \end{aligned} \tag{7.109}$$

Solve this problem through geometric reasoning. What is the minimum objective value? What is the predicted objective based on Exercise 7.7?

Exercise 7.9 Consider the problem

$$\begin{aligned} \underset{x,y}{\text{minimize}} \quad & x^2 + (y-2)^2 \\ \text{s.t.} \quad & x - y - 1 \leq 0 \end{aligned} \tag{7.110}$$

Geometrically argue whether the constraint is active. Then solve for the local minimum and Lagrange multiplier. What is the corresponding objective value?

Exercise 7.10 Consider the problem

$$\begin{aligned} \underset{x,y}{\text{minimize}} \quad & x^2 + (y-2)^2 + (z-1)^2 \\ \text{s.t.} \quad & -x - y + z + 1 \leq 0 \\ & -x + 2y - z + 6 \leq 0 \end{aligned} \tag{7.111}$$

Solve this problem using geometric arguments. Set up the KKT equations and verify that the solution satisfies all the necessary conditions. What is the minimum objective value?

Exercise 7.11 Solve the problem in Exercise 7.9 using `fmincon`; observe that the constraints are linear.

Exercise 7.12 Solve the problem in Exercise 7.10 using `fmincon`; observe that the constraints are linear.

Exercise 7.13 Consider the problem

$$\begin{aligned} \underset{x,y}{\text{minimize}} \quad & e^x(x^2 - xy) \\ \text{s.t.} \quad & x^2 + y^2 - 100 \leq 0 \\ & y - xy \leq 0 \end{aligned} \tag{7.112}$$

Write a MATLAB class that contains an objective function and a constraint function; these two functions must optionally return the gradients as well. Then solve this problem using `fmincon` under the following conditions:

(a) without explicit use of gradients
(b) with explicit use of objective gradients
(c) with explicit use of objective and constraint gradients.

Compare the three results (solution, function calls and time taken), with (1, 1) as an initial guess.

Exercise 7.14 Consider the problem

$$\begin{aligned} \underset{x,y}{\text{minimize}} \quad & \frac{\sin^3(2\pi x)\sin(2\pi y)}{x^3(x+y)} \\ \text{s.t.} \quad & x^2 - y + 1 \leq 0 \\ & 1 - x + (y-4)^2 \leq 0 \\ & 0 \leq x \leq 10 \\ & 0 \leq y \leq 10 \end{aligned} \tag{7.113}$$

Write a MATLAB class that contains an objective function and a constraint function to capture the non-linear constraints; these two functions must optionally return the gradients as well. Then solve the problem using `fmincon` under the following conditions:

(a) without explicit use of gradients
(b) with explicit use of objective gradients
(c) with explicit use of objective and constraint gradients.

Compare the results (solution, function calls and time taken), with (1, 4) as an initial guess.

Exercise 7.15 Consider the single equality-constrained problem

$$\begin{aligned} \underset{\mathbf{x}}{\text{minimize}} \quad & f(\mathbf{x}) \\ \text{s.t.} \quad & h(\mathbf{x}) = 0 \end{aligned} \tag{7.114}$$

The *penalty method* is a popular method to solve this numerically. The idea is to convert the constrained problem into an unconstrained problem by using a penalty parameter α as follows:

$$\underset{\mathbf{x}}{\text{minimize}} \quad f(\mathbf{x}) + \alpha h^2(\mathbf{x}) \tag{7.115}$$

One can show that, for large values of α, Equation (7.115) is equivalent to Equation (7.114). With this as motivation, the penalty method proceeds as follows:

1. Let $\mathbf{x}_0$ be the initial guess; initialize α as follows:

$$\alpha = \frac{f(\mathbf{x}_0)}{h^2(\mathbf{x}_0)} \tag{7.116}$$

To avoid numerical issues (for example, the numerator or denominator being close to zero), the initial value of α is constrained to be between 0.001 and 1000.

2. Starting at the current solution, solve the unconstrained problem (using methods such as `fminunc`) in Equation (7.115) to find the optimal solution.
3. If the solution $\mathbf{x}$ has converged, exit. Otherwise increase α by a factor of 10 and go back to step 2.

Write a MATLAB function `fminPenaltySingle.m` with the following syntax:

```
[xMin,fMin] = fminPenaltySingle(fun,x0,constraintFun)
```

You can assume that the constraint function returns an empty inequality constraint and one equality constraint. Test your method on various single equality-constrained problems.

Exercise 7.16 The penalty method can be generalized to handle multiple constraints:

$$\begin{aligned} &\underset{\mathbf{x}}{\text{minimize}} \; f(\mathbf{x}) \\ &\text{s.t.} \qquad h_i(\mathbf{x}) = 0; \; i = 1, 2, ..., M \end{aligned} \tag{7.117}$$

The penalized unconstrained problem is

$$\underset{\mathbf{x}}{\text{minimize}} \quad f(\mathbf{x}) + \sum_{i=1,2}^{M} \alpha_i h_i^2(\mathbf{x}) \tag{7.118}$$

Here, each of the parameters is initialized as follows:

$$\alpha_i = \frac{f(\mathbf{x}_0)}{h_i^2(\mathbf{x}_0)} \tag{7.119}$$

Further, each parameter is increased by a factor of 10 during each iteration.

Write a MATLAB function `fminPenalty.m` with the following syntax:

```
[xMin,fMin] = fminPenalty(fun,x0,constraintFun)
```

You can assume that the constraint function returns an empty inequality constraint and one or more equality constraints. Test your method on various multiple equality-constrained problems.

Exercise 7.17 The *augmented Lagrangian* is a variation of the penalty method where one also estimates the Lagrangian. Consider the problem

$$\begin{aligned} &\underset{\mathbf{x}}{\text{minimize}} \; f(\mathbf{x}) \\ &\text{s.t.} \qquad h(\mathbf{x}) = 0 \end{aligned} \tag{7.120}$$

The augmented Lagrangian is defined as follows:

$$\underset{\mathbf{x}}{\text{minimize}} \; f(\mathbf{x}) + \alpha h^2(\mathbf{x}) + \mu h(\mathbf{x}) \tag{7.121}$$

The method proceeds as follows:

1. Let $\mathbf{x}_0$ be the initial guess; initialize α as follows:

$$\alpha = \frac{f(\mathbf{x}_0)}{h^2(\mathbf{x}_0)} \tag{7.122}$$

To avoid numerical issues (for example, the numerator or denominator being close to zero), the initial value of α is constrained to be between 0.001 and 1000.

2. Initialize the Lagrange multiplier, $\mu = 0$.
3. Starting at the current solution and the current values of α and μ, solve the unconstrained problem (using methods such as `fminunc`) in Equation (7.121) to find the optimal solution.
4. Update the Lagrange multiplier as follows (see Reference [8]):

$$\mu^{i+1} = \mu^i + \alpha^i 2h(\mathbf{x}^i) \tag{7.123}$$

5. If the solution has converged, exit. Otherwise increase α by a factor of 10 and go back to step 3.

Write a MATLAB function `fminAugmentedLagrangianSingle` with the following syntax:

```
[xMin,fMin] =
fminAugmentedLagrangianSingle(fun,x0,constraintFun)
```

You can assume that the constraint function returns an empty inequality constraint and one equality constraint. Test your method on various single equality-constrained problems.

Exercise 7.18 The *active set* method is often used for solving inequality-constrained problems. Specifically, consider the problem

$$\begin{aligned} &\underset{\mathbf{x}}{\text{minimize}} && f(\mathbf{x}) \\ &\text{s.t.} && g(\mathbf{x}) \leq 0 \end{aligned} \tag{7.124}$$

The method proceeds as follows:

1. Assume that the constraint is *inactive*; the problem therefore reduces to an unconstrained problem. Solve the unconstrained problem (numerically) to obtain all stationary points $\mathbf{x}^*$ (with $\lambda^* = 0$). Now, for each stationary point, verify whether the KKT conditions are satisfied. Reject points that do not satisfy these equations.
2. Next, assume that the constraint is *active*; the problem reduces to an equality-constrained problem. Solve the equality problem to obtain all stationary points $\lambda^* = 0$ and Lagrange multipliers λ^*. For each stationary point, verify whether the KKT conditions are satisfied; if they are not satisfied, reject this stationary point.
3. Valid solutions are all solutions from both step 1 and step 2.

Write a MATLAB function `fminActiveSetSingle` with the following syntax:

```
[xMin,fMin] = fminActiveSetSingle(fun,x0,constraintFun)
```

You can assume that the constraint function returns one inequality constraint and an empty equality constraint. Test your method on various single inequality-constrained problems.

Exercise 7.19 The active set method may be generalized to multiple inequality-constrained problems. Specifically, consider the problem

$$\begin{aligned} &\underset{\mathbf{x}}{\text{minimize}} \ f(\mathbf{x}) \\ &\text{s.t.} \qquad g_j(\mathbf{x}) \leq 0; \ j = 1, 2, ..., L \end{aligned} \tag{7.125}$$

One must now consider all possible combinations of active constraints. Given L inequality constraints, one must consider 2^L combinations, namely

$$\begin{aligned} &\text{No constraint} : \{\phi\} \\ &\text{One constraint} : \{1\}, \{2\}, \ldots, \{L\} \\ &\text{Two constraints} : \{1,2\}, \{1,3\}, \ldots, \{L-1, L\} \\ &\ldots \\ &\text{All constraints} : \{1, 2, 3, \ldots, L\} \end{aligned} \tag{7.126}$$

For each of the combinations in Equation (7.126), let A be the set of active constraints. For example, $A = \{1, 3, 7\}$ implies constraints 1, 3 and 7 are active, while the remaining constraints are inactive. Next, for each set A, solve the equality problem

$$\begin{aligned} &\underset{\mathbf{x}}{\text{minimize}} \ f(\mathbf{x}) \\ &\text{s.t.} \qquad g_k(\mathbf{x}) = 0; \ k \in A \end{aligned} \tag{7.127}$$

Let a stationary point be $\mathbf{x}^*$, and let the Lagrange multiplier corresponding to the active constraints be $\lambda_k^*, k \in A$. Only the points that satisfy the KKT conditions are retained, and the rest are rejected.

Write a MATLAB function `fminActiveSetMultiple` with the following syntax:

```
[xMin,fMin] = fminActiveSetMultiple(fun,x0,constraintFun)
```

You can assume that the constraint function returns one or more inequality constraints and an empty equality constraint. Test your method on various inequality-constrained problems.

Exercise 7.20 For the three-spring system in Figure 7.8, add an additional constraint, $v - u^2 = 0$. Add the constraint to the class `threeSpring.m` that you developed in Exercise 6.17. Then use `fmincon` to find the equilibrium point for node 1. Find the corresponding Lagrange multiplier.

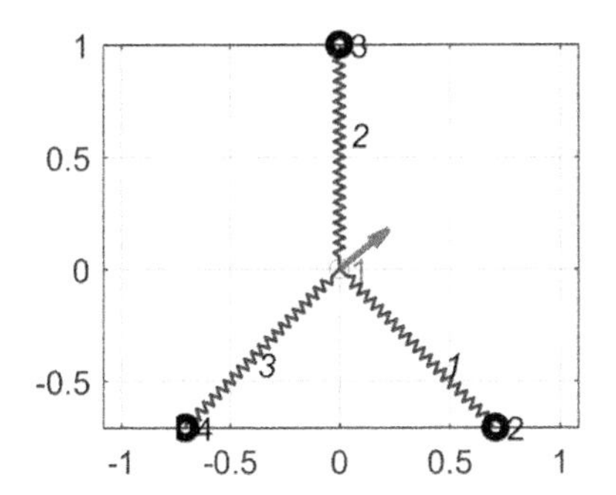

Figure 7.8 A three-spring system.

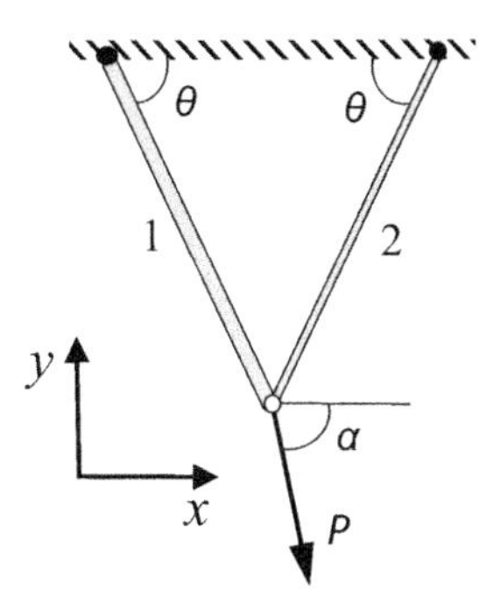

Figure 7.9 A truss system.

Exercise 7.21 Repeat Exercise 7.20 using `fminsearchcon`.

Exercise 7.22 Consider the problem posed in Figure 7.9, where a two-bar truss system is subject to a load. The objective is to optimize the cross-sectional areas A_1 and A_2 of the two bars so as to minimize the total volume, subject to both tensile and compressive stress constraints on each of the two bars.

We will study the formulations of this problem in Chapter 10. For now, it is sufficient to note that the optimization problem can be posed as follows, with four inequality constraints, where σ_Y is the yield strength of the material.

$$
\begin{aligned}
\underset{\{A_1, A_2\}}{\text{minimize}} \quad & (A_1 l_1 + A_2 l_2) \\
\text{s.t.} \quad & \frac{P \sin(\alpha + \theta)}{A_1 \sin 2\theta} - \sigma_Y \leq 0 \\
& -\frac{P \sin(\alpha + \theta)}{A_1 \sin 2\theta} - \sigma_Y \leq 0
\end{aligned}
$$

$$\frac{P \sin(\alpha - \theta)}{A_2 \sin 2\theta} - \sigma_Y \leq 0$$
$$-\frac{P \sin(\alpha - \theta)}{A_2 \sin 2\theta} - \sigma_Y \leq 0 \tag{7.128}$$

For a particular configuration, let us assume that the Lagrange multipliers with these four constraints were numerically computed as 0.05, 1×10^{-6}, 1×10^{-6} and 0.065. From this information, can you deduce the active and inactiveconstraints?

Exercise 7.23 For the problem in Exercise 7.22, suppose the optimal volume computed was 5 units. If the yield strength were to be increased by 1 unit, what would be the expected optimal volume?

8 Special Classes of Problems

Highlights

1. In this chapter, we discuss certain special types of optimization problems. One of the primary reasons to do so is that they often require the use of specialized numerical methods.
2. First, we will consider linear-programming problems, of the form

$$\begin{aligned} \underset{\mathbf{x}}{\text{minimize}} \quad & \mathbf{f}^T\mathbf{x} \\ \text{s.t.} \quad & \mathbf{A}_{eq}\mathbf{x} = \mathbf{b}_{eq} \\ & \mathbf{A}\mathbf{x} \le \mathbf{b} \\ & \mathbf{x}_{\min} \le \mathbf{x} \le \mathbf{x}_{\max} \end{aligned}$$

   where the objective and constraints are linear functions of the design variables.
3. A special case of such problems is mixed integer linear-programming problems, where one or more design variables are required to be integers:

$$\mathbf{x}_I \subseteq \mathbf{x} \in Z \text{ (integer solutions)}$$

4. We will also consider constrained least-squares problems, of the form

$$\begin{aligned} \underset{\mathbf{x}}{\text{minimize}} \quad & \sum \left(r_i(\mathbf{x})\right)^2 \\ \text{s.t.} \quad & h_i(\mathbf{x}) = 0; \; i = 1, 2, ..., M \\ & g_j(\mathbf{x}) \le 0; \; j = 1, 2, ..., L \\ & \mathbf{x}_{\min} \le \mathbf{x} \le \mathbf{x}_{\max} \end{aligned}$$

   where the objective is the sum of squares of (residual) functions.
5. Finally, we will consider multi-objective problems, of the form

$$\begin{aligned} \underset{\mathbf{x}}{\text{minimize}} \quad & \{f_1(\mathbf{x}) \quad f_2(\mathbf{x}) \quad \ldots \quad f_N(\mathbf{x})\} \\ \text{s.t.} \quad & h_i(\mathbf{x}) = 0; \; i = 1, 2, ..., M \\ & g_j(\mathbf{x}) \le 0; \; j = 1, 2, ..., L \\ & \mathbf{x}_{\min} \le \mathbf{x} \le \mathbf{x}_{\max} \end{aligned}$$

8.1 Linear Programming

First, we consider a special class of optimization problems where the objective and constraints are linear, i.e., of the form

$$\begin{aligned} \underset{\mathbf{x}}{\text{minimize}} \quad & \mathbf{f}^T\mathbf{x} \\ \text{s.t.} \quad & \mathbf{A}_{eq}\mathbf{x} = \mathbf{b}_{eq} \\ & \mathbf{A}\mathbf{x} \leq \mathbf{b} \\ & \mathbf{x}_{\min} \leq \mathbf{x} \leq \mathbf{x}_{\max} \end{aligned} \tag{8.1}$$

Such problems are referred to as *linear-programming* problems, and often occur in resource allocation.

Example 8.1 A company makes three types of chocolate, namely, light, medium and dark, with varying amounts of milk, sugar and cocoa. At its disposal the company has 50.5 lb of milk, 28.5 lb of sugar and 10 lb of cocoa. Further, each pound of

- light chocolate requires 0.6 lb of milk, 0.3 lb of sugar and 0.1 lb of cocoa
- medium chocolate requires 0.5 lb of milk, 0.25 lb of sugar and 0.25 lb of cocoa
- dark chocolate requires 0.4 lb of milk, 0.2 lb of sugar and 0.4 lb of cocoa.

The company can sell each pound of chocolate as follows:

- light chocolate: $10.00
- medium chocolate: $12.50
- dark chocolate: $15.00.

How many pounds of each chocolate should the company manufacture to maximize its profit?

Solution: Let the number of pounds of each type of chocolate be L, M and D, respectively. Since the goal is to maximize profit, the objective function is given by

$$f = -(10L + 12.5M + 15D)$$

Observe the negative sign (since we will be minimizing the objective).

There are several constraints. First, we are limited to 50.5 lb of milk, i.e.,

$$0.6L + 0.5M + 0.4D \leq 50.5$$

Observe that this is an inequality since we may not be able to use all the milk. Similarly, we are constrained by the amount of sugar available:

$$0.3L + 0.25M + 0.2D \leq 28.5$$

and by the amount of cocoa available:

$$0.1L + 0.25M + 0.4D \leq 10$$

Finally, we will not allow negative values for the unknowns:

$$\begin{aligned} L &\geq 0 \\ M &\geq 0 \\ D &\geq 0 \end{aligned}$$

(cont.)

To summarize:

$$\begin{aligned}\underset{L,M,D}{\text{minimize}} \quad & -(10L + 12.5M + 15D)\\ \text{s.t.} \quad & 0.6L + 0.5M + 0.4D \le 50.5\\ & 0.3L + 0.25M + 0.2D \le 28.5\\ & 0.1L + 0.25M + 0.4D \le 10\\ & L \ge 0\\ & M \ge 0\\ & D \ge 0\end{aligned}$$

There are no equality constraints in this problem. It can be posed in the standard form

$$\begin{aligned}\underset{\mathbf{x}}{\text{minimize}} \quad & -\{10 \quad 12.5 \quad 15\}\mathbf{x}\\ \text{s.t.} \quad & \begin{bmatrix}0.6 & 0.5 & 0.4\\ 0.3 & 0.25 & 0.2\\ 0.1 & 0.25 & 0.4\end{bmatrix}\mathbf{x} \le \begin{Bmatrix}50.5\\ 28.5\\ 10\end{Bmatrix}\\ & 0 \le \mathbf{x}\end{aligned}$$

Simplex methods are often used for solving such problems. We will not be discussing the theory behind these methods; instead, we will illustrate the use of MATLAB's `linprog` to solve such problems. The syntax for `linprog` is

```
>> X = linprog(f,A,b,Aeq,beq,LB,UB);
```

For this problem, we have

```
>> f = -[10 12.5 15];
>> A = [0.6 0.5 0.4; 0.3 0.25 0.2; 0.1 0.25 0.4];
>> b = [50.5 28.5 10];
>> Aeq = [];
>> beq = [];
>> LB = [0 0 0];
>> UB = [];
```

Passing these arguments to `linprog`, we obtain the solution:

```
>> X = linprog(f,A,b,Aeq,beq,LB,UB)
Optimal solution found.
X =
   76.2500
    9.5000
         0
```

(cont.)

In other words, to maximize its profit, the company must make 76.25 lb of light chocolate, 9.5 lb of medium chocolate and 0 lb of dark chocolate. Observe that the amount of milk, sugar and cocoa used in this production plan can be determined via

```
>> A*X
ans =
    50.5000
    25.2500
    10.0000
```

Thus, the company has used up all the milk and cocoa, but it is left with 3.25 lb of sugar.

8.2 Mixed Integer Linear Programming

A *mixed integer linear-programming* (MILP) problem is a special case of a linear-programming problem where one or more of the design variables is restricted to be an integer (see reference [22]). To illustrate, we modify Example 8.1 as follows.

Example 8.2 A company makes three types of chocolate, namely, light, medium and dark, with varying amounts of milk, sugar and cocoa. At its disposal the company has 50.5 lb of milk, 28.5 lb of sugar and 10 lb of cocoa. Further, *each piece* of

- light chocolate requires 0.06 lb of milk, 0.03 lb of sugar and 0.01 lb of cocoa
- medium chocolate requires 0.05 lb of milk, 0.025 lb of sugar and 0.025 lb of cocoa
- dark chocolate requires 0.04 lb of milk, 0.02 lb of sugar and 0.04 lb of cocoa.

The company can sell the chocolates *per piece* as follows:

- light chocolate: $1.00
- medium chocolate: $1.25
- dark chocolate: $1.50.

How many *pieces* of each type of chocolate should the company manufacture to maximize its profit?

Solution: Let the number of pieces of each type of chocolate be L, M and D, respectively (note that L, M and D are now integers). As before, we pose the problem as

$$\begin{aligned} \underset{\mathbf{x}}{\text{minimize}} \quad & -\{1 \quad 1.25 \quad 1.5\}\mathbf{x} \\ \text{s.t.} \quad & \begin{bmatrix} 0.06 & 0.05 & 0.04 \\ 0.03 & 0.025 & 0.02 \\ 0.01 & 0.025 & 0.04 \end{bmatrix} \mathbf{x} \leq \begin{Bmatrix} 50.5 \\ 28.5 \\ 10 \end{Bmatrix} \\ & 0 \leq \mathbf{x} \\ & \mathbf{x}: \text{integers} \end{aligned}$$

(cont.)

We will illustrate the use of MATLAB's `intlinprog` to solve such problems. The syntax for `intlinprog` is

```
>> X = intlinprog(f,intcon,A,b,Aeq,beq,LB,UB);
```

The main difference is the second variable, `intcon` (short for integer constraints), which specifies which of the design variables are constrained to be integers. In this problem, all three are required to be integers. Thus,

```
>> f = -[1 1.25 1.5];
>> intcon = [1 2 3]; % all 3 are required to be integers
>> A = [0.06 0.05 0.04; 0.03 0.025 0.02; 0.01 0.025 0.04];
>> b = [50.5 28.5 10];
>> Aeq = [];
>> beq = [];
>> LB = [0 0 0];
```

```
>> UB = [];
```

Passing these arguments to `intlinprog`, we obtain the solution:

```
>> X = intlinprog(f,intcon,A,b,Aeq,beq,LB,UB)
Optimal solution found.
X =
   X =
  762.0000
   95.0000
         0
```

In other words, to maximize its profit, the company must make 762 pieces of light chocolate, 95 pieces of medium chocolate and 0 pieces of dark chocolate. Observe that the amount of milk, sugar and cocoa used in this production plan can be determined via

```
>> A*X
ans =
   50.4700
   25.2350
    9.9950
```

Now some amount of all three ingredients remains unused.

8.3 Least-squares Problems

Recall that in Chapter 2 we considered fitting a straight line to a series of data points (x_i, y_i). This led to the optimization problem

$$\underset{m,c}{\text{minimize}} \sum_{i=1,2,\ldots}^{N} (mx_i + c - y_i)^2 \tag{8.2}$$

As noted in Chapter 2, this is a quadratic problem, and can be solved using, say, `fminunc`. However, more efficient methods can be used if one observes that the objective is the sum of squares of simpler functions, i.e., of the form

$$f(\mathbf{x}) = \sum \left(r_k(\mathbf{x})\right)^2, \quad \text{where } r_k = (mx_k + c - y_k); \ \mathbf{x} = \{m, c\} \tag{8.3}$$

Such problems are referred to as *least-squares* problems. They are of particular interest for two reasons: (1) they occur often in all forms of data-fitting scenarios ([23]), and (2) one can develop specialized methods that are more efficient and robust than generic non-linear optimization methods discussed thus far.

Typically there are constraints that need to be considered:

$$\begin{aligned} &\underset{\mathbf{x}}{\text{minimize}} \ f(\mathbf{x}) = \sum \left(r_k(x)\right)^2 \\ &\text{s.t.} \qquad h_i(\mathbf{x}) = 0; \ i = 1, 2, \ldots, M \\ &\qquad\quad\ g_j(\mathbf{x}) \leq 0; \ j = 1, 2, \ldots, L \end{aligned} \tag{8.4}$$

Such problems are referred to as *constrained* least-squares problems.

The individual functions $r_i(\mathbf{x})$ in least-squares problems are typically referred to as *residuals*. Further, if all the residuals are linear functions of the optimization variables, then the problem is referred to as a *linear* least-squares problem. The problem posed above is thus a linear least-squares problem.

On the other hand, the problem

$$\underset{m,c}{\text{minimize}} \sum_{k=1,2,\ldots}^{N} (m^2 x_k + mc - c^3 y_k)^2 \tag{8.5}$$

is a *non-linear* least-squares problem since the residuals are not linear functions of the design variables $\{m, c\}$.

8.3.1 Linear Least Squares

Consider a linear least-squares problem with the objective function

$$\underset{\mathbf{x}}{\text{minimize}} f(\mathbf{x}) = \frac{1}{2} \sum_{i=1,2}^{R} \left(r_i(\mathbf{x})\right)^2 \tag{8.6}$$

where each r_i is a linear function of $\mathbf{x}$:

$$r_i = \mathbf{c}_i \cdot \mathbf{x} - d_i \tag{8.7}$$

It is convenient to group these linear equations as

$$\begin{Bmatrix} r_1 \\ r_2 \\ \vdots \\ r_R \end{Bmatrix} = \begin{bmatrix} \mathbf{c}_1 \\ \mathbf{c}_2 \\ \vdots \\ \mathbf{c}_R \end{bmatrix} \mathbf{x} - \begin{Bmatrix} d_1 \\ d_2 \\ \vdots \\ d_R \end{Bmatrix} \tag{8.8}$$

In matrix form, we have

$$\mathbf{r} = \mathbf{C}\mathbf{x} - \mathbf{d} \tag{8.9}$$

Therefore, the least-squares problem may be posed as

$$\underset{\mathbf{x}}{\text{minimize}} f(\mathbf{x}) = \frac{1}{2}\mathbf{r}(\mathbf{x})^T \mathbf{r}(\mathbf{x}) \tag{8.10}$$

Since this is a quadratic problem, one can exploit the gradient identities to yield

$$\nabla f(\mathbf{x}) = \frac{1}{2} 2\nabla \mathbf{r}(\mathbf{x}) \mathbf{r}(\mathbf{x})^T = \mathbf{C}^T(\mathbf{C}\mathbf{x} - \mathbf{d}) \tag{8.11}$$

Thus, the stationary point is given by

$$\mathbf{C}^T\mathbf{C}\mathbf{x} = \mathbf{C}^T\mathbf{d} \tag{8.12}$$

It also follows that the stationary point is a global minimum if the matrix $\mathbf{C}^T\mathbf{C}$ is positive definite. Of course, this is nothing more than a convenient interpretation of methods to minimize quadratic functions.

Example 8.3 Solve the problem

$$\underset{x_1,x_2}{\text{minimize}} \ \frac{1}{2}[(3x_1 + x_2 - 4)^2 + \ (-x_1 + 2x_2 - 2)^2] \tag{8.13}$$

Solution: First, observe that the objective can be expressed as a sum of squares of linear residual functions; thus, it is a linear least-squares problem. We now express the residuals in a matrix form:

$$\begin{Bmatrix} r_1 \\ r_2 \end{Bmatrix} = \begin{Bmatrix} 3x_1 + x_2 - 4 \\ -x_1 + 2x_2 - 2 \end{Bmatrix} = \begin{bmatrix} 3 & 1 \\ -1 & 2 \end{bmatrix} \begin{Bmatrix} x_1 \\ x_2 \end{Bmatrix} - \begin{Bmatrix} 4 \\ 2 \end{Bmatrix}$$

Thus,

$$\mathbf{C} = \begin{bmatrix} 3 & 1 \\ -1 & 2 \end{bmatrix}; \ \mathbf{d} = \begin{Bmatrix} 4 \\ 2 \end{Bmatrix}$$

(cont.)

From Equation (8.12), the stationary point satisfies

$$\begin{bmatrix} 3 & 1 \\ -1 & 2 \end{bmatrix}^T \begin{bmatrix} 3 & 1 \\ -1 & 2 \end{bmatrix} \mathbf{x} = \begin{bmatrix} 3 & 1 \\ -1 & 2 \end{bmatrix}^T \begin{Bmatrix} 4 \\ 2 \end{Bmatrix}$$

i.e.,

$$\begin{bmatrix} 10 & 1 \\ 1 & 5 \end{bmatrix} \mathbf{x} = \begin{Bmatrix} 10 \\ 8 \end{Bmatrix}$$

Thus $\mathbf{x} = \{0.8571,\ 1.4286\}^T$; further, the matrix $C^T C$ is positive definite. So the stationary point is a local minimum.

In MATLAB, one can use the `lsqlin` function to solve such problems. The syntax for this method is

```
>> X = lsqlin(C,d,A,b,Aeq,beq,LB,UB,X0);
```

For this problem,

```
>> C=[3 1;-1 2]; d =[4 2]';
```

Since the other variables are null, we have

```
>> [x,f] = lsqlin(C,d)
x =
    0.8571
    1.4286
f =
  1.9722e-031
```

One can generalize the use of `lsqlin` to problems with linear constraints of the form

$$\begin{aligned} &\underset{\mathbf{x}}{\text{minimize}}\, f(\mathbf{x}) = \frac{1}{2}\mathbf{r}(\mathbf{x})^T \mathbf{r}(\mathbf{x}) \\ &\text{s.t.} \qquad \mathbf{A}\mathbf{x} = \mathbf{b} \\ &\qquad\qquad \mathbf{B}\mathbf{x} \le \mathbf{c} \end{aligned} \tag{8.14}$$

Example 8.4 Solve the constrained linear least-squares problem

$$\begin{aligned} &\underset{x_1, x_2}{\text{minimize}} \quad \frac{1}{2}\left[(3x_1 + x_2 - 4)^2 + \ (-x_1 + 2x_2 - 2)^2\right] \\ &\text{s.t.} \qquad x_1 \le 0.5 \end{aligned} \tag{8.15}$$

Solution: In standard form, this is

(cont.)

$$\begin{aligned} &\underset{x_1,x_2}{\text{minimize}} \quad \frac{1}{2}(\mathbf{Cx}-\mathbf{d})^T(\mathbf{Cx}-\mathbf{d}) \\ &\text{s.t.} \qquad \mathbf{Ax} \le \mathbf{b} \end{aligned} \tag{8.16}$$

where

$$\mathbf{C} = \begin{bmatrix} 3 & 1 \\ -1 & 2 \end{bmatrix}; \ \mathbf{d} = \begin{Bmatrix} 4 \\ 2 \end{Bmatrix}$$

$$\mathbf{A} = [1 \quad 0]; \ \mathbf{b} = \{0.5\}$$

In MATLAB, one can solve this as follows:

```
>> C=[3 1;-1 2]; d =[4 2]';
>> A = [1 0]; b =[0.5];
>> [x,f] = lsqlin(C,d,A,b)
...
x =
    0.5000
    1.5000
f =
    1.2500
```

8.3.2 Non-linear Least Squares

Next consider the unconstrained non-linear least-squares problem

$$f(\mathbf{x}) = \sum \left(r_i(\mathbf{x})\right)^2 \tag{8.17}$$

One can once again exploit any of the non-linear methods discussed earlier to solve such problems. However, the popular *Levenberg–Marquardt* algorithm offers an efficient blend of the first-order steepest-descent method and the second-order Gauss–Newton method, specifically for such problems.

We will not discuss the theory behind this algorithm. Instead, we will demonstrate the use of MATLAB's `lsqnonlin` function, which is an implementation of this algorithm. The syntax for this method is

```
>> X = lsqnonlin(FUN,X0,LB,UB,OPTIONS)
```

where the objective returns multiple residuals, as the example below illustrates.

Example 8.5 Solve the following problem:

$$\underset{\mathbf{x}}{\text{minimize}} \;\; (r_1(\mathbf{x}))^2 + (r_2(\mathbf{x}))^2 \tag{8.18}$$

where

$$\begin{aligned} r_1(\mathbf{x}) &= (x_1 - 1)e_{x_1} + x_2 \\ r_2(\mathbf{x}) &= x_1 - (x_2 - 2)^2 \end{aligned} \tag{8.19}$$

Solution: First we create a MATLAB function `test_lsnonlin.m`, with the following contents:

```
function r = test_lsnonlin(x)
r(1) = (x(1)-1)*exp(x(1)) + x(2);
r(2) = x(1) - (x(2)-2)^2;
```

Then the solution can be found as follows:

```
>> x0 = [0 0];
>> [x,resnorm] = lsqnonlin(@test_lsnonlin,x0)
...
x =
    0.4172    1.2103
resnorm =
    0.1487
```

8.4 Multi-objective Optimization

To understand multi-objective problems, consider two independent single-objective problems:

$$\underset{\mathbf{x}}{\text{minimize}} \;\; (x-1)^2 + (x-3)^2 \tag{8.20}$$

and

$$\underset{\{x,y\}}{\text{minimize}} \;\; (x-2)^2 + (y-5)^2 \tag{8.21}$$

By inspection, we note that the minimum for the first problem is (1, 3) and the minimum for the second problem is (2, 5).

Now consider the problem of minimizing both objectives, i.e., a multi-objective problem:

$$\underset{\{x,y\}}{\text{minimize}} \;\; \left\{(x-1)^2 + (y-3)^2 \quad (x-2)^2 + (y-5)^2\right\} \tag{8.22}$$

Multi-objective problems typically exhibit an infinite number of solutions. In this case, the solution consists of all points that lie on the line segment joining (1, 3) and (2, 5). The underlying reason lies in the concept of Pareto optimality.

A point $\mathbf{x} = (x, y)$ is considered a solution to the above multi-objective problem if it satisfies the *non-dominated* or *Pareto-optimal* property, i.e., if none of the objectives can be improved, without degrading the other objective (see reference [24]).

For example, consider a point $\mathbf{x} = (1.25, 3.5)$ which lies on the line segment. Observe that at this point, the objectives take the values $\mathbf{f} = \{0.3125, 2.8125\}$. There are an infinite number of points that have the same objective of $f_1 = 0.3125$ (exercise for the reader). However, for all such points other than $\mathbf{x} = (1.25, 3.5)$, the second objective, f_2, is greater than 2.8125 (exercise for the reader). In other words, the point $\mathbf{x} = (1.25, 3.5)$ is non-dominated or Pareto-optimal, i.e., it is a solution for the multi-objective problem.

A typical approach to finding Pareto-optimal solutions to such a problem is to convert it to a single-objective problem, via weights. Specifically, given a two-objective problem

$$\underset{\mathbf{x}}{\text{minimize}} \ \{f_1(\mathbf{x}) \quad f_2(\mathbf{x})\} \tag{8.23}$$

we pick a positive weight $0 \leq w \leq 1$, and solve the single-objective problem

$$\underset{\mathbf{x}}{\text{minimize}} \ wf_1(\mathbf{x}) + (1 - w)f_2(\mathbf{x}) \tag{8.24}$$

Once the solution is numerically computed, the weight is modified and a new solution is found. This method is simple but does not guarantee that all solutions can be found. There are other strategies for solving multi-objective problems, including goal-attainment and minimax; the reader is referred to MATLAB resources for their description.

A generic multi-objective problem may have numerous objectives and constraints:

$$\begin{aligned} \underset{\mathbf{x}}{\text{minimize}} \quad & \{f_1(\mathbf{x}) \quad f_2(\mathbf{x}) \quad \ldots \quad f_N(\mathbf{x})\} \\ \text{s.t.} \quad & \mathbf{A}_{eq}\mathbf{x} = \mathbf{b}_{eq} \\ & \mathbf{A}\mathbf{x} \leq \mathbf{b} \\ & h_i(\mathbf{x}) = 0; \ i = 1, 2, \ldots, M \\ & g_j(\mathbf{x}) \leq 0; \ j = 1, 2, \ldots, L \\ & \mathbf{x}_{\min} \leq \mathbf{x} \leq \mathbf{x}_{\max} \end{aligned} \tag{8.25}$$

A similar strategy can be adopted to find Pareto-optimal solutions. In MATLAB, we will rely on an implementation of this method, namely `paretosearch`, whose syntax is

```
>> X =
paretosearch(FUN,NVARS,A,b,Aeq,beq,lb,ub,NONLCON,options)
```

To illustrate its use, we first construct two functions that correspond to the ones discussed above:

```
>>f1 = @(x) (x(1)-1)^2+(x(2)-3)^2;
>>f2 = @(x) (x(1)-3)^2+(x(2)-5)^2;
```

We then combine them into a multi-objective function:

```
>>fMulti = @(x) [f1(x), f2(x)];
```

Since there are no constraints, we can solve for the Pareto-optimal solutions as follows:

```
>>X = paretosearch(fMulti,2);
```

By default, 60 different solutions are computed for this problem. These solutions lie (approximately) on the line segment joining (1, 3) and (2, 5), as the reader can verify.

8.5 EXERCISES

Exercise 8.1 In Example 8.1, suppose the price for dark chocolate increases to \$17.50 per pound (everything else remaining the same), What should be the company production plan?

Exercise 8.2 In Example 8.1, suppose only 45 lb of milk is available (everything else remaining the same). What should be the company production plan?

Exercise 8.3 In Example 8.1, suppose the company has 25 lb of sugar, and wants to use up all the sugar, but is willing to accept left-over milk and cocoa. How would you reformulate the problem, and what should be the company production plan?

Exercise 8.4 In Example 8.1, the company wants to limit light chocolate to 60 lb (because of limited market demand). How would you reformulate the problem, and what should be the company production plan?

Exercise 8.5 In Example 8.2, suppose the price for dark chocolate increases to \$1.75 per piece (everything else remaining the same). What should be the company production plan?

Exercise 8.6 In Example 8.2, suppose only 40 lb of milk is available (everything else remaining the same). What should be the company production plan?

Exercise 8.7 In Example 8.2, the company wants to limit light chocolate to 600 pieces and medium chocolate to 100 pieces (on account of limited market demand). How would you reformulate the problem, and what should be the company production plan?

Exercise 8.8 Suppose you are given the following set of data points:

x	0.0975	0.2785	0.5469	0.9575	0.9649	0.1576	0.9706	0.9572
y	0.4100	0.5103	0.6871	0.9672	0.9701	0.4323	0.9831	0.9676

Using the linear least-squares method `lsqlin`, find a straight line that best fits these data points, using the y-distance between the point and the straight line as the measure of error (residual). Plot these data points, along with the constructed straight line over $x \in [0, 1]$. Find the y-intercept, i.e., where the straight line intersects the y-axis.

Exercise 8.9 Repeat Exercise 8.8, but use the x-distance between the point and the straight line as the measure of error.

Exercise 8.10 Repeat Exercise 8.8, but use the shortest distance between the point and the straight line as the measure of error. However, this is now a *non-linear* least-squares problem.

Exercise 8.11 Suppose you are given the following set of data points:

x	0.0975	0.2785	0.5469	0.9575	0.9649	0.1576	0.9706	0.9572
y	0.1712	0.7060	0.0318	0.2769	0.0462	0.0971	0.8235	0.6948
z	0.1790	−0.297	0.5775	0.5097	0.6947	0.2525	−0.084	0.0682

Using the linear least-squares formulation `lsqlin`, find a plane that best fits these data points using the z-distance between the point and the plane as the measure of error (residual). Plot the data points, and the constructed plane over the range $x, y \in [0, 1]$. Find the z-intercept, i.e., where the plane intersects the z-axis.

Exercise 8.12 Using the data points in Exercise 8.11, fit a plane using the shortest distance as the measure of error; note that this will be a non-linear least-squares problem. Find the z-intercept, i.e., where the plane intersects the z-axis.

Exercise 8.13 Suppose you are given the following set of data points:

x	3.5243	3.7332	1.9981	3.9643	2.4575	2.1396	3.1674	3.0115
y	1.8554	1.6765	1.0127	0.7466	1.8456	0.4852	1.9880	0.0031

Using the least-squares formulation `lsqnonlin`, find a circle of the form $(x - x_c)^2 + (y - y_c)^2 = r^2$ that best fits these data points, using the shortest distance between the point and the circle as the measure of error.

Exercise 8.14 Repeat Exercise 8.13 with the constraint $x_c \geq 3$ and $y_c \leq 1$.

Exercise 8.15 For the multi-objective problem

$$\underset{\{x,y\}}{\text{minimize}} \left\{ (x-1)^2 + (y-3)^2 \quad (x-2)^2 + (y-5)^2 \right\}$$

prove analytically that all points on the line segment joining (1, 3) and (2, 5) are Pareto-optimal.

Exercise 8.16 Recall that a typical approach to finding Pareto-optimal solutions to multi-objective problems is to convert them into single-objective problems, via weights. Specifically, given a two-objective problem

$$\underset{\mathbf{x}}{\text{minimize}} \ \{f_1(\mathbf{x}) \quad f_2(\mathbf{x})\}$$

we pick a positive weight between 0 and 1, and solve the single-objective problem

$$\underset{\mathbf{x}}{\text{minimize}} \ w f_1(\mathbf{x}) + (1-w) f_2(\mathbf{x})$$

Once the solution is numerically computed, the weight is modified, and a new solution is found.

Write a MATLAB code that takes as its argument a two-objective function; it should return 100 Pareto-optimal solutions. You can use `fminunc` within your code. For example, your code should be able to handle examples such as these:

```
>> f1 = @(x) (x(1)-1)^2+(x(2)-3)^2;
>> f2 = @(x) (x(1)-10)^2+(x(2)-12)^2;
>> fMulti = @(x) [f1(x), f2(x)];
>> X = myParetoMethod(fMulti);
plot(X(:,1),X(:,2),'.');
axis equal;
```

Exercise 8.17 Consider the multi-objective problem

$$\underset{\{x,y,z\}}{\text{minimize}} \ \{f_1 \quad f_2 \quad f_3\}$$

where

$$\begin{aligned} f_1 &= (x-1)^2 + (y-2)^2 + (z-3)^2 \\ f_2 &= (x-4)^2 + (y+3)^2 + (z-1)^2 \\ f_3 &= (x-2)^2 + (y-8)^2 + (z)^2 \end{aligned}$$

What do you expect the solution set to be? Confirm this numerically using `paretosearch`.

9 Truss Analysis

Highlights

1. A truss system is a collection of bars connected at their ends by pin joints. Truss systems are the simplest structural systems to analyze and optimize.
2. In this chapter, we consider the analysis of such truss systems, i.e., given an external force, compute the displacements at the nodes and the stresses in the bars.
3. Further, through this study we will explore the deep connection between engineering principles and mathematical optimization.
4. To analyze a truss, we can apply either the principle of force balance or the principle of minimum potential energy. The former typically leads to a set of linear equations governing the displacements at the nodes, while the latter leads to a minimization of a quadratic function. The equivalence between the two is established, and the equivalence is related to the first-order optimality conditions.
5. A MATLAB implementation for posing and analyzing truss systems is also presented. This will be employed in Chapter 10 to optimize such systems.

9.1 Overview

A truss is a collection of bars connected by pins at their endpoints. Since pin joints cannot transfer moments, the bars can only carry axial forces (either tension or compression). A simple truss is illustrated in Figure 9.1, where the two nodes at the top are not allowed to move, while a horizontal force is applied at the third node. Observe that the free node can move in both x- and y-directions; we will denote these displacements by u and v, respectively.

The truss in Figure 9.1 is statically *determinate* in that the internal force in each bar can be computed using only the equations of force balance; this can then be followed by displacement calculations (see Section 9.4.1). Thus, for this simple determinate structure, one can easily solve for the displacements using force balance.

On the other hand, consider the truss in Figure 9.2, consisting of seven bars and five nodes. Nodes 4 and 5 are fixed, while the remaining nodes are free to move. A force is applied to nodes 1 and 2. Our objective, once again, is to compute the

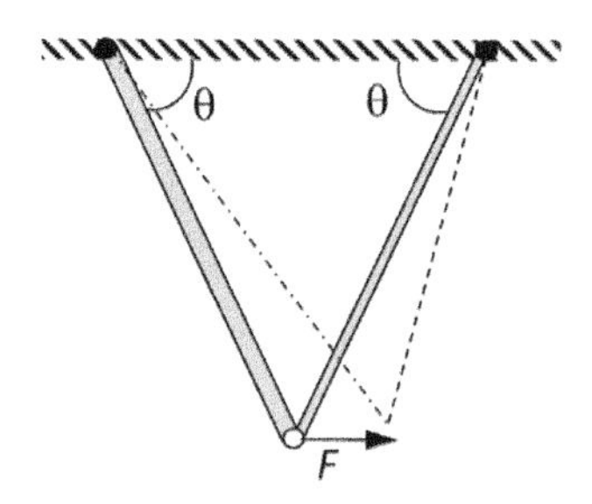

Figure 9.1 A determinate truss.

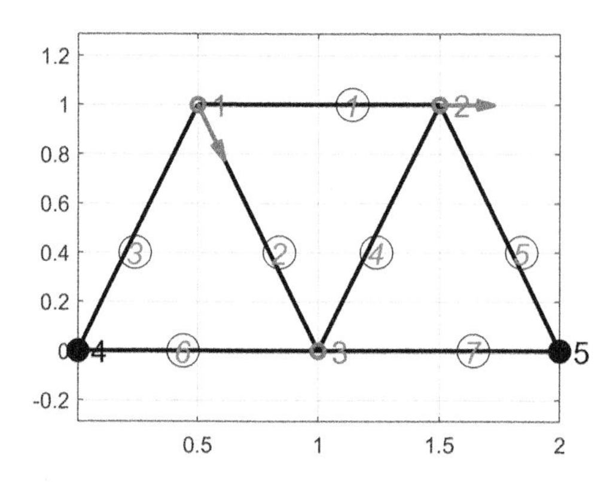

Figure 9.2 An indeterminate truss system.

displacements of the free nodes. Further, given the displacement of the free nodes, one can also determine the stresses in each of the bars. This is an *indeterminate* truss in that the forces cannot be determined using only force balance (see Section 9.4.2). The reader is referred to reference [25] for a comprehensive treatment of such structural systems.

The primary reason for studying the analysis of truss systems is to establish the deep connection between engineering principles and optimization.

9.2 Conventions

To make the analysis easy to follow, we will adopt the following conventions:

- The initial location of node i will be denoted by $\{x_i, y_i\}$.
- The x- and y-displacements of node i will be denoted by u_i and v_i, respectively.
- Thus the final position of node i is given by $\{x_i + u_i, y_i + v_i\}$.
- The length of a bar prior to deformation will be denoted by l_{ij}, i.e.,

$$(l_{ij})^2 = (x_j - x_i)^2 + (y_j - y_i)^2 \tag{9.1}$$

- After deformation, the lengths will be denoted by L_{ij}; recall that the position of node i is given by $\{x_i + u_i, y_i + v_i\}$ and the position of node j is given by $\{x_j + u_j, y_j + v_j\}$. Thus the length of the bar after deformation is given by

$$(L_{ij})^2 = [(x_j + u_j) - (x_i + u_i)]^2 + [(y_j + v_j) - (y_i + v_i)]^2 \tag{9.2}$$

9.3 Small-Displacement Assumption

A key assumption that we will make is that the node displacements are small compared to the lengths of the bars. Such an assumption is reasonable for most engineering structures. Under this assumption, one can obtain a simple expression for the deformation of a bar.

Consider Equation (9.2):

$$(L_{ij})^2 = [(x_j - x_i) + (u_j - u_i)]^2 + [(y_j - y_i) - (v_j - v_i)]^2 \tag{9.3}$$

Expanding, we have

$$(L_{ij})^2 = \begin{bmatrix} (x_j - x_i)^2 + 2(u_j - u_i)(x_j - x_i) + (u_j - u_i)^2 \\ + (y_j - y_i)^2 + 2(v_j - v_i)(y_j - y_i) + (v_j - v_i)^2 \end{bmatrix} \tag{9.4}$$

Exploiting Equation (9.1),

$$(L_{ij})^2 = \begin{bmatrix} (l_{ij})^2 + 2(u_j - u_i)(x_j - x_i) + 2(v_j - v_i)(y_j - y_i) \\ + (u_j - u_i)^2 + (v_j - v_i)^2 \end{bmatrix} \tag{9.5}$$

Denote the change in length by

$$\Delta L_{ij} = L_{ij} - l_{ij} \tag{9.6}$$

Observe that ΔL_{ij} can be positive or negative. Since $\Delta L_{ij} = L_{ij} - l_{ij}$, we have

$$(l_{ij} + \Delta L_{ij})^2 = \begin{bmatrix} (l_{ij})^2 + 2(u_j - u_i)(x_j - x_i) + 2(v_j - v_i)(y_j - y_i) \\ + (u_j - u_i)^2 + (v_j - v_i)^2 \end{bmatrix} \tag{9.7}$$

Since the displacements are small, i.e.,

$$\max(|u_i|, |u_j|, |v_i|, |v_j|) \ll l_{ij} \tag{9.8}$$

one can neglect the quadratic terms $(\Delta L_{ij})^2$, $(u_j - u_i)^2$ and $(v_j - v_i)^2$ in Equation (9.7), resulting in

$$2l_{ij}\Delta L_{ij} \approx 2(u_j - u_i)(x_j - x_i) + 2(v_j - v_i)(y_j - y_i) \tag{9.9}$$

i.e.,

$$\Delta L_{ij} \approx (u_j - u_i)\frac{(x_j - x_i)}{l_{ij}} + (v_j - v_i)\frac{(y_j - y_i)}{l_{ij}} \tag{9.10}$$

Let the bar from node i to node j make an angle of α_{ij} with respect to the x-axis; note that $\alpha_{ji} = \alpha_{ij} - \pi$. Because of the small-displacement assumption, we conclude that α_{ij} will remain unchanged after deformation, i.e.,

$$\begin{Bmatrix} \cos \alpha_{ij} \\ \sin \alpha_{ij} \end{Bmatrix} = \frac{1}{l_{ij}} \begin{Bmatrix} x_j - x_i \\ y_j - y_i \end{Bmatrix} \tag{9.11}$$

Exploiting Equation (9.11), we have

$$\Delta L_{ij} \approx (u_j - u_i)\cos \alpha_{ij} + (v_j - v_i)\sin \alpha_{ij} \tag{9.12}$$

In Section 9.4 we will exploit Equation (9.12) to compute the force within each truss bar.

9.4 Force Balance Method

As noted earlier, each truss bar is under either tension or compression. From basic strength of materials, recall that the deformation and the force running through the bar are related via

$$\Delta L = \frac{F}{(EA/l)} \tag{9.13}$$

where E is the Young's modulus, A is the cross-sectional area of the bar and l is its length. Equation (9.13) is often expressed as

$$F = k\Delta L \tag{9.14}$$

where k is the stiffness of the bar, given by

$$k = \frac{EA}{l} \tag{9.15}$$

Each bar can exhibit a different stiffness. The stiffness of the bar joining node i and node j is denoted by

$$k_{ij} = \frac{E_{ij}A_{ij}}{l_{ij}} \tag{9.16}$$

Therefore, the unknown deformation is related to the unknown force via

$$F_{ij} = k_{ij}\Delta L_{ij} \tag{9.17}$$

Since the angle with respect to the x-axis made by the bar from node i to node j is α_{ij}, the direction of the force is given by $(\cos \alpha_{ij}, \sin \alpha_{ij})$, i.e., the force vector is given by

$$\mathbf{F}_{ij} = k_{ij}\Delta L_{ij}\begin{Bmatrix} \cos\alpha_{12} \\ \sin\alpha_{12} \end{Bmatrix} \tag{9.18}$$

We will first consider a force balance approach to compute the displacements for the determinate truss in Figure 9.3, and then generalize this to indeterminate trusses.

The *law of force balance* states that *for a structural system at rest, the sum of all internal forces is equal to the external force at each node:*

$$\sum \mathbf{F}_i^{\text{int}} + \mathbf{F}_i^{\text{ext}} = 0 \tag{9.19}$$

where the internal forces are those forces acting through the bar attached to node i.

9.4.1 Determinate Truss System

Consider the truss problem in Figure 9.3 where P is the external force. Let the internal forces in bars 1 and 2 be F_1 and F_2, respectively. Then a force balance yields

$$\begin{aligned} P\cos\beta &= F_1\cos\theta - F_2\cos\theta \\ P\sin\beta &= F_1\sin\theta + F_2\sin\theta \end{aligned} \tag{9.20}$$

i.e.,

$$\begin{aligned} F_1 &= \frac{P\sin(\beta+\theta)}{\sin 2\theta} \\ F_2 &= \frac{P\sin(\beta-\theta)}{\sin 2\theta} \end{aligned} \tag{9.21}$$

Therefore the deformation of the two bars is given by

$$\begin{aligned} \Delta L_1 &= \frac{F_1}{(EA_1/l)} = \frac{P\sin(\beta+\theta)l}{EA_1\sin 2\theta} \\ \Delta L_2 &= \frac{F_2}{(EA_2/l)} = \frac{P\sin(\beta-\theta)l}{EA_2\sin 2\theta} \end{aligned} \tag{9.22}$$

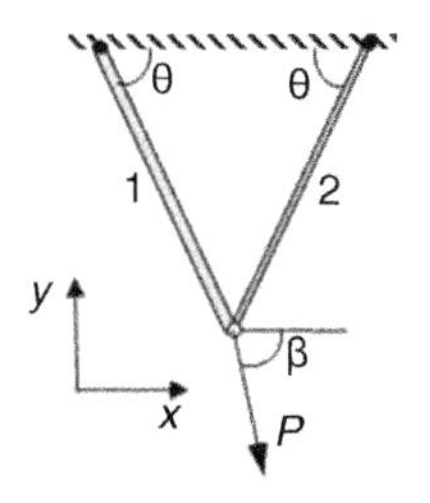

Figure 9.3 A determinate truss system.

Let the displacement at the free node be (u, v). Using the generic expression in Equation (9.12), one can show that

$$\begin{aligned}\Delta L_1 &= u \cos\theta - v \sin\theta \\ \Delta L_2 &= -u \cos\theta - v \sin\theta\end{aligned} \tag{9.23}$$

In other words

$$\begin{aligned}u &= \frac{1}{2\cos\theta}(\Delta L_1 - \Delta L_2) \\ v &= \frac{-1}{2\sin\theta}(\Delta L_1 + \Delta L_2)\end{aligned} \tag{9.24}$$

After substituting, one can show that

$$\begin{aligned}u &= \frac{Pl}{2E\cos\theta}\left(\frac{\sin(\beta+\theta)}{A_1 \sin 2\theta} - \frac{\sin(\beta-\theta)}{A_2 \sin 2\theta}\right) \\ v &= \frac{-Pl}{2E\sin\theta}\left(\frac{\sin(\beta+\theta)}{A_1 \sin 2\theta} + \frac{\sin(\beta-\theta)}{A_2 \sin 2\theta}\right)\end{aligned} \tag{9.25}$$

These expressions yield a closed-form solution, and this will be particularly useful in verifying numerical implementation.

Example 9.1 Let $\beta = \theta = 45°$. For each bar let the Young's modulus be 2×10^{11} N/m^2, its length 1 m and its cross-sectional area 10^{-6} m^2. Let $P = 1$ N. Find the displacements.

Solution: Observe that when $\beta = \theta$ Equation (9.25) reduces to

$$\begin{aligned}u &= \frac{Pl}{2EA_1 \cos\theta} = 3.536 \times 10^{-6} \text{ m} \\ v &= \frac{-Pl}{2EA_1 \sin\theta} = -3.536 \times 10^{-6} \text{ m}\end{aligned} \tag{9.26}$$

Note that the displacements are independent of the cross-sectional area of the second bar.

Example 9.2 Suppose $\theta = 45°, \beta = 60°$. For each bar let the Young's modulus be 2×10^{11} N/m^2, its length 1 m and its cross-sectional area 10^{-6} m^2. Let $P = 1$ N. Find the displacements.

Solution: Equation (9.25) reduces to

$$\begin{aligned}u &= \frac{Pl}{2E(0.7071)}\left(\frac{0.9659}{A_1} - \frac{0.2588}{A_2}\right) = 2.50 \times 10^{-6} \text{ m} \\ v &= \frac{-Pl}{2E(0.7071)}\left(\frac{0.9659}{A_1} + \frac{0.2588}{A_2}\right) = -4.3301 \times 10^{-6} \text{ m}\end{aligned} \tag{9.27}$$

These analytical solutions will be used later to verify our MATLAB implementation.

9.4.2 Indeterminate Truss System

We now consider the indeterminate truss in Figure 9.2. Consider a free node, say, node 1 in Figure 9.2, and the bar joining node 1 and node 2.

As mentioned earlier, the force experienced by this bar is given by

$$\mathbf{F}_{12} = k_{12}\Delta L_{12}\begin{Bmatrix} \cos\alpha_{12} \\ \sin\alpha_{12} \end{Bmatrix} \tag{9.28}$$

Substituting from Equation (9.12), we have

$$\mathbf{F}_{12} \approx \frac{E_{12}A_{12}}{l_{12}}[(u_2 - u_1)\cos\alpha_{12} + (v_2 - v_1)\sin\alpha_{12}]\begin{Bmatrix} \cos\alpha_{12} \\ \sin\alpha_{12} \end{Bmatrix} \tag{9.29}$$

i.e.,

$$\mathbf{F}_{12} \approx \mathbf{k}_{12}\left[\begin{Bmatrix} u_2 \\ v_2 \end{Bmatrix} - \begin{Bmatrix} u_1 \\ v_1 \end{Bmatrix}\right] \tag{9.30}$$

where

$$\mathbf{k}_{12} = \frac{E_{12}A_{12}}{l_{12}}\begin{bmatrix} \cos^2\alpha_{12} & \sin\alpha_{12}\cos\alpha_{12} \\ \sin\alpha_{12}\cos\alpha_{12} & \sin^2\alpha_{12} \end{bmatrix} \tag{9.31}$$

is the 2×2 stiffness matrix associated with the bar joining nodes 1 and 2. Observe that Equation (9.30) can be expressed as

$$\mathbf{F}_{12} \approx -[\mathbf{k}_{12} \quad -\mathbf{k}_{12} \quad \mathbf{0} \quad \mathbf{0} \quad \mathbf{0}]\begin{Bmatrix} u_1 \\ v_1 \\ u_2 \\ v_2 \\ u_3 \\ v_3 \\ u_4 \\ v_4 \\ u_5 \\ v_5 \end{Bmatrix} \tag{9.32}$$

where $\mathbf{0}$ denotes a 2×2 zero matrix. In short,

$$\mathbf{F}_{12} \approx -[\mathbf{k}_{12} \quad -\mathbf{k}_{12} \quad \mathbf{0} \quad \mathbf{0} \quad \mathbf{0}]\mathbf{d} \tag{9.33}$$

where the entire set of displacements is denoted as

$$\mathbf{d} = \begin{Bmatrix} u_1 \\ v_1 \\ u_2 \\ v_2 \\ u_3 \\ v_3 \\ u_4 \\ v_4 \\ u_5 \\ v_5 \end{Bmatrix} \tag{9.34}$$

Similarly,

$$\begin{aligned} \mathbf{F}_{13} &\approx -[\mathbf{k}_{13} \quad \mathbf{0} \quad -\mathbf{k}_{13} \quad \mathbf{0} \quad \mathbf{0}]\mathbf{d} \\ \mathbf{F}_{14} &\approx -[\mathbf{k}_{14} \quad \mathbf{0} \quad \mathbf{0} \quad -\mathbf{k}_{14} \quad \mathbf{0}]\mathbf{d} \end{aligned} \tag{9.35}$$

Thus, the total *internal* force acting on node 1 is

$$\mathbf{F}_1 = -[\mathbf{k}_{12} + \mathbf{k}_{13} + \mathbf{k}_{14} \quad -\mathbf{k}_{12} \quad -\mathbf{k}_{13} \quad -\mathbf{k}_{14} \quad \mathbf{0}]\mathbf{d} \tag{9.36}$$

Finally, if the external force on node 1 is $\mathbf{f}_1$, then

$$-[\mathbf{k}_{12} + \mathbf{k}_{13} + \mathbf{k}_{14} \quad -\mathbf{k}_{12} \quad -\mathbf{k}_{13} \quad -\mathbf{k}_{14} \quad \mathbf{0}]\mathbf{d} + \mathbf{f}_1 = 0 \tag{9.37}$$

or

$$[\mathbf{k}_{12} + \mathbf{k}_{13} + \mathbf{k}_{14} \quad -\mathbf{k}_{12} \quad -\mathbf{k}_{13} \quad -\mathbf{k}_{14} \quad \mathbf{0}]\mathbf{d} = \mathbf{f}_1 \tag{9.38}$$

Equation (9.38) is the force balance equation for node 1 and constitutes a pair of scalar equations, governing the unknown displacements. We must establish the remaining force balance equations in order to solve for all the displacements.

Now consider node 2, which is attached to nodes 1, 3 and 5. Using the pattern established above, the force balance yields

$$[-\mathbf{k}_{21} \quad \mathbf{k}_{21} + \mathbf{k}_{23} + \mathbf{k}_{25} \quad -\mathbf{k}_{23} \quad \mathbf{0} \quad -\mathbf{k}_{25}]\mathbf{d} = \mathbf{f}_2 \tag{9.39}$$

Finally, the force balance at node 3 yields

$$[-\mathbf{k}_{31} \quad -\mathbf{k}_{32} \quad \mathbf{k}_{31} + \mathbf{k}_{32} + \mathbf{k}_{34} + \mathbf{k}_{35} \quad -\mathbf{k}_{34} \quad -\mathbf{k}_{35}]\mathbf{d} = \mathbf{f}_3 \tag{9.40}$$

We shall not derive the force equations for nodes 4 and 5 since they are fixed.

Next, observe that the stiffness matrices are symmetric (see Equation(9.31)), i.e.,

$$\mathbf{k}_{ij} = \mathbf{k}_{ji} \tag{9.41}$$

Combining Equations (9.38), (9.39) and (9.40) results in six equations:

$$\begin{bmatrix} \mathbf{k}_{12} + \mathbf{k}_{13} + \mathbf{k}_{14} & -\mathbf{k}_{12} & -\mathbf{k}_{13} & -\mathbf{k}_{14} & \mathbf{0} \\ -\mathbf{k}_{21} & \mathbf{k}_{21} + \mathbf{k}_{23} + \mathbf{k}_{25} & -\mathbf{k}_{23} & \mathbf{0} & -\mathbf{k}_{25} \\ -\mathbf{k}_{31} & -\mathbf{k}_{32} & \mathbf{k}_{31} + \mathbf{k}_{32} + \mathbf{k}_{34} + \mathbf{k}_{35} & -\mathbf{k}_{34} & -\mathbf{k}_{35} \end{bmatrix} \mathbf{d} = \begin{Bmatrix} \mathbf{f}_1 \\ \mathbf{f}_2 \\ \mathbf{f}_3 \end{Bmatrix} \tag{9.42}$$

Further, observe that $u_4 = v_4 = u_5 = v_5 = 0$.

Consequently, we can define a reduced set of unknowns:

$$\hat{\mathbf{d}} = \begin{Bmatrix} u_1 \\ v_1 \\ u_2 \\ v_2 \\ u_3 \\ v_3 \end{Bmatrix} \tag{9.43}$$

Therefore, we can get rid of the corresponding columns, resulting in

$$\begin{bmatrix} \mathbf{k}_{12}+\mathbf{k}_{13}+\mathbf{k}_{14} & -\mathbf{k}_{12} & -\mathbf{k}_{13} \\ -\mathbf{k}_{21} & \mathbf{k}_{21}+\mathbf{k}_{23}+\mathbf{k}_{25} & -\mathbf{k}_{23} \\ -\mathbf{k}_{31} & -\mathbf{k}_{32} & \mathbf{k}_{31}+\mathbf{k}_{32}+\mathbf{k}_{34}+\mathbf{k}_{35} \end{bmatrix} \hat{\mathbf{d}} = \begin{Bmatrix} \mathbf{f}_1 \\ \mathbf{f}_2 \\ \mathbf{f}_3 \end{Bmatrix} \tag{9.44}$$

Equation (9.44) is a linear system of the form

$$\hat{\mathbf{K}}\hat{\mathbf{d}} = \hat{\mathbf{f}} \tag{9.45}$$

The linear system happens to be symmetric since the individual stiffness matrices are symmetric.

Example 9.3 Consider the truss in Figure 9.4. The Young's modulus of each bar is 2×10^{11} N/m^2, cross-sectional area is 10^{-6} m^2 and length is 1 m. If a force of $(4, 2)$ N is applied to node 1, establish the linear system of equations and find the displacement of the node.

Solution: Observe that the only degrees of freedom are (u, v) of node 1. The stiffness matrices are

$$\mathbf{k}_{12} = \frac{2 * 10^{11} * 10^{-6}}{1} \begin{bmatrix} (0)^2 & 0 * 1 \\ 0 * 1 & (1)^2 \end{bmatrix} = 10^5 \begin{bmatrix} 0 & 0 \\ 0 & 2 \end{bmatrix} \tag{9.46}$$

$$\mathbf{k}_{13} = \frac{2 * 10^{11} * 10^{-6}}{1} \begin{bmatrix} (-1/\sqrt{2})^2 & (-1/\sqrt{2})(-1/\sqrt{2}) \\ Sym & (-1/\sqrt{2})^2 \end{bmatrix} = 10^5 \begin{bmatrix} 1 & 1 \\ 1 & 1 \end{bmatrix} \tag{9.47}$$

$$\mathbf{k}_{14} = \frac{2 * 10^{11} * 10^{-6}}{1} \begin{bmatrix} (1/\sqrt{2})^2 & (1/\sqrt{2})(-1/\sqrt{2}) \\ Sym & (-1/\sqrt{2})^2 \end{bmatrix} = 10^5 \begin{bmatrix} 1 & -1 \\ -1 & 1 \end{bmatrix} \tag{9.48}$$

where "*Sym*" indicates a symmetric matrix. The global linear system (after eliminating all the fixed degrees of freedom) is given by

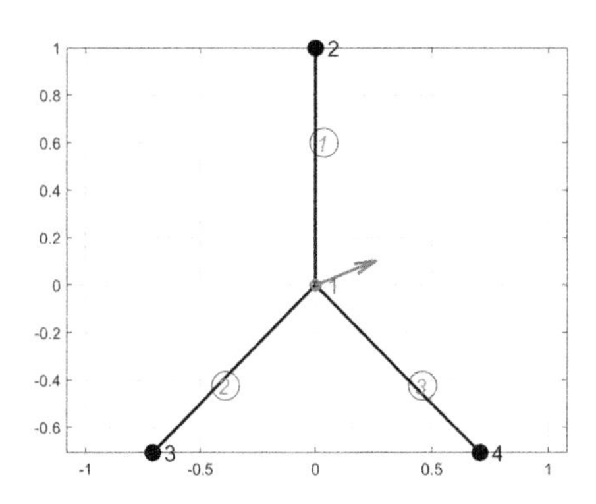

Figure 9.4 A three-bar truss.

(cont.)

$$[\mathbf{k}_{12} + \mathbf{k}_{13} + \mathbf{k}_{14}]\begin{Bmatrix} u \\ v \end{Bmatrix} = \begin{Bmatrix} 4 \\ 2 \end{Bmatrix} \tag{9.49}$$

i.e.,

$$10^5\begin{bmatrix} 2 & 0 \\ 0 & 4 \end{bmatrix}\begin{Bmatrix} u \\ v \end{Bmatrix} = \begin{Bmatrix} 4 \\ 2 \end{Bmatrix} \tag{9.50}$$

i.e.,

$$\begin{Bmatrix} u \\ v \end{Bmatrix} = 10^{-5}\begin{Bmatrix} 2 \\ 0.5 \end{Bmatrix} \tag{9.51}$$

9.5 Potential Energy Method

The strategy studied in the previous section was based on a force balance at the nodes. We now consider the potential energy approach to arrive at the same result.

From Section 2.5, recall the statement of the *principle of minimum potential energy*. This states that *a structural system subject to an external force will come to rest when its potential energy is a minimum*. The potential energy is defined as

$$\Pi = U - 2W \tag{9.52}$$

where U is the internal elastic energy and W is the quasi-static external work done by the force. The elastic energy associated with any truss bar is defined as

$$U = \frac{1}{2}\frac{EA}{l}(\Delta L)^2 \tag{9.53}$$

Consider the elastic energy of the bar joining node 1 and node 2. Exploiting Equation (9.12) we have

$$U_{12} = \frac{1}{2}\frac{E_{12}A_{12}}{l_{12}}(\Delta L_{12})^2 \approx \frac{1}{2}\frac{E_{12}A_{12}}{l_{12}}[(u_2 - u_1)\cos\alpha_{12} + (v_2 - v_1)\sin\alpha_{12}]^2 \tag{9.54}$$

Upon expansion, one can show that this can be expressed as

$$U_{12} = \frac{1}{2}\{u_2 \quad v_2\}\mathbf{k}_{12}\begin{Bmatrix} u_2 \\ v_2 \end{Bmatrix} + \frac{1}{2}\{u_1 \quad v_1\}\mathbf{k}_{12}\begin{Bmatrix} u_1 \\ v_1 \end{Bmatrix} - \{u_2 \quad v_2\}\mathbf{k}_{12}\begin{Bmatrix} u_1 \\ v_1 \end{Bmatrix} \tag{9.55}$$

Grouping terms, we have

$$U_{12} = \frac{1}{2}\{u_1 \quad v_1 \quad u_2 \quad v_2\}\begin{bmatrix} \mathbf{k}_{12} & -\mathbf{k}_{12} \\ -\mathbf{k}_{12} & \mathbf{k}_{12} \end{bmatrix}\begin{Bmatrix} u_1 \\ v_1 \\ u_2 \\ v_2 \end{Bmatrix} \tag{9.56}$$

Observe that this is equivalent to

$$U_{12} = \frac{1}{2}\{u_1 \quad v_1 \quad u_2 \quad v_2 \quad u_3 \quad v_3\}\begin{bmatrix} \mathbf{k}_{12} & -\mathbf{k}_{12} & \mathbf{0} \\ -\mathbf{k}_{12} & \mathbf{k}_{12} & \mathbf{0} \\ \mathbf{0} & \mathbf{0} & \mathbf{0} \end{bmatrix}\begin{Bmatrix} u_1 \\ v_1 \\ u_2 \\ v_2 \\ u_3 \\ v_3 \end{Bmatrix} \tag{9.57}$$

Repeating these expressions for other bars, and observing that the deformations of the fixed nodes are zero, we obtain the total elastic energy:

$$U = \frac{1}{2}\hat{\mathbf{d}}^T\hat{\mathbf{K}}\hat{\mathbf{d}} \tag{9.58}$$

where $\hat{\mathbf{K}}$ and $\hat{\mathbf{d}}$ are defined as in Section 9.4.2. The total quasi-static work done is given by

$$W = \frac{1}{2}\hat{\mathbf{d}}^T\hat{\mathbf{f}} \tag{9.59}$$

Finally, the total potential energy is the difference between the elastic energy and twice the work done:

$$\Pi = \frac{1}{2}\hat{\mathbf{d}}^T\hat{\mathbf{K}}\hat{\mathbf{d}} - \hat{\mathbf{d}}^T\hat{\mathbf{f}} \tag{9.60}$$

Not surprisingly, setting the gradient of the potential energy to zero yields the force balance equation:

$$\begin{aligned} &\nabla\Pi = 0 \\ &\Rightarrow \hat{\mathbf{K}}\hat{\mathbf{d}} = \hat{\mathbf{f}} \end{aligned} \tag{9.61}$$

9.6 Assembly of Truss Linear System

Whether we consider the force balance approach or the potential energy approach, we observe that one must assemble the **K** matrix and the **f** vector.

Given an arbitrary truss with M free nodes (for example, $M = 1$ in Figure 9.4), one creates a $2M\times 2M$ global stiffness matrix $\mathbf{K}$ and a $2M\times 1$ force vector $\mathbf{f}$, as follows. For the bar joining nodes i and j:

1. We first construct the 2×2 stiffness matrices $\mathbf{k}_{ij}$ via

$$\mathbf{k}_{ij} = \frac{E_{ij}A_{ij}}{l_{ij}} \begin{bmatrix} \cos^2\alpha_{ij} & \sin\alpha_{ij}\cos\alpha_{ij} \\ \sin\alpha_{ij}\cos\alpha_{ij} & \sin^2\alpha_{ij} \end{bmatrix} \tag{9.62}$$

2. The sub-matrix $\mathbf{k}_{ij}$ is *appended* to the $\mathbf{K}$ matrix at the locations $(2i-1, 2i-1)$ and $(2j-1, 2j-1)$; then $-\mathbf{k}_{ij}$ is *inserted* into the $\mathbf{K}$ matrix at the locations $(2i-1, 2j-1)$ and $(2j-1, 2i-1)$.

The right-hand side of $\mathbf{f}$ is simply the linear assembly of all external forces. Once $\mathbf{K}$ and $\mathbf{f}$ are assembled, the linear system is solved to obtain the displacements (after removing the columns and rows corresponding to fixed degrees of freedom).

Note that once the displacements are computed, the force and stress in each of the bars can be computed as follows. From Equation (9.12), given the displacements at the two nodes of a bar, the deformation of the bar is given by

$$\Delta L_{ij} \approx (u_j - u_i)\cos\alpha_{ij} + (v_j - v_i)\sin\alpha_{ij} \tag{9.63}$$

Once the deformation has been computed, the force can be determined as

$$F_{ij} = k_{ij}\Delta L_{ij} \tag{9.64}$$

Finally, the stress is given by

$$\sigma_{ij} = \frac{F_{ij}}{A_{ij}} \tag{9.65}$$

9.7 Truss Modeling Using MATLAB

We now consider the MATLAB implementation of the algorithm in Section 9.6, captured via a `truss2d` class for analyzing truss systems:

```
classdef truss2d
    properties(GetAccess = 'public', SetAccess = 'private')
        % public read access, but private write access.
        myNodeLocations; % (x,y) node locations, (2, N)
        myConnectivity; % (startnode, endnode) (2, M),
        ...
    end
    methods
```

```
function obj = truss2d(nodeXY,connectivity)
    % constructor for a truss
    % nodeXY: 2 x N, N is the number of nodes
    % connectivity: 2 x M, M is the number of bars
    ...
end
function obj = assignE(obj,E,members)
    % obj = assignE(obj,E,members)
    % assign E to one or more members
    % if members is not given, then assign E to all
    if (nargin == 2)
        members = 1:obj.myNumTrussBars;
    else
        assert(max(members) <= obj.myNumTrussBars);
        assert(min(members) >=  1);
    end
    obj.myE(members) = E;
end
function obj = assignA(obj,A,members)
    ...
end
function obj = fixXofNodes(obj,nodes)
    %fix the x locations of the nodes
    obj.myDOFFree(2*nodes-1) = 0;
end
function obj = fixYofNodes(obj,nodes)
    ...
end
function obj = applyForce(obj,node,force)
    % apply force at specified nodes
    obj.myForceExternal(2*node-1,1) = force(1);% x
    obj.myForceExternal(2*node,1) = force(2);% y
end
function plot(obj,withNumbering)
        ...
end
function plotDeformed(obj)
    ...
end
```

```
        function obj = assemble(obj)
            % Node based
            ...
            % k = EA/L of each member
            Ke = (obj.myE).*(obj.myArea)./(obj.myL);
            ...
        end
        function obj = assembleElemBased(obj)
            % more efficient means of assembly
            % Create the K matrix
            ...
        end
        function obj = solve(obj)
            ...
        end
        function vol = getVolume(obj)
            vol = sum(obj.myArea.*obj.myL);
        end
        function obj = computeStresses(obj)
            ...
        end
    end
end
```

The reader is encouraged to study the code accompanying this text, and to observe the following:

1. The constructor `truss2d` takes two arguments: the (x, y) coordinates of all the nodes, and their connectivity. For convenience, the length and orientation of each bar is computed within the constructor.
2. Two implementations of the assembly of the stiffness matrix are included: node-based assembly (default) and element-based assembly; the latter is more efficient. The reader is encouraged to study these two functions carefully.
3. The truss is solved in the `solve` function; since the problem reduces to solving a linear system of equations, explicit minimization of the quadratic function is not required.
4. Once the displacements have been computed, the stresses are also computed, as described in Equation (9.65).
5. Functions are provided to: (1) assign material properties and cross-sectional areas, (2) fix specific nodes and (3) apply external forces on free nodes.
6. Functions are included to plot the undeformed and deformed truss, with and without numbering.

We will check our code using problems with analytical solutions.

Example 9.4 Solve Example 9.1 using the `truss2d` class.

Solution: The construction of a truss object requires the nodal coordinates and their connectivity. From Figure 9.3, we gather the following information:

```
theta = pi/4;
xy = [0 -cos(theta) cos(theta); 0 sin(theta) sin(theta)];%
nodes
connectivity = [1 2; 1 3]';
t = truss2d(xy,connectivity);
```

We then assign cross-sectional areas and Young's modulus:

```
t = t.assignE(2e11); % for all bars
t = t.assignA(1e-6); % for all bars
```

Assignment of boundary conditions and forces are as below:

```
t = t.fixXofNodes([2 3]);
t = t.fixYofNodes([2 3]);
beta = pi/4;
t = t.applyForce(1,[cos(beta); -sin(beta)]);
```

The truss (prior to deformation) can be displayed as follows:

```
t = t.plot();
```

We then assemble the stiffness matrix and force vector as follows:

```
t = t.assemble();
```

Finally, one can solve for the unknown displacements via

```
t = t.solve();
```

The displacements at all three nodes is given by

```
t.myUV
ans =
   1.0e-05 *
    0.3536         0         0
   -0.3536         0         0
```

Observe that this matches the expected answer in Equation (9.26).

Example 9.5 Solve Example 9.2 using the `truss2d` class.

Solution: The truss is constructed as before; the only difference is the force direction.

```
beta = pi/3;
t = t.applyForce(1,[cos(beta); -sin(beta)]);
```

(cont.)

We assemble and solve:

```
t = t.assemble();
t = t.solve();
t.myUV
ans =
   1.0e-05 *
    0.2500         0         0
   -0.4330         0         0
```

This matches the expected answer in Equation (9.27).

Example 9.6 Solve Example 9.3 using the `truss2d` class.

Solution: The truss is constructed as follows:

```
xy = [0 0 -0.707 0.707; 0 1 -0.707 -0.707]; % nodes
connectivity = [1 2; 1 3; 1 4]';
t = truss2d(xy,connectivity);
```

We then assign cross-sectional areas and Young's modulus.

```
t = t.assignE(2e11); % for all bars
t = t.assignA(1e-6); % for all bars
```

Assignment of boundary conditions and forces are as below.

```
t = t.fixXofNodes([2 3 4]);
t = t.fixYofNodes([2 3 4]);
t = t.applyForce(1,[4; 2]);
```

We assemble and solve:

```
t = t.assemble();
t = t.solve();
t.myUV
ans =
   1.0e-04 *
    0.2000         0         0         0
    0.0500         0         0         0
```

This matches the expected answer in Equation (9.51).

We now consider other examples.

Example 9.7 For the truss in Figure 9.5, the Young's modulus of each bar is 2×10^{11} N/m^2, and its cross-sectional area is 10^{-6} m^2. A force of (1, −2) N is

(cont.)

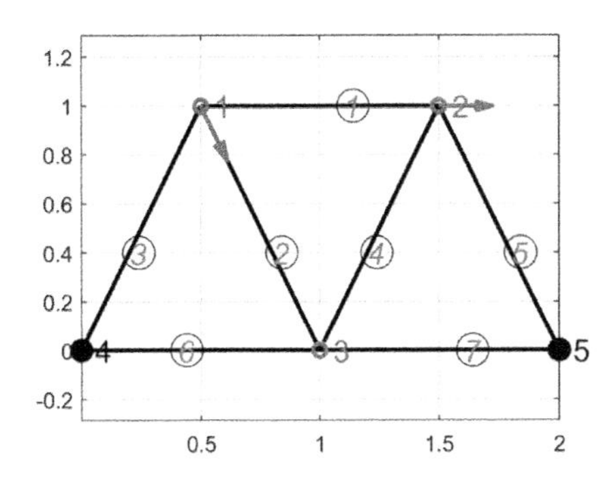

Figure 9.5 A seven-bar truss.

applied at node 1, and (2, 0) N is applied at node 2; nodes 4 and 5 are fixed. Using the `truss2d` implementation, find the displacements of all nodes.

Solution: The construction of a truss object requires the nodal coordinates and their connectivity. From Figure 9.5, we gather the following information:

```
xy = [0.5 1.5 1.0 0 2.0; 1 1 0 0 0];% nodes
connectivity = [1 2; 1 3; 1 4; 2 3; 2 5; 3 4; 3 5]';
t = truss2d(xy,connectivity);
```

We then assign cross-sectional areas and Young's modulus.

```
t = t.assignE(2e11); % for all members
t = t.assignA(1e-6); % for all members
```

Assignment of boundary conditions and forces are as below.

```
t = t.fixXofNodes([4 5]);
t = t.fixYofNodes([4 5]);
t = t.applyForce(1,[1;-2]);
t = t.applyForce(2,[2;0]);
```

The truss (prior to deformation) can be displayed as follows:

```
t = t.plot();
```

We then assemble the stiffness matrix and force vector as follows:

```
t = t.assemble();
```

Finally, one can solve for the unknown displacements via

```
t = t.solve();
```

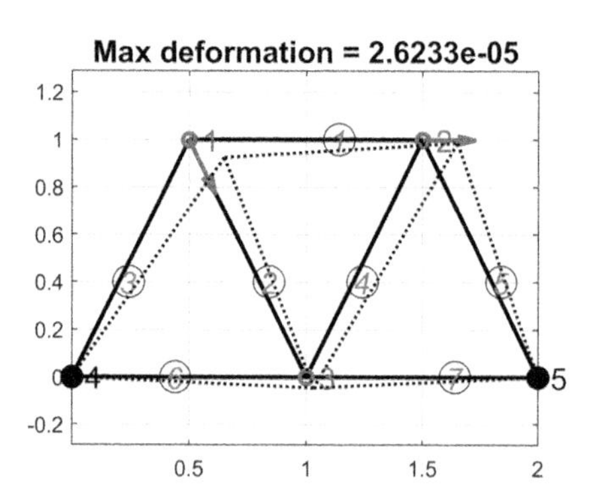

Figure 9.6 Deformed truss; deformation is exaggerated for illustration.

The displacements at all five nodes are given by

```
t.myUV
ans =
  1.0e-004 *
    0.2346    0.2346    0.0500         0         0
   -0.1173   -0.0224   -0.0699         0         0
```

The deformed truss is displayed via

```
t.plotDeformed();
```

(see Figure 9.6; deformation is exaggerated for illustrative purposes).

Example 9.8 The truss problem in Example 9.7 is modified as follows: node 5 is fixed only in the y-direction, but free to move in the x-direction (node 4 is fixed both in x and y); see Figure 9.7. Using the `truss2d` implementation, find the displacement of all the nodes.

Solution: As before, the truss is constructed as follows:

```
xy = [0.5 1.5 1.0 0 2.0; 1 1 0 0 0];% nodes
connectivity = [1 2; 1 3; 1 4; 2 3; 2 5; 3 4; 3 5]';
t = truss2d(xy,connectivity);
```

We then assign cross-sectional areas and Young's modulus.

```
t = t.assignE(2e11); % for all members
t = t.assignA(1e-6); % for all members
```

Observe that for node 5 only the x-direction is not fixed, and the forces are as before:

(cont.)

```
t = t.fixXofNodes(4);
t = t.fixYofNodes([4 5]);
t = t.applyForce(1,[1;-2]);
t = t.applyForce(2,[2;0]);
```

The truss (prior to deformation) can be displayed as follows:

```
t = t.plot();
```

We then assemble the stiffness matrix and force vector as follows:

```
t = t.assemble();
```

Finally, one can solve for the unknown displacements via

```
t = t.solve();
```

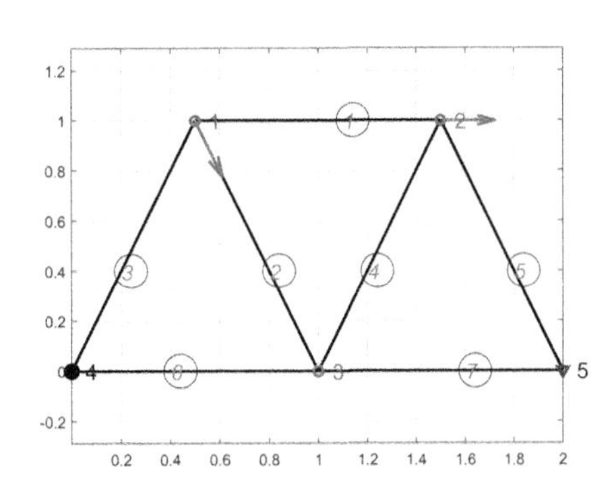

Figure 9.7 A seven-bar truss with node 5 allowed to slide.

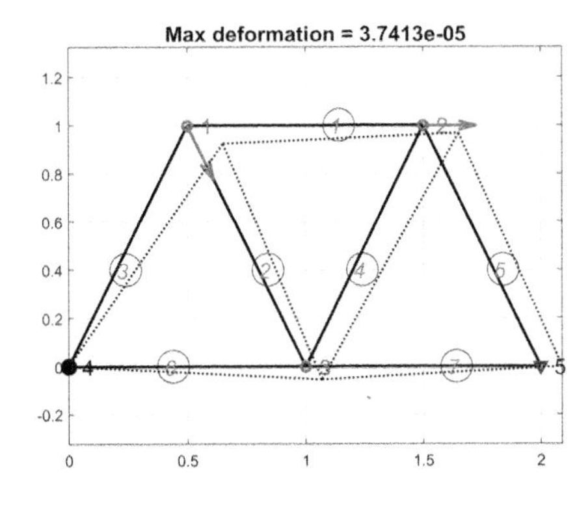

Figure 9.8 A deformed truss with node 5 allowed to slide; deformation is exaggerated for illustration.

(cont.)

The displacements at all five nodes are given by

```
t.myUV
ans =
   1.0e-04 *
  0.3346    0.3346    0.1500         0    0.2000
 -0.1673   -0.0724   -0.1199         0         0
```

The deformed truss is displayed via

```
t.plotDeformed();
```

(see Figure 9.8; deformation is exaggerated for illustrative purposes). Observe that node 5 slides along x, but does not move along y.

Example 9.9 Consider the truss in Figure 9.9. Assume that the Young's modulus for all bars is 2×10^{11} N/m^2, and that the cross-sectional area of every bar except bar 4 is 10^{-6} m^2, while the cross-sectional area of bar 4 is 2×10^{-6} m^2. A force of 100 N is applied vertically downward at node 4, as illustrated. Node 1 is fixed, while node 3 is allowed to slide along x. Using the `truss2d` implementation, find the stress in bar 1.

Solution: As before, the truss is constructed as follows:

```
xy = [0  1.25 2 1.25; 0 0 0 1];% nodes
connectivity = [1 2; 1 4; 2 3; 2 4; 3 4]';
t = truss2d(xy,connectivity);
```

We then assign cross-sectional areas and Young's modulus.

```
t = t.assignE(2e11); % for all bars
t = t.assignA(1e-6); % for all bars
```

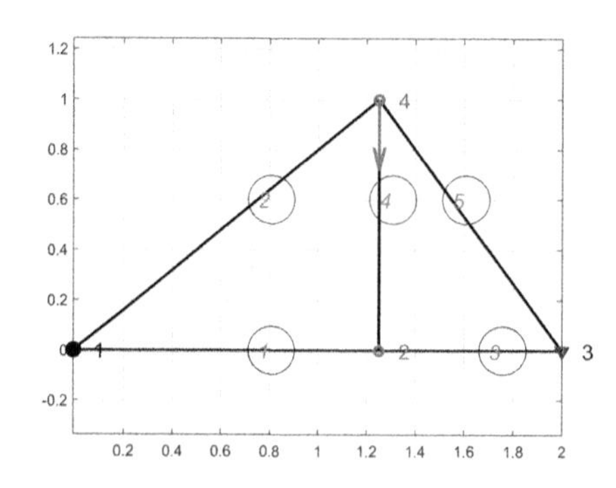

Figure 9.9 A five-bar truss with cross-sectional area bars.

(cont.)

Then we override the cross-sectional area for bar 4:

```
t = t.assignA(2*1e-6,4);
```

We apply the restraints and forces:

```
t = t.fixXofNodes(1);
t = t.fixYofNodes([1 3]);
t = t.applyForce(4,[0; -100]);
```

We then assemble and solve as follows:

```
t = t.assemble();
t = t.solve();
```

The stress in bar 1 is extracted as follows

```
t.myStress(1)
ans =
   4.6875e+07
```

The deformed truss (see Figure 9.10) is displayed via

```
t.plotDeformed();
```

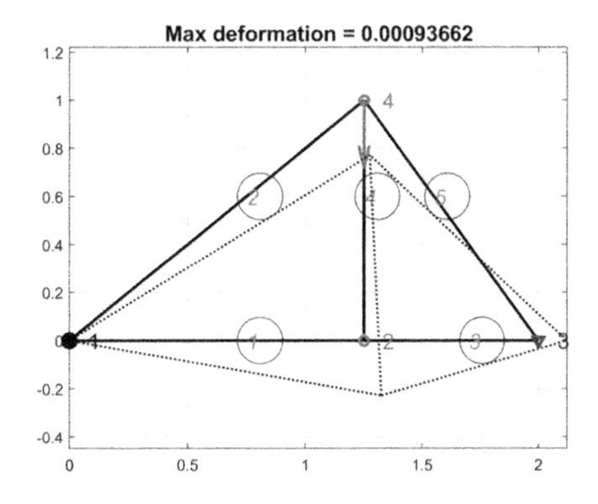

Figure 9.10 A deformed truss with node 3 allowed to slide; deformation is exaggerated for illustration.

9.8 EXERCISES

Exercise 9.1 Consider the truss in Figure 9.11, where each bar makes an angle of 30° with the vertical, and the nodes at the top are separated by a distance d. A force P is applied as shown in the figure. Assume that the Young's modulus and cross-sectional area of all bars are E and A, respectively.

Using force balance, find an expression for the vertical deflection of the bottom node as a function of P, d, E and A. Then find the numerical value of the deflection when $d = 0.1$ m, $E = 2 \times 10^{11}$ N/m^2, $A = 10^{-6}$ m^2 and $P = 1$ N.

Hint: First find the forces in bars 1 and 2, and then the forces in the remaining bars. Then find the deformation of each truss member, and finally the displacements u and v at the tip.

Exercise 9.2 Set up and solve Exercise 9.1 using the `truss2d` class.

Exercise 9.3 Consider the truss in Figure 9.12. Assume that the Young's modulus of each bar is 2×10^{11} N/m^2, and the cross-sectional area of each bar is 1×10^{-6} m^2. A force of 100 N is applied vertically to node 2. Nodes 1 and 4 are fixed. Using the `truss2d` implementation, find the vertical displacement at node 2.

Exercise 9.4 For the truss problem in Exercise 9.3, we will now allow node 1 to slide along y (see Figure 9.13). Everything else being the same, find the vertical displacement at node 2.

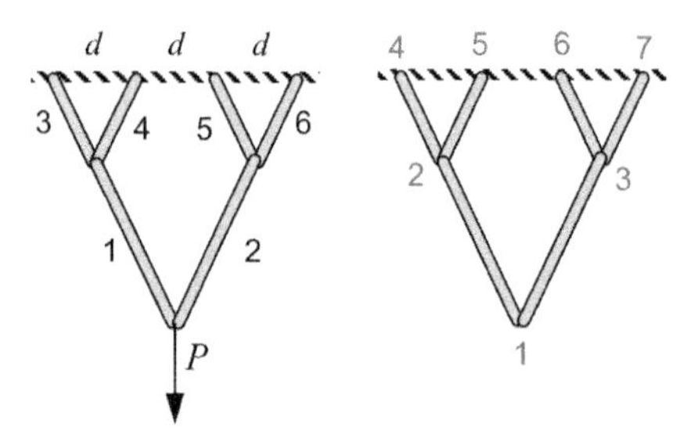

Figure 9.11 A chain truss.

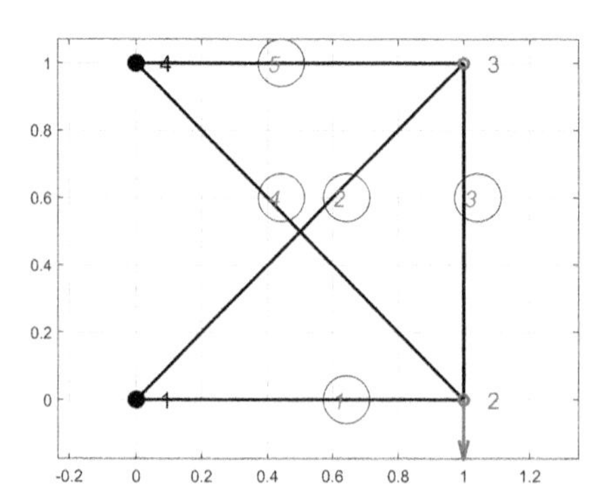

Figure 9.12 A cantilevered truss.

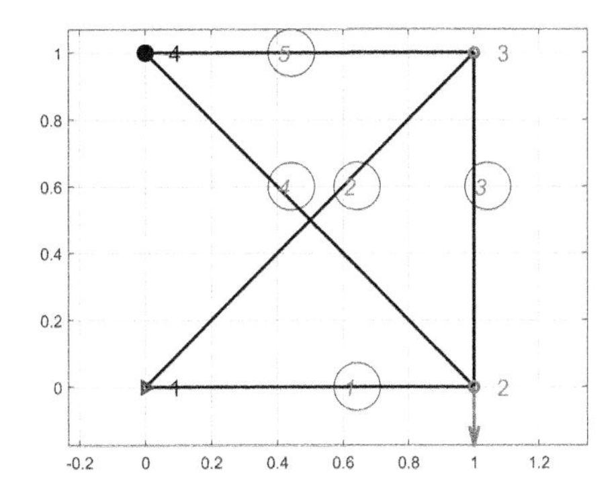

Figure 9.13 A cantilevered truss with sliding node.

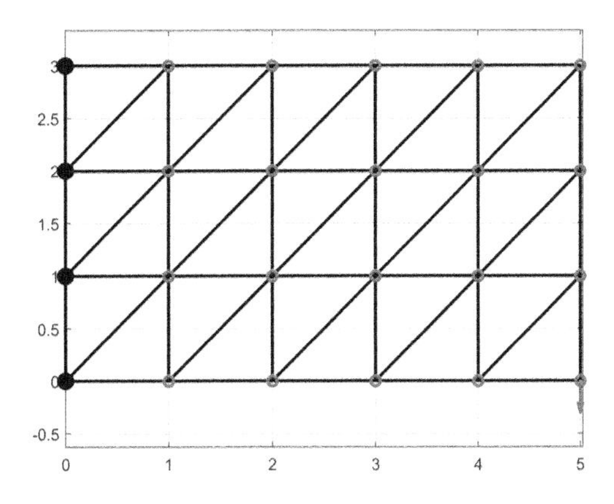

Figure 9.14 Truss grid.

Exercise 9.5 Implement a *function* `truss2dCantileverGrid.m` for creating a grid of bars, as illustrated in Figure 9.14.

You would call this function as follows:

```
t = truss2DCantileverGrid(Nx,Ny,L,F);
```

`Nx` and `Ny` are the number of unit cells (these are respectively equal to 5 and 3 in the figure), `L` is the length of each unit cell (this is 1 in the figure) and `F` is the force. `truss2dCantileverGrid.m` should return a `truss2d` object, where all nodes on the far left side are fixed, and `F` is applied to the node at the bottom right, as shown in the figure. Assume that Young's modulus is 2×10^{11} N/m^2 and the cross-sectional area is 10^{-6} m^2 for all bars.

The recommended numbering of nodes and bars is illustrated in Figure 9.15. Observe that there are no truss bars at the wall (since the nodes are fixed, the bars are not needed).

Exercise 9.6 Using the function created in Exercise 9.5, create a truss with `Nx` = 5, `Ny` = 3, `L` =1 and `F` = 100. Find the maximum deformation.

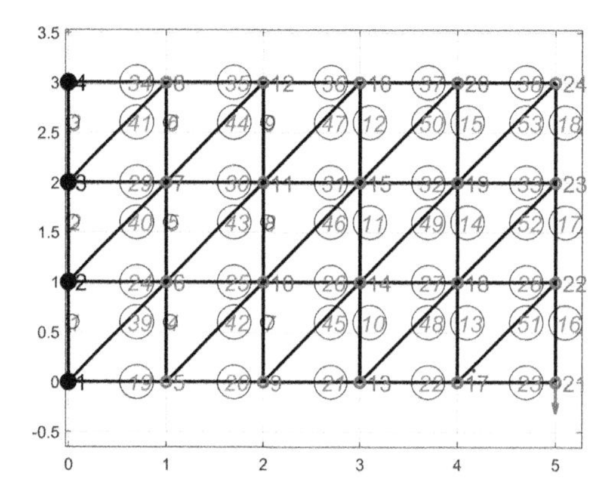

Figure 9.15 Truss grid numbering.

Exercise 9.7 Recall that the elastic energy in a bar is given by

$$U = \frac{1}{2}\frac{EA}{l}(\Delta L)^2$$

The method `elasticEnergyInBar(k)` in the `truss2d` class uses this equation to compute the elastic energy in any bar `k`. For the problem in Exercise 9.6, find the bar with the highest elastic energy and its corresponding elastic energy.

Exercise 9.8 For the truss in Exercise 9.6, assign a volume (and therefore an area) to each bar that is proportional to the elastic energy in that bar, such that the total volume is unchanged. This is a simple approach to the problem of optimal distribution of material. Now find the maximum deformation, and compare this with the deformation computed for that exercise, in which all areas were equal.

Exercise 9.9 Consider the determinate 3D truss illustrated in Figure 9.16, where the four nodes are located symmetrically as follows:

$$(0,0,0);\ (1/\sqrt{2},\ 0,\ 1/\sqrt{2});\ \left(\frac{-1}{2\sqrt{2}}, \frac{\sqrt{3}}{2\sqrt{2}}, \frac{1}{\sqrt{2}}\right);\ \left(\frac{-1}{2\sqrt{2}}, \frac{-\sqrt{3}}{2\sqrt{2}}, \frac{1}{\sqrt{2}}\right)$$

Observe that the lengths of all bars are 1, and connectivity is as illustrated. A unit force is applied along the x-direction, as shown. Let the Young's modulus of all bars be E, and the cross-sectional area be A. Using force balance, find the displacement of node 1 along the x-direction analytically.

Exercise 9.10 Generalize the `truss2d` class to 3D, i.e., implement a `truss3d` class. The constructor for the class should be as follows:

```
a = 1/sqrt(2); b = sqrt(3);
xy = [0 0 0; a 0 a; -a/2 a*b/2 a; -a/2 -a*b/2 a]';
connectivity = [1 2; 1 3;  1 4]'; % connectivity
t = truss3d(xy,connectivity);
```

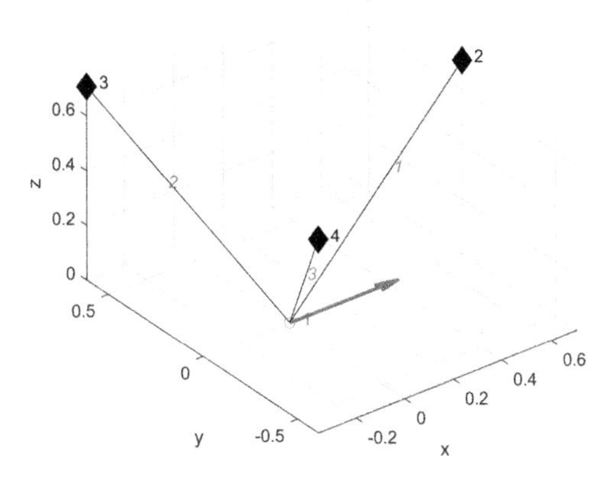

Figure 9.16 A determinate 3D truss.

You should be able to assign Young's modulus and cross-sectional area:

```
t = t.assignE(2e11);
t = t.assignA(1e-6);
```

Finally, you should be able to apply restraints and force:

```
t = t.fixXofNodes([2 3 4]);
t = t.fixYofNodes([2 3 4]);
t = t.fixZofNodes([2 3 4]);
t = t.applyForce(1,[1;0; 0]);
```

Also, implement the 3D plot function:

```
t.plot();
```

This should display Figure 9.16. Note that in this exercise we are merely constructing and displaying the truss.

Exercise 9.11 As a continuation of the previous exercises, your `truss3d` class should implement various methods such as `assemble`, `solve`, etc. It is sufficient if you implement either the node-based or the element-based assembly.

```
t = t.assemble();
t = t.solve();
t.plotDeformed();
```

Compute the vertical deformation at node 1, and compare against the analytical solution obtained in Exercise 9.9.

Exercise 9.12 Note that the `truss3d` class is capable of solving 2D truss problems as well. Verify your `truss3d` class by solving Example 9.7, where the z-coordinates of all nodes are set to zero.

Exercise 9.13 Using the `truss3d` class of Exercise 9.12, construct the problem illustrated in Figure 9.17. Find the displacements u, v and w of node 5.

```
xyz = [-1 1 1 -1 0; -1 -1 1 1 0; 0 0 0 0 2];% (x,y,z) of
nodes
connectivity = [1 5; 2 5; 3 5; 4 5]'; % connectivity
t = truss3d(xyz,connectivity);
t = t.assignE(2e11); % for all members
t = t.assignA(1e-6); % for all members
t = t.fixXofNodes([1 2 3 4]);
t = t.fixYofNodes([1 2 3 4]);
t = t.fixZofNodes([1 2 3 4]);
t = t.applyForce(5,[0; 25; 0]);
t.plot();
t = t.assemble();
t = t.solve();
t.plotDeformed();
```

Exercise 9.14 Consider the truss in Figure 9.18, which is duplicate of the problem in Figure 9.2 except that there is a "hanging" node (node 6) that is not connected to any bar. Pose and attempt to solve this problem using the `truss2d` class. Explain the outcome.

Exercise 9.15 In Exercise 9.14, create a bar joining node 1 and node 6. Pose and attempt to solve this problem using the `truss2d` class. Explain the outcome.

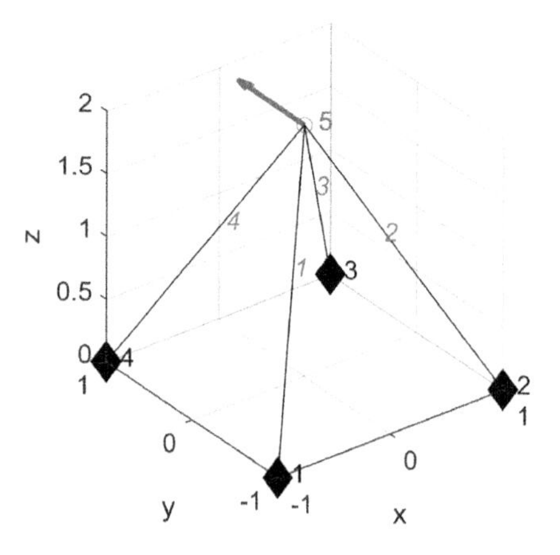

Figure 9.17 A four-bar 3D truss.

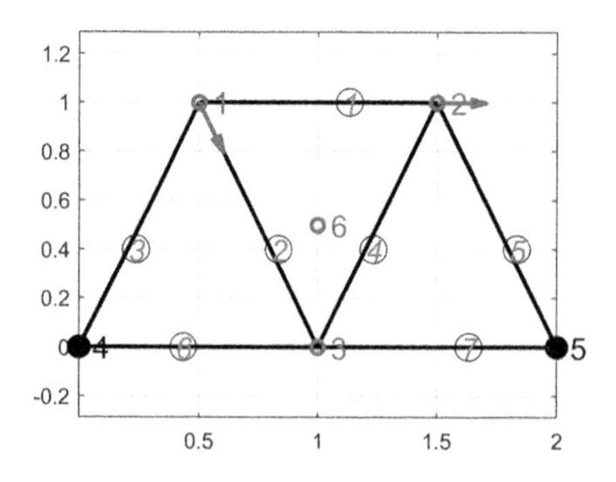

Figure 9.18 A truss with "hanging" node.

10 Size Optimization of Trusses

Highlights

1. In this chapter, we consider the size optimization of truss systems, i.e., we will optimize the cross-sectional areas of truss bars to minimize certain objectives.
2. There are various types of size optimization problems one can pose. The simplest is to minimize the compliance, subject to a volume constraint.
3. We will consider several examples of compliance minimization, some of which have analytical solutions.
4. We will then attempt to solve these using numerical methods. From a numerical perspective, the most important conclusion we will draw is that *proper scaling of the optimization variables, objective and constraints is crucial for numerical convergence.*
5. Other types of size optimization problems, such as volume minimization subject to compliance and/or stress constraint, are also considered.

Consider the truss illustrated in Figure 10.1. In Chapter 9, we discussed methods to *analyze* such systems. In this chapter, we consider structural *optimization* ([26]); specifically, optimizing the cross-sectional areas of the bars. For example, we will minimize the *compliance* (a popular term in structural optimization, to be explained in Section 10.1), subject to a volume constraint. Alternatively, we will minimize the volume, subject to compliance constraints.

10.1 Compliance Minimization

For simplicity, consider the two-bar truss in Figure 10.2. An engineer may be interested, for example, in minimizing the displacement at the free node, for a fixed mass of truss material. In other words, how should the material be distributed between the two truss bars such that the displacement is minimized? Thus, the goal is to *maximize the stiffness* of the truss for a fixed amount of material used. The design variables are the cross-sectional areas of the bars.

The truss in Figure 10.2 has only one free node, and therefore minimizing the displacement of that node is meaningful. For a more complex truss such as

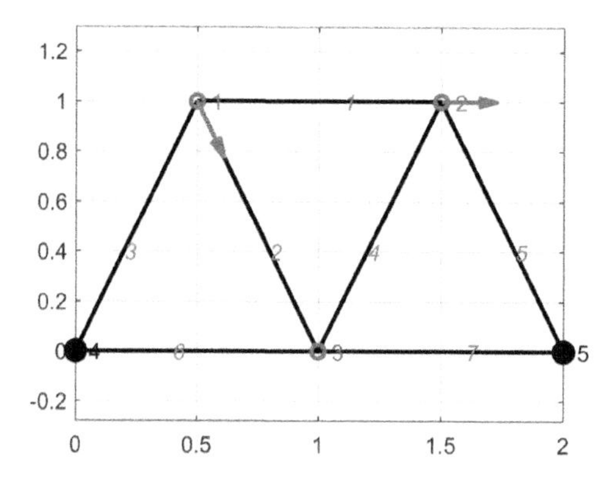

Figure 10.1 A truss system.

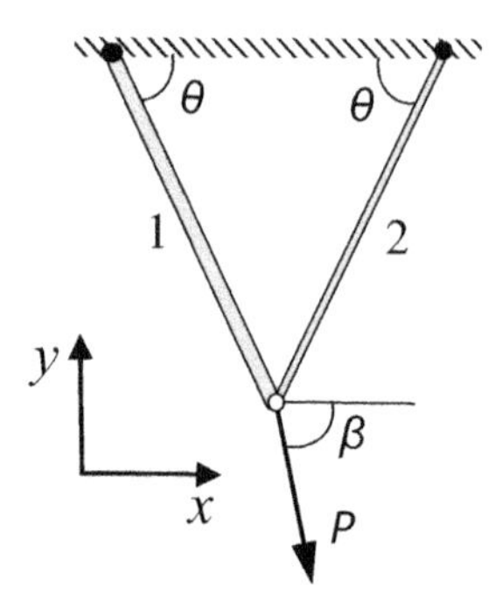

Figure 10.2 A determinate truss system.

the one in Figure 10.1, there are many free nodes. Thus, one should consider the displacements at all nodes as part of the optimization objective. While there are many ways of doing this, a simple and elegant approach is via the compliance, defined as

$$\text{Compliance: } J = \mathbf{f}^T\mathbf{d} \tag{10.1}$$

where **f** is the applied force and **d** is the displacement vector. Observe that the compliance is simply the dot product of the force and the displacement at all nodes. For the problem in Figure 10.2, where a single force is applied at a node, the compliance is the product of the force vector and the displacement vector at that node. For the problem in Figure 10.1, the compliance accounts for the displacement at the two nodes where forces are being applied (and where the displacements are going to be large). By minimizing the compliance, we are essentially minimizing (in an average sense) the displacement at all free nodes where forces are being applied.

The reader may wonder: *Why not just minimize the maximum displacement?* Unfortunately, this is much harder than it may appear. Specifically, since the location of the maximum displacement is not known a priori, it will turn out to be mathematically and numerically more challenging. In other words, minimizing the compliance is a much simpler problem to begin with. Once we have a firm understanding of compliance minimization, other objectives can also be considered.

Now that we have defined an objective (compliance) and the design variables (cross-sectional areas), we will also need to impose a constraint on the volume or the material used (otherwise the problem is ill-posed). For simplicity, given an initial volume V^0, we will require that the total volume during optimization be constrained by the expression

$$\sum A_i l_i - V^0 \leq 0 \tag{10.2}$$

where

$$\begin{aligned} &A_i\text{: cross-sectional area of bar } i \\ &l_i\text{: length of bar } i \\ &V^0\text{: initial volume of the truss} \end{aligned} \tag{10.3}$$

Finally, to avoid physically meaningless solutions, we will require that the cross-sectional areas remain positive, i.e., we impose a lower bound:

$$A_i > 0 \tag{10.4}$$

Thus, the compliance minimization problem may be posed as

$$\begin{aligned} \underset{\{A_1,A_2,\ldots\}}{\text{minimize}} \quad & \mathbf{f}^T\mathbf{d} \\ \text{s.t.} \quad & \sum A_i l_i - V^0 \leq 0 \\ & A_i > 0 \\ & \mathbf{K}\mathbf{d} = \mathbf{f} \end{aligned} \tag{10.5}$$

10.1.1 Determinate Truss System

We will now solve Equation (10.5) analytically for the two-bar problem in Figure 10.2. Since there are only two design variables, the problem can be posed as

$$\begin{aligned} \underset{\{A_1,A_2\}}{\text{minimize}} \quad & \mathbf{f}^T\mathbf{d} \\ \text{s.t.} \quad & A_1 l + A_2 l - V^0 \leq 0 \\ & A_i > 0 \\ & \mathbf{K}\mathbf{d} = \mathbf{f} \end{aligned} \tag{10.6}$$

where

$$\begin{aligned}&A_1, A_2\text{: cross-sectional areas of the two bars}\\&l\text{: length of each bar}\\&V^0\text{: initial volume of the truss}\\&\mathbf{f} = \{P\cos\beta,\ -P\sin\beta\}\end{aligned} \tag{10.7}$$

Since the truss problem is determinate, there is no need to set up and solve the stiffness matrix: the unknown displacement at the free node can be computed analytically. Therefore, the optimal cross-sectional areas can also be computed analytically.

From the previous chapter, recall that the displacement at the free node is given by

$$\begin{aligned}u &= \frac{Pl}{2E\cos\theta}\left(\frac{\sin(\beta+\theta)}{A_1\sin 2\theta} - \frac{\sin(\beta-\theta)}{A_2\sin 2\theta}\right)\\v &= \frac{-Pl}{2E\sin\theta}\left(\frac{\sin(\beta+\theta)}{A_1\sin 2\theta} + \frac{\sin(\beta-\theta)}{A_2\sin 2\theta}\right)\end{aligned} \tag{10.8}$$

Since the compliance is given by

$$J = \mathbf{f}^T\mathbf{d} = (P\cos\beta)u - (P\sin\beta)v \tag{10.9}$$

we have, upon simplification,

$$J = \frac{P^2 l}{E}\left[\frac{\sin^2(\beta+\theta)}{A_1\sin^2 2\theta} + \frac{\sin^2(\beta-\theta)}{A_2\sin^2 2\theta}\right] \tag{10.10}$$

Thus, the optimization problem may be posed as

$$\begin{aligned}&\underset{\{A_1,A_2\}}{\text{minimize}} \quad \frac{P^2 l}{E}\left[\frac{\sin^2(\beta+\theta)}{A_1\sin^2 2\theta} + \frac{\sin^2(\beta-\theta)}{A_2\sin^2 2\theta}\right]\\&\text{s.t.} \quad A_1 l + A_2 l - V^0 \le 0\\&\qquad A_i > 0\end{aligned} \tag{10.11}$$

To find numerical solutions, we consider special cases.

Example 10.1 For the truss problem in Figure 10.2, let $P = 10$ N, $E = 2\times10^{11}$ N/m^2, $l = 1.0$ m and $V^0 = 2\times10^{-6}$ m^3. Find the optimal cross-sections if $\beta = \theta = 45°$.

Solution: Observe that, when $\beta = \theta$, Equation (10.11) reduces to

$$\begin{aligned}&\underset{\{A_1,A_2\}}{\text{minimize}} \quad \frac{P^2 l}{EA_1}\\&\text{s.t.} \quad A_1 + A_2 - \frac{V^0}{l} \le 0\\&\qquad A_i > 0\end{aligned} \tag{10.12}$$

(cont.)

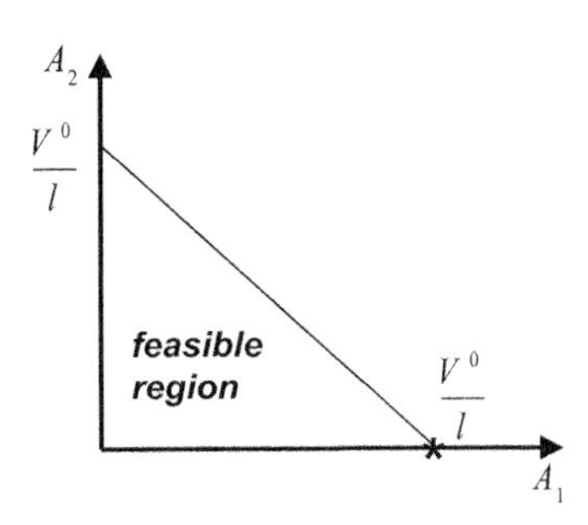

Figure 10.3 The feasible space for Equation (10.12).

The feasible region is illustrated in Figure 10.3.

Further, observe that the objective is independent of A_2, and is minimized when A_1 is maximum. Thus, the optimal solution occurs at

$$\begin{aligned} A_1 &= \frac{V^0}{l} = 2 \times 10^{-6}\ \text{m}^2 \\ A_2 &\approx 0 \end{aligned} \tag{10.13}$$

The minimum compliance is

$$J^* = \frac{P^2 l^2}{EV^0} = 2.5 \times 10^{-4}\ \text{Nm}$$

Example 10.2 In Example 10.1, suppose $\theta = 45°$ and $\beta = 60°$, and find the optimal cross-sections.

Solution: Equation (10.11) reduces to

$$\begin{aligned} &\underset{\{A_1, A_2\}}{\text{minimize}} && \frac{P^2 l}{E}\left\{\frac{\sin^2(\beta+\theta)}{A_1 \sin^2 2\theta} + \frac{\sin^2(\beta-\theta)}{A_2 \sin^2 2\theta}\right\} \\ &\text{s.t.} && A_1 + A_2 \leq \frac{V^0}{l} \\ & && A_i > 0 \end{aligned} \tag{10.14}$$

Since the constant multiplier does not play a role in the minimization, we have

$$\begin{aligned} &\underset{\{A_1, A_2\}}{\text{minimize}} && \left\{\frac{0.933}{A_1} + \frac{0.067}{A_2}\right\} \\ &\text{s.t.} && A_1 + A_2 - \frac{V^0}{l} \leq 0 \\ & && A_i > 0 \end{aligned} \tag{10.15}$$

(cont.)

The feasible space is exactly as before (see Figure 10.3). However, the objective depends on both A_1 and A_2, and is minimized when A_1 and A_2 are maximized. We will therefore make a reasonable assumption that the constraint is active, i.e., we will use as much material as allowed with $A_1 + A_2 = V^0/l$.

Since the constraint is linear, we can eliminate one of the variables and reduce the problem to an unconstrained minimization problem:

$$\underset{\{A_1,A_2\}}{\text{minimize}} \frac{P^2 l}{E} \left\{ \frac{0.933}{A_1} + \frac{0.067}{V^0/l - A_1} \right\}$$

Let

$$\overline{A} \equiv V_{max}/l = 2 \times 10^{-6} \text{ m}^2$$

Setting the derivative to zero, we have

$$-\frac{0.933}{(A_1)^2} + \frac{0.067}{(\overline{A} - A_1)^2} = 0$$

i.e.,

$$0.866(A_1/\overline{A})^2 - 1.866(A_1/\overline{A}) + 0.933 = 0$$

The two solutions are: $A_1 = 0.7887\overline{A}$ and $A_1 = 1.366\overline{A}$; the second solution is rejected since this leads to a negative value for A_2. Thus, the optimal solution is

$$\begin{aligned} A_1 &= 0.7887\overline{A} = 1.5774 \times 10^{-6} \text{ m}^2 \\ A_2 &= 0.2113\overline{A} = 4.226 \times 10^{-7} \text{ m}^2 \end{aligned} \tag{10.16}$$

Further, the optimal compliance is

$$J^* = 3.75 \times 10^{-4} \text{ Nm}$$

10.2 Compliance Minimization Using MATLAB

For indeterminate truss systems such as the one in Figure 10.1, one must numerically solve the optimization problem posed in Equation (10.5). We therefore develop an optimization class as a sub-class of the analysis class `truss2d`, introduced in Chapter 9. As an important side-note, readers are referred to specialized structural optimization methods such as the method of moving asymptotes ([27]), which are often more efficient than the generic methods discussed here. However, considering the focus of this text, we will restrict ourselves to optimization methods supported by MATLAB.

10.2.1 Direct Implementation

We will start with a direct implementation of Equation (10.5) via the baseline class `truss2dMinCompliance_0.m`. Observations include:

1. The class is a child class of the `truss2d` class; we will rely on the parent `truss2d` class for analysis.
2. Additional properties such as `myInitialVolume` are included for convenience.
3. The class constructor calls the parent constructor.
4. The optimization routine `optimize` relies on MATLAB's `fmincon` routine for optimization. `fmincon` is passed the following arguments:
   a. the `complianceObjective` function
   b. the initial cross-sectional areas
   c. a lower bound of 10^{-12} imposed on the areas

   Since the volume constraint is linear with respect to the design variables, we also pass the corresponding `A` matrix and `B` vector.
5. The message from `fmincon` is displayed and will be useful for numerical debugging.
6. Note that gradient information is not provided; finite difference is assumed by default. Explicit computation of gradients is discussed in Chapter 11.

The reader is encouraged to study the class carefully before proceeding.

```
classdef truss2dMinCompliance_0 < truss2d
    % This class minimizes the compliance of a 2d truss
    properties(GetAccess = 'public', SetAccess = 'private')
        myInitialVolume;
        myInitialArea;
        myInitialCompliance;
        myFinalVolume;
        myFinalArea;
        myFinalCompliance;
    end
    methods
        function obj =
truss2dMinCompliance_0(xy,connectivity)
            obj = obj@truss2d(xy,connectivity);
        end
        function J = complianceObjective(obj,Area)
            obj = obj.assignA(Area);
            obj = obj.assemble();
            obj = obj.solve();
            J = obj.mySol'*obj.myForceExternal(:);
```

```
        end
        function obj = optimize(obj)
            obj.myInitialArea = obj.myArea;
            obj.myInitialVolume = sum(obj.myArea.*obj.myL);
            LB = 1e-12*ones(1,obj.myNumTrussBars);
            AinEq = obj.myL;
            BinEq = obj.myInitialVolume;
            obj = obj.assemble();
            obj = obj.solve();
            obj.myInitialCompliance =
obj.mySol'*obj.myForceExternal(:);

[AMin,~,~,Output]=fmincon(@obj.complianceObjective, obj.
myInitialArea, AinEq,BinEq,[],[],LB);
            obj = obj.assignA(AMin);
            obj = obj.assemble();
            obj = obj.solve();
            obj.myFinalArea = obj.myArea.*obj.myL;
            obj.myFinalCompliance =
obj.mySol'*obj.myForceExternal(:);
            obj.myFinalVolume = sum(obj.myArea.*obj.myL);
            disp(Output);
        end
    end
end
```

We are now ready to use and test this class.

Example 10.3 Solve Example 10.1 using the MATLAB class `truss2dMinCompliance_0.m`.

Solution: Following Examples 9.4 and 9.5, the truss is initialized as follows:

```
theta = pi/4; % bar orientation
beta = pi/4; %angle of force
xy = [0 -cos(theta) cos(theta); 0 sin(theta) sin(theta)];
connectivity = [1 2; 1 3]';
t = truss2dMinCompliance_0(xy,connectivity);
t = t.assignE(2e11);
t = t.assignA(1e-6);
t = t.fixXofNodes([2 3]);
t = t.fixYofNodes([2 3]);
t = t.applyForce(1,[10*cos(beta); -10*sin(beta)]);
```

Then the truss is optimized via the command

```
t = t.optimize();
```

The result is summarized below.

(cont.)

```
Optimization completed because at the initial point, the
objective function is non-decreasing in feasible directions
to within the default value of the optimality tolerance,
and  constraints are satisfied to within the default value
of the constraint tolerance.
           iterations: 16
           funcCount: 59
           constrviolation: 0
           stepsize: 1.044357910506142e-11
           algorithm: 'interior-point'
           firstorderopt: 7.703355208832363e-06
           cgiterations: 0
            message: 'Local minimum found that satisfies the
constraints.…'
```

We can inspect the results via

```
>>t
myInitialArea: [1.0000e-06 1.0000e-06]
        myInitialVolume: 2.0000e-06
    myInitialCompliance: 5.0000e-04
            myFinalArea: [1.9690e-06 1.5503e-08]
      myFinalCompliance: 2.5394e-04
          myFinalVolume: 1.9845e-06
```

The compliance is fairly close to the correct answer, and so are the final cross-sectional areas; see Example 10.1.

Before we draw strong conclusions, let us consider one more example.

Example 10.4 Solve Example 10.2 using `truss2dMinCompliance_0.m`.

Solution: The minor change compared to Example 10.3 is

```
…
beta = pi/3; %angle of force
…
```

After optimization, the result is summarized below:

```
>>t
          myInitialArea: [1.0000e-06 1.0000e-06]
        myInitialVolume: 2.0000e-06
    myInitialCompliance: 5.0000e-04
            myFinalArea: [1.4082e-06 4.3964e-07]
      myFinalCompliance: 4.0745e-04
          myFinalVolume: 1.8479e-06
```

The computed solution is unfortunately incorrect; see Example 10.2.

In the next section, the reasons for failure are explained and corrected.

10.2.2 Scaling the Design Variables

As noted in Example 10.4, optimization failed. *The underlying problem is that the initial areas are relatively small, i.e., they are of the same order as the step size used by* `fmincon`, *resulting in early termination.* To avoid early termination, we must scale the design variables, as follows.

We introduce dimensionless variables, defined as

$$x_i = A_i / A_i^0 \tag{10.17}$$

where $x_i = A_i / A_i^0$ is the initial area of bar i. These dimensionless variables are then used as design variables:

$$\begin{aligned} &\underset{\{x_1, x_2, \ldots\}}{\text{minimize}} && \mathbf{f}^T \mathbf{d} \\ &\text{s.t.} && \sum x_i A_i^0 l_i - V^0 \leq 0 \\ & && x_i > 0 \\ & && \mathbf{Kd} = \mathbf{f} \end{aligned} \tag{10.18}$$

The use of dimensionless design variables is crucial in numerical optimization. Equation (10.18) is implemented below via the class `truss2dMinCompliance_1.m`. The reader is encouraged to study the class, especially the important differences, relative to the earlier implementation.

```
classdef truss2dMinCompliance_1 < truss2d
    ...
        function J = complianceObjective(obj,x)
            Area = x.*obj.myInitialArea;
            obj = obj.assignA(Area);
            ...
        end
        function obj = optimize(obj)
            ...
            AinEq = obj.myInitialArea.*obj.myL;
            BinEq = obj.myInitialVolume;
            x0 = ones(1,obj.myNumTrussBars); % unitless
            ...
            obj = obj.assignA(xMin.*obj.myInitialArea);

        end
    end
end
```

Observations include:

1. The design variables are denoted by $\mathbf{x}$ and initialized to unity.
2. In the objective, each design variable x_i is multiplied by the initial area A_i^0 to obtain the current cross-sectional area A_i.

3. Similarly, the volume constraint matrix `AinEq` is scaled by the initial areas.
4. The optimal solution is scaled to recover the correct areas.

Example 10.5 Solve Example 10.1 using the MATLAB class `truss2dMinCompliance_1.m`.

Solution: We have

```
>> t = truss2dMinCompliance_1(xy,connectivity);
```

The result is summarized below.

```
>>t
        myInitialVolume: 2.0000e-06
          myInitialArea: [1.0000e-06 1.0000e-06]
    myInitialCompliance: 5.0000e-04
            myFinalArea: [1.9987e-06 6.5220e-10]
      myFinalCompliance: 2.5016e-04
          myFinalVolume: 1.9993e-06
```

The answers are consistent with analytical results.

Example 10.6 Solve Example 10.2 using `truss2dMinCompliance_1.m`.

Solution: The result is summarized below.

```
>>t
        myInitialVolume: 2.0000e-06
          myInitialArea: [1.0000e-06 1.0000e-06]
    myInitialCompliance: 5.0000e-04
            myFinalArea: [1.5743e-06 4.2272e-07]
      myFinalCompliance: 3.7557e-04
          myFinalVolume: 1.9970e-06
```

The computed solutions are consistent with analytical results.

Next, we will attempt to improve on the accuracy.

10.2.3 Scaling the Constraints and Objective

Although we scaled the variables, the objective and constraint values are relatively small, i.e., of the same order as the termination error, and the optimization can terminate early, resulting in an unnecessary loss of accuracy. To avoid this numerical issue, (1) we must scale the objective, relative to the initial compliance value, and (2) the constraint must be scaled to be dimensionless. We can accomplish this as follows:

$$\begin{aligned} &\underset{\{x_1,x_2,\ldots\}}{\text{minimize}} && (\mathbf{f}^T\mathbf{d})/J^0 \\ &\text{s.t.} && \frac{\sum x_i A_i^0 l_i}{V^0} - 1 \leq 0 \\ &&& x_i > 0 \\ &&& \mathbf{K}\mathbf{d} = \mathbf{f} \end{aligned} \tag{10.19}$$

where the initial compliance is given by

$$J^0 = \mathbf{f}^T\mathbf{d}^0 \tag{10.20}$$

This ensures, for example, that the objective starts at 1 and therefore a small termination criterion is meaningful. Equation (10.19) is implemented via the class `truss2dMinComplianceVolumeConstraint.m`, which the reader is encouraged to study; the important differences, relative to the earlier implementation, are highlighted below.

```
classdef truss2dMinComplianceVolumeConstraint < truss2d
    ...
        function obj = initialize(obj)
            obj.myInitialArea = obj.myArea;
            obj.myInitialVolume = sum(obj.myArea.*obj.myL);
            obj = obj.assemble();
            obj = obj.solve();
            obj.myInitialCompliance =
obj.mySol'*obj.myForceExternal(:);
        end

        function JRelative = complianceObjective(obj,x)
            ...
            J = sol'*obj.myForceExternal(:);
            JRelative = J/obj.myInitialCompliance;
        end
        function obj = optimize(obj)
            ...
            t = t.initialize();
            x0 = ones(1,obj.myNumTrussBars); % unitless
            ...
            obj = obj.assignA(xMin.*obj.myInitialArea);
            ...
        end
    end
end
```

Observations include:

1. Scaling constants are initialized once at the start of optimization.
2. The objective function is of the order of unity, because of scaling.
3. The constraint is scaled such that the initial value is zero.

Example 10.7 Solve Example 10.1 using the MATLAB class `truss2dMinCompliance VolumeConstraint.m`.

Solution: We have

```
>>t=truss2dMinComplianceVolumeConstraint(xy,connectivity);
```

The result is summarized below.

```
>>t
        myInitialVolume: 2.0000e-06
          myInitialArea: [1.0000e-06 1.0000e-06]
    myInitialCompliance: 5.0000e-04
            myFinalArea: [2.0000e-06 3.2250e-13]
      myFinalCompliance: 2.5000e-04
          myFinalVolume: 2.0000e-06
```

We now converge to the expected solution; see Example 10.1.

Example 10.8 Solve Example 10.2 using MATLAB.

Solution: The result is summarized below.

```
>>t
        myInitialVolume: 2.0000e-06
          myInitialArea: [1.0000e-06 1.0000e-06]
    myInitialCompliance: 5.0000e-04
            myFinalArea: [1.5774e-06 4.2265e-07]
      myFinalCompliance: 3.7500e-04
          myFinalVolume: 2.0000e-06
```

The computed solution perfectly matches the analytical results; see Example 10.2.

The important conclusions are as follows:

1. Analytical solutions are extremely helpful when developing optimization routines. They help us identify mistakes in our formulation and/or implementation.
2. Scaling is important for numerical convergence; in particular, the scaling should be such that, at the start of the optimization cycle,
   (a) the optimization variables are scaled to unity
   (b) the objective must return a value close to unity
   (c) the constraints must be dimensionless.

3. Further, the scaling constants used for the variables, objective and constraints must be determined at the start of the optimization process and should not be modified during optimization.

10.2.4 Indeterminate Truss System

We can now use the implementation `truss2dMinComplianceVolumeConstraint.m`, discussed in Section 10.2.3, to optimize indeterminate trusses.

Example 10.9 Consider the problem illustrated in Figure 10.4, where a force of $(1, -2)$ N is applied to node 1 and $(2, 0)$ N is applied to node 2. The Young's modulus is 2×10^{11} N/m^2 and the initial cross-sectional area of all bars is 10^{-6} m^2. Find the truss structure with minimal compliance, but with the same volume as the initial structure.

Solution: We have

```
xy = [0.5 1.5 1.0 0 2.0; 1 1 0 0 0];
connectivity = [1 2; 1 3; 1 4; 2 3; 2 5; 3 4; 3 5]';
t = truss2dMinComplianceVolumeConstraint(xy,connectivity);
t = t.assignE(2e11);
t = t.assignA(1e-6); % for all members
t = t.fixXofNodes([4 5]);
t = t.fixYofNodes([4 5]);
t = t.applyForce(1,[1;-2]);
t = t.applyForce(2,[2;0]);
```

Before optimizing, let us solve and plot the deformed truss:

```
t = t.assemble();
t = t.solve();
t.plotDeformed();
```

The deformation of the initial truss is illustrated in Figure 10.5; the deformation has been exaggerated for visual clarity.

```
t = t.optimize();
```

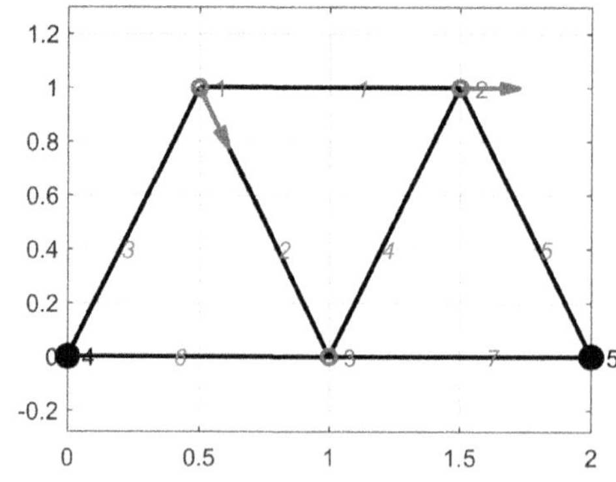

Figure 10.4 An indeterminate truss system.

(cont.)

The result is summarized below:

```
>>t
Initial Area:
   1.0e-06 *
    1.0000     1.0000     1.0000     1.0000     1.0000
 1.0000    1.0000
Final Area:
   1.0e-05 *
    0.0000     0.1759     0.0000     0.1759     0.1759     0.0787
 0.0787
Initial Volume:
   7.4721e-06
Final Volume:
   7.4721e-06
Initial Compliance:
   9.3853e-05
Final Compliance:
   6.0391e-05
```

Observe that the compliance dropped by almost 36 percent, for the same volume used. This illustrates the power of optimization. The deformed truss can be visualized via

```
t.plotDeformed();
```

The deformation of the optimized truss is illustrated in Figure 10.6. Observe that the maximum deformation is 36 percent lower than the initial design (with the same volume), i.e., the optimized truss is much stiffer.

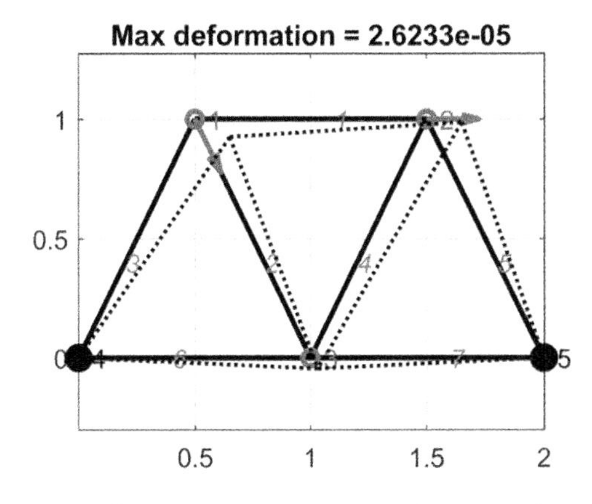

Figure 10.5 Deformation of the initial truss.

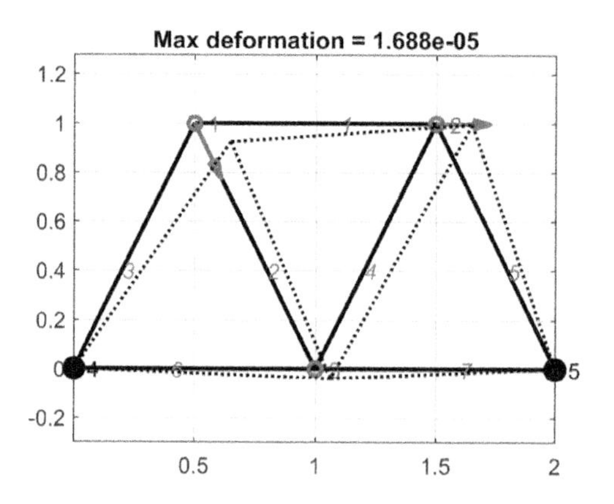

Figure 10.6 Deformation of the optimized truss.

We will consider additional examples in the exercises at the end of this chapter.

10.3 Compliance-Constrained Volume Minimization

We now switch the constraint and objective, i.e., we will minimize the volume with a constraint on the compliance. This may be posed as

$$\begin{aligned}
&\underset{\{A_1,A_2,\ldots\}}{\text{minimize}} && \sum A_i l_i \\
&\text{s.t.} && \mathbf{f}^T\mathbf{d} - J^0 \leq 0 \\
& && A_i > 0 \\
& && \mathbf{Kd} = \mathbf{f}
\end{aligned} \tag{10.21}$$

For the two-bar problem in Figure 10.2, since there are only two design variables, the problem can be posed as

$$\begin{aligned}
&\underset{\{A_1,A_2\}}{\text{minimize}} && A_1 l + A_2 l \\
&\text{s.t.} && \mathbf{f}^T\mathbf{d} - J^0 \leq 0 \\
& && A_i > 0 \\
& && \mathbf{Kd} = \mathbf{f}
\end{aligned} \tag{10.22}$$

Using the closed-form solution for the compliance (derived earlier), we have

$$\begin{aligned}
&\underset{\{A_1,A_2\}}{\text{minimize}} && A_1 l + A_2 l \\
&\text{s.t.} && \frac{P^2 l}{E}\left[\frac{\sin^2(\beta+\theta)}{A_1\sin^2 2\theta} + \frac{\sin^2(\beta-\theta)}{A_2\sin^2 2\theta}\right] - J^0 \leq 0 \\
& && A_i > 0
\end{aligned} \tag{10.23}$$

To find numerical solutions, we consider special cases.

Example 10.10 Repeat Example 10.1, but now minimize the volume subject to compliance constraint.

Solution: Equation (10.11) reduces to

$$\begin{aligned}
&\underset{\{A_1,A_2\}}{\text{minimize}} && A_1 l + A_2 l \\
&\text{s.t.} && \frac{P^2 l}{EA_1} - J^0 \leq 0 \\
& && A_i > 0
\end{aligned} \tag{10.24}$$

Thus, the optimal solution occurs at

$$\begin{aligned}
A_1 &= \frac{P^2 l}{EJ^0} = 2 \times 10^{-6}\ \text{m}^2 \\
A_2 &\approx 0
\end{aligned} \tag{10.25}$$

Note that $J^0 = 5 \times 10^{-4}$. Therefore,

$$\begin{aligned}
A_1 &= \frac{P^2 l}{EJ^0} = 1 \times 10^{-6}\ \text{m}^2 \\
A_2 &\approx 0
\end{aligned} \tag{10.26}$$

and the optimal volume is $V^* = A_1 l + A_2 l = 1 \times 10^{-6}\ \text{m}^3$.

We now consider the numerical implementation of volume minimization. Specifically, Equation (10.19) is modified as follows:

$$\begin{aligned}
&\underset{\{x_1,x_2,\ldots\}}{\text{minimize}} && \frac{\sum x_i A_i^0 l_i}{V_0} \\
&\text{s.t.} && \frac{(\mathbf{f}^T \mathbf{d})}{J^0} - 1 \leq 0 \\
& && x_i > 0 \\
& && \mathbf{K}\mathbf{d} = \mathbf{f}
\end{aligned} \tag{10.27}$$

Observe that the fundamental rules of scaling are adhered to in Equation (10.27), i.e., the scaling constants are such that, *at the start of optimization*,

1. the design variables are initialized to a value of 1
2. the objective returns a value of 1
3. the constraint is dimensionless and returns a value of 0.

A synopsis of the MATLAB implementation is provided below.

```
classdef truss2dMinVolumeComplianceConstraint < truss2d
    properties(GetAccess = 'public', SetAccess = 'private')
        ...
    end
    methods
        ...
        function volRelative = volumeObjective(obj,x)
```

```
            ...
            volRelative = vol/obj.myInitialVolume;
        end
        function [cineq,ceq] = complianceConstraint(obj,x)
            ...
            cineq = J/obj.myInitialCompliance - 1;
        end
        function obj = initialize(obj)
            ...
        end
        function obj = optimize(obj)
            ...
        end
    end
end
```

Observe that the ideas are borrowed from `truss2dMinCompliance VolumeConstraint.m`, except that the objective and constraint are switched. To illustrate its use, consider the following example.

Example 10.11 Consider the problem illustrated in Figure 10.4, where a force of $(1, -2)$ is applied to node 1 and $(2, 0)$ is applied to node 2. The Young's modulus is 2×10^{11} N/m^2, and the initial area of all bars is 10^{-6}. Find the truss structure with minimal volume, but with the same compliance as the initial structure.

Solution: The key difference is

```
t = truss2dMinVolumeComplianceConstraint(xy,connectivity);
```

The deformation of the initial truss is as before; see Figure 10.5. We now optimize:

```
t = t.optimize();
```

The result is summarized below.

```
myInitialVolume: 7.4721e-06
myInitialArea: [1.0000e-06 1.0000e-06 1.0000e-06 1.0000e-06
1.0000e-06 1.0000e-06 1.0000e-06]
myInitialCompliance: 9.3853e-05
myFinalArea: [1.4944e-13 1.1317e-06 1.3367e-13 1.1317e-06
1.1317e-06 5.0611e-07 5.0611e-07]
myFinalCompliance: 9.3853e-05
myFinalVolume: 4.8081e-06
```

Observe that the volume dropped by almost 36 percent, for the same compliance. *What do you expect the maximum deformation to be? Why?*

10.4 Stress-Constrained Volume Minimization

Thus far, we have disregarded stresses in our optimization formulation. Stresses are extremely important in engineering. In this section, we will explore including stress constraints in our formulation.

10.4.1 Stress Constraints: Determinate Truss

For simplicity, we will start with the determinate two-bar problem illustrated in Figure 10.2. For determinate trusses, the stresses can be computed analytically. Indeed, a simple force balance will show that the forces in the two bars are given by

$$\begin{aligned} F_1 &= \frac{P \sin(\beta + \theta)}{\sin 2\theta} \\ F_2 &= \frac{P \sin(\beta - \theta)}{\sin 2\theta} \end{aligned} \tag{10.28}$$

Therefore, the stresses are given by

$$\begin{aligned} \sigma_1 &= \frac{P \sin(\beta + \theta)}{A_1 \sin 2\theta} \\ \sigma_2 &= \frac{P \sin(\beta - \theta)}{A_2 \sin 2\theta} \end{aligned} \tag{10.29}$$

In order to impose stress constraints, we will assume for now that the two bars are in tension. This essentially amounts to assuming that $\beta \geq \theta$.

Let the allowable tensile stress be σ_Y (this denotes the yield strength of the material). For example, the yield strength of steel can vary from 400 MPa to 1000 MPa. Then the stress-constrained volume-minimization problem may be posed as

$$\begin{aligned} &\underset{\{A_1, A_2\}}{\text{minimize}} \; (A_1 l_1 + A_2 l_2) \\ &\text{s.t.} \quad \frac{P \sin(\beta + \theta)}{A_1 \sin 2\theta} \leq \sigma_Y \\ &\qquad \frac{P \sin(\beta - \theta)}{A_2 \sin 2\theta} \leq \sigma_Y \end{aligned} \tag{10.30}$$

Observe that we have disregarded compliance constraints for now. Equivalently, this formulation can be posed as

$$\begin{aligned} &\underset{\{A_1, A_2\}}{\text{minimize}} \; (A_1 l_1 + A_2 l_2) \\ &\text{s.t.} \quad A_1 \geq \frac{P \sin(\beta + \theta)}{\sigma_Y \sin 2\theta} \\ &\qquad A_2 \geq \frac{P \sin(\beta - \theta)}{\sigma_Y \sin 2\theta} \end{aligned} \tag{10.31}$$

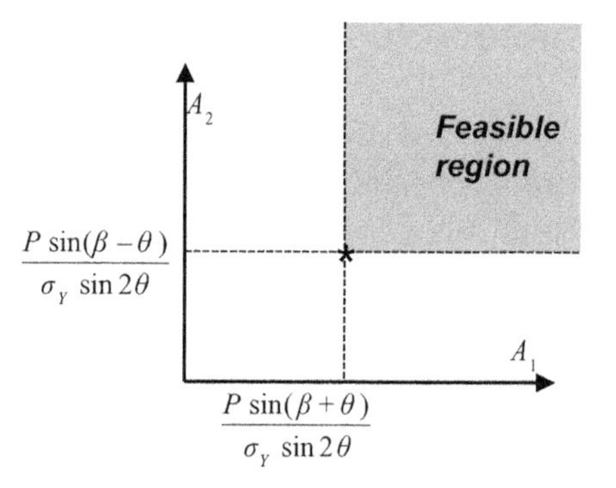

Figure 10.7 Feasible region for bars in pure tension.

The feasible region is illustrated in Figure 10.7, and the optimal solution (closest point to the origin) is

$$A_1^* = \frac{P\sin(\beta + \theta)}{\sigma_Y \sin 2\theta}$$
$$A_2^* = \frac{P\sin(\beta - \theta)}{\sigma_Y \sin 2\theta} \tag{10.32}$$

Example 10.12 Suppose $\theta = \beta = 45°$, $l = 1$ m, $P = 1$ N, $E = 2 \times 10^{11}$ N/m^2 and $\sigma_Y = 100$ MPa. Find the optimal cross-sections by considering the stress constraints.

Solution: Observe that, when $\beta = \theta$, Equation (10.32) reduces to

$$A_1^* = \frac{P}{\sigma_Y}$$
$$A_2^* = 0 \tag{10.33}$$

This is to be expected since bar 1 takes the entire load in this scenario. Thus, the optimal areas are

$$A_1^* = 10^{-8} \text{ m}^2$$
$$A_2^* = 0 \tag{10.34}$$

Example 10.13 Suppose $\theta = 45°$, $\beta = 60°$, $l = 1$ m, $P = 1$ N, $E = 2 \times 10^{11}$ N/m^2 and $\sigma_Y = 100$ MPa. Find the optimal cross-sections.

Solution: From Equation (10.32), we have

$$A_1^* = \frac{P\sin(\beta + \theta)}{\sigma_Y \sin 2\theta} = 9.659 \times 10^{-9} \text{ m}^2$$

$$A_2^* = \frac{P\sin(\beta - \theta)}{\sigma_Y \sin 2\theta} = 2.588 \times 10^{-9} \text{ m}^2 \tag{10.35}$$

In Example 10.12, $\beta \geq \theta$, therefore, both stresses were positive, i.e., both bars were in tension. On the other hand, if $\beta < \theta$, the assumption that both bars are in tension is not true anymore. One must consider failure due to both tension and compression. Assuming the failure strength is the same in tension and compression, we generalize Equation (10.30) to

$$\begin{aligned} &\underset{\{A_1,A_2\}}{\text{minimize}} \ (A_1 l_1 + A_2 l_2) \\ &\text{s.t.} \quad \left|\frac{P \sin(\beta+\theta)}{A_1 \sin 2\theta}\right| \leq \sigma_Y \\ &\qquad \left|\frac{P \sin(\beta-\theta)}{A_2 \sin 2\theta}\right| \leq \sigma_Y \end{aligned} \tag{10.36}$$

i.e.,

$$\begin{aligned} &\underset{\{A_1,A_2\}}{\text{minimize}} \ (A_1 l_1 + A_2 l_2) \\ &\text{s.t.} \quad -\sigma_Y \leq \frac{P \sin(\beta+\theta)}{A_1 \sin 2\theta} \leq \sigma_Y \\ &\qquad -\sigma_Y \leq \frac{P \sin(\beta-\theta)}{A_2 \sin 2\theta} \leq \sigma_Y \end{aligned} \tag{10.37}$$

Equivalently,

$$\begin{aligned} &\underset{\{A_1,A_2\}}{\text{minimize}} \ (A_1 l_1 + A_2 l_2) \\ &\text{s.t.} \quad A_1 \geq \frac{P \sin(\beta+\theta)}{\sigma_Y \sin 2\theta} \\ &\qquad A_1 \geq -\frac{P \sin(\beta+\theta)}{\sigma_Y \sin 2\theta} \\ &\qquad A_2 \geq \frac{P \sin(\beta-\theta)}{\sigma_Y \sin 2\theta} \\ &\qquad A_2 \geq -\frac{P \sin(\beta-\theta)}{\sigma_Y \sin 2\theta} \end{aligned} \tag{10.38}$$

Only two of the constraints will be active; the specific two will depend on the values of β and θ. Consider the following example.

Example 10.14 Suppose $\theta = 45°$, $\beta = 30°$, $l = 1$ m, $P = 1$ N, $E = 2 \times 10^{11}$ N/m^2 and $\sigma_Y = 100$ MPa. Find the optimal cross-sections.

Solution: From Equation (10.38), we have

$$\begin{aligned} &\underset{\{A_1,A_2\}}{\text{minimize}} \ (A_1 l_1 + A_2 l_2) \\ &\text{s.t.} \quad A_1 \geq 9.659 \times 10^{-9} \\ &\qquad A_1 \geq -9.659 \times 10^{-9} \\ &\qquad A_2 \geq -2.588 \times 10^{-9} \\ &\qquad A_2 \geq 2.588 \times 10^{-9} \end{aligned} \tag{10.39}$$

Observe that only two of the constraints are active, i.e.,

(cont.)

$$\begin{aligned} &\underset{\{A_1,A_2\}}{\text{minimize}} \ (A_1 l_1 + A_2 l_2) \\ &\text{s.t.} \qquad A_1 \geq 9.659 \times 10^{-9} \\ &\qquad\qquad A_2 \geq 2.588 \times 10^{-9} \end{aligned} \tag{10.40}$$

Thus, the optimal areas are

$$\begin{aligned} A_1 &= 9.659 \times 10^{-9} \text{ m}^2 \\ A_2 &= 2.588 \times 10^{-9} \text{ m}^2 \end{aligned} \tag{10.41}$$

These analytical solutions will be very useful in verifying numerical implementation of stress-constrained optimization, considered next.

10.4.2 Stress Constraints: MATLAB Implementation

We now consider indeterminate trusses such as the one illustrated in Figure 10.1. For such systems, one can pose the generic stress-constrained volume-minimization problem as

$$\begin{aligned} &\underset{\{A_1,A_2,\ldots\}}{\text{minimize}} \sum A_i l_i \\ &\text{s.t.} \qquad \sigma_i \leq \sigma_Y \\ &\qquad\qquad \sigma_i \geq -\sigma_Y \\ &\qquad\qquad \mathbf{Kd} = \mathbf{f} \end{aligned} \tag{10.42}$$

Observe that we have imposed two constraints (tension and compression) on each bar. In standard form, we have

$$\begin{aligned} &\underset{\{A_1,A_2,\ldots\}}{\text{minimize}} \sum A_i l_i \\ &\text{s.t.} \qquad \sigma_i - \sigma_Y \leq 0 \\ &\qquad\qquad -\sigma_i - \sigma_Y \leq 0 \\ &\qquad\qquad \mathbf{Kd} = \mathbf{f} \end{aligned} \tag{10.43}$$

Recall that, for numerical implementation, one must scale the design variables, the objective and constraints. We therefore replace Equation (10.43) with

$$\begin{aligned} &\underset{\{x_1,x_2,\ldots\}}{\text{minimize}} \frac{\sum x_i A_i^0 l_i}{V^0} \\ &\text{s.t.} \qquad \frac{\sigma_i}{\sigma_Y} - 1 \leq 0 \\ &\qquad\qquad \frac{-\sigma_i}{\sigma_Y} - 1 \leq 0 \\ &\qquad\qquad x_i > 0 \\ &\qquad\qquad \mathbf{Kd} = \mathbf{f} \end{aligned} \tag{10.44}$$

A snippet of the MATLAB implementation of Equation (10.44) is provided below.

```
classdef truss2dMinVolumeStressConstraint < truss2d
    % stress-constrained volume minimization
    properties(GetAccess = 'public', SetAccess = 'private')
        ..
        myYieldStress;
    end
    methods
        ...
        function obj =
assignYieldStress(obj,yieldStress,members)
            % assign yieldStress to one or more members
        ...
            obj.myYieldStress(members) = yieldStress;
        end
        ...
        function [cineq,ceq] = stressConstraint(obj,x)
            Area = x.*obj.myInitialArea;
            obj = obj.assignA(Area);
            obj = obj.assemble();
            obj = obj.solve();
            nConstrants = 2*obj.myNumTrussBars;
            cineq = zeros(1,nConstrants);
            constraint = 1;
            for m = 1:obj.myNumTrussBars
                cineq(constraint) =
obj.myStress(m)/obj.myYieldStress(m)-1;% tension
                cineq(constraint+1) = -
obj.myStress(m)/obj.myYieldStress(m) -1;%compression
                constraint = constraint+2; % increment
            end
            ceq = [];
        end
        function obj = optimize(obj)
            ...
            [xMin,~,~,~] =
fmincon(@obj.volumeObjective,x0, ...
[],[],[],[],LB,[],@obj.stressConstraint);
            ...
        end
    end
end
```

The main differences with respect to `truss2dMinVolumeCompliance Constraint` are the following:

1. An additional property, namely `myYieldStress`, is needed for this class.
2. An additional utility function, `assignYieldStress`, is needed.

3. The constraint function `complianceConstraint` is replaced by `stressConstraint`; observe that, unlike in `complianceConstraint`, two constraints are enforced per truss bar.

We will now verify the implementation by comparing the results against analytical solutions.

Example 10.15 Solve Example 10.12 using MATLAB.

Solution: See the solution to Example 10.3 to define the truss. However, we now assign the yield stress as well, and optimize:

```
...
t = t.assignYieldStress(100e6);
t = truss2dMinVolumeStressConstraint(xy,connectivity);
t = t.optimize();
```

The areas computed are

```
t.myFinalArea
ans =
   1.0e-07 *
    0.1000    0.0000
```

This is the expected solution!

Example 10.16 Solve Example 10.13 using the MATLAB class `truss2dMinVolumeStressConstraint.m`.

Solution: We leave it to the reader to define the truss problem. The computed areas are

```
...
    1.0e-08 *
  0.965929026280051   0.258822245062461
```

This is the expected solution (to within numerical precision).

Example 10.17 Solve Example 10.14 using `truss2dMinVolumeStressConstraint.m`.

Solution: We leave it to the reader to define the truss problem. The computed areas are

(cont.)

```
>> t.myFinalArea
ans =
   1.0e-08 *
    0.9659     0.2588
```

This is the expected solution (to within numerical precision).

Having gained confidence in our implementation, we can now address indeterminate trusses such as the one in Figure 10.1.

Example 10.18 Solve Example 10.11 with the exception that, instead of imposing a compliance constraint, we impose a stress constraint with yield stress of 100 MPa.

Solution: We leave it to the reader to define the truss problem. The final results are

```
...
myInitialVolume: 7.4721e-06
myInitialCompliance: 9.3853e-05
myFinalCompliance: 0.0047
myFinalVolume: 9.5004e-08
```

Observe that the volume has decreased by 98 percent! However, also observe that the compliance has increased significantly (we did not impose a constraint on the compliance). In other words, the computed truss meets the stress constraints (verify!), but it is not stiff. We will leave it as an exercise for the reader to impose both compliance and stress constraints.

10.5 Buckling Constraints

Yet another constraint often imposed on structural systems is buckling. Specifically, when a bar is in compression, it must be prevented from buckling. Recall that if a bar that is pinned at both ends (as in a truss) and is in compression, then the critical load for the first mode of buckling is given by

$$P_{cr} = \frac{\pi^2 EI}{l^2} \tag{10.45}$$

(see reference [10]), where I is the moment of inertia. Thus, one must also impose the following constraint on each bar:

$$F \geq -\frac{\pi^2 EI}{l^2} \tag{10.46}$$

Observe the negative sign in Equation (10.46). The equation imposes a constraint on the compressive stress; tensile stresses are not of concern in buckling.

We shall restrict our attention to bars with circular cross-sections. For such bars, the moment of inertia and cross-sectional areas are related as follows:

$$I = \frac{\pi r^4}{4} = \frac{\pi^2 r^4}{4\pi} = \frac{A^2}{4\pi} \tag{10.47}$$

Therefore,

$$F \geq -\frac{\pi E A^2}{4l^2} \tag{10.48}$$

To illustrate, we will once again consider the determinate two-bar truss problem illustrated in Figure 10.2.

10.5.1 Buckling Constraints: Determinate Truss

A simple force balance will show that the forces in the two bars are given by

$$\begin{aligned} F_1 &= \frac{P\sin(\beta+\theta)}{\sin 2\theta} \\ F_2 &= \frac{P\sin(\beta-\theta)}{\sin 2\theta} \end{aligned} \tag{10.49}$$

Therefore, the buckling constraints are given by

$$\begin{aligned} \frac{P\sin(\beta+\theta)}{\sin 2\theta} &\geq -\frac{\pi E (A_1)^2}{4(l_1)^2} \\ \frac{P\sin(\beta-\theta)}{\sin 2\theta} &\geq -\frac{\pi E (A_2)^2}{4(l_2)^2} \end{aligned} \tag{10.50}$$

Equivalently,

$$\begin{aligned} (A_1)^2 &\geq -\frac{P\sin(\beta+\theta)}{\sin 2\theta}\frac{4(l_1)^2}{\pi E} \\ (A_2)^2 &\geq -\frac{P\sin(\beta-\theta)}{\sin 2\theta}\frac{4(l_2)^2}{\pi E} \end{aligned} \tag{10.51}$$

The constraints are active only when the right-hand side is positive.

Example 10.19 Evaluate Equation (10.51) when $\theta = 45°$, $\beta = 30°$, $l = 1$ m, $P = 1$ N and $E = 2 \times 10^{11}$ N/m^2.

Solution: Substituting, we have

$$\begin{aligned} (A_1)^2 &\geq -6.1493 \times 10^{-12} \\ (A_2)^2 &\geq 1.6477 \times 10^{-12} \end{aligned} \tag{10.52}$$

As one can observe, only the constraint on the second bar (that is, in compression) is active. In practice, additional compliance and/or stress constraints must be imposed (see Exercise 10.10).

10.6 EXERCISES

Exercise 10.1 Consider the truss in Figure 10.8. Assume that the Young's modulus of each bar is 2×10^{11} N/m^2 and the initial cross-sectional area of each bar is 10^{-6} m^2. A force of 100 N is applied vertically to node 2; nodes 1 and 4 are fixed. Minimize the compliance subject to the (initial) volume constraint. What is the reduction in compliance achieved?

Exercise 10.2 For the problem in Figure 10.8, minimize the volume subject to the (initial) compliance constraint. What is the reduction in volume achieved?

Exercise 10.3 Consider the truss illustrated in Figure 10.9 (see Chapter 9 on the creation of such trusses). A force of 100 N is applied to the node at the bottom right, as illustrated. Assume that the Young's modulus is 2×10^{11} N/m^2 and the cross-sectional area is 10^{-6} m^2 for all bars. Minimize the compliance subject to a volume constraint. What is the reduction in compliance achieved? Note that you will need to change the truss class in `truss2dCantileverGrid`.

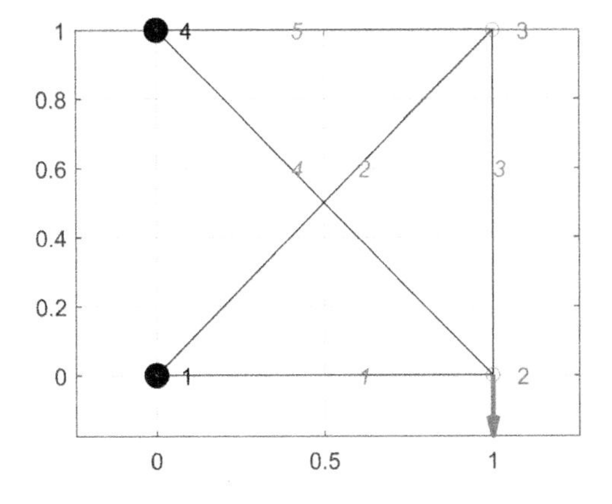

Figure 10.8 A cantilevered truss.

Exercise 10.4 For the problem illustrated in Figure 10.9, minimize the volume subject to an (initial) compliance constraint. What is the reduction in volume achieved?

Exercise 10.5 This exercise is a study in the computational cost of optimization. The truss illustrated in Figure 10.9 is made of up of five horizontal cells and three vertical cells. Create trusses with the following grid sizes and minimize the compliance subject to a volume constraint: (a) (6, 3), (b) (8, 4), (c) (10, 5) and (d) (12, 6). Tabulate the following results for each case: (i) number of design variables (i.e., the number of truss bars), (ii) percentage reduction in compliance achieved and (iii) time taken.

Exercise 10.6 For the truss problem in Figure 10.10, suppose $\theta = 45°$, $\alpha = 70°$, $l = 1$ m, $P = 100$ N, $E = 2 \times 10^{11}$ N/m^2, $\sigma_Y = 100$ MPa (yield strength) and initial cross-sectional areas of 10^{-6} m^2. Find the optimal cross-sections (analytically), with the constraints that the stress should not exceed the yield strength and that the compliance should not exceed the initial compliance.

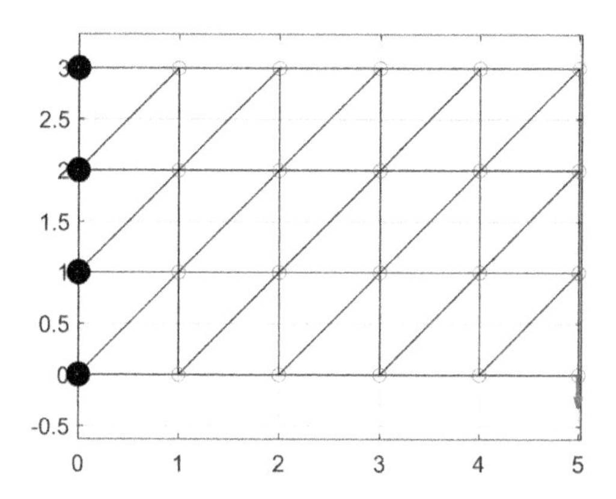

Figure 10.9 Truss grid.

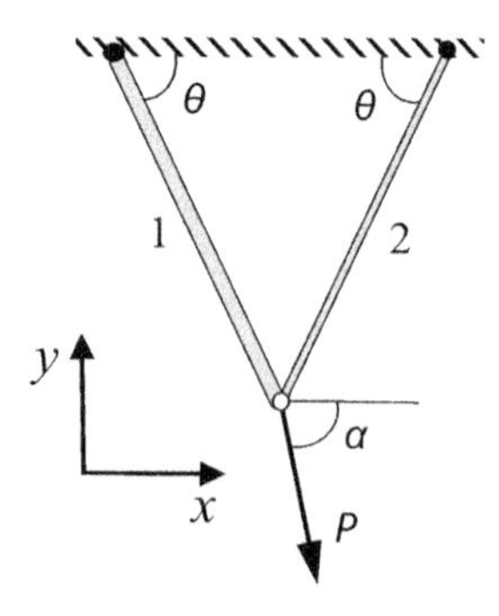

Figure 10.10 A determinate truss system.

Exercise 10.7 Consider the truss illustrated in Figure 10.11 (all length units are in meters). A force of (100, −20) N is applied at node 1. The Young's modulus is 2×10^{11} N/m^2, the cross-sectional area of each bar is 10^{-6} m^2 and the maximum stress allowed is 100 MPa.

(a) Minimize the compliance subject to volume constraint.
(b) Minimize the volume subject to the initial compliance constraint.

Exercise 10.8 Find analytical solutions to Example 10.12, with the additional constraint that the compliance must not exceed the initial compliance.

Exercise 10.9 Recall that Equation (10.27) captures the problem of minimizing volume subject to compliance constraint, while Equation (10.44) captures the problem of minimizing volume subject to stress constraints. Now consider the problem of minimizing volume subject to compliance and stress constraints, i.e.,

$$\begin{aligned}
&\underset{\{x_1, x_2, \ldots\}}{\text{minimize}} && \frac{\sum x_i A_i^0 l_i}{V^0} \\
&\text{s.t.} && \frac{(\mathbf{f}^T \mathbf{d})}{J^0} - 1 \leq 0 \\
& && \frac{\sigma_i}{\sigma_Y} - 1 \leq 0 \\
& && \frac{-\sigma_i}{\sigma_Y} - 1 \leq 0 \\
& && x_i > 0 \\
& && \mathbf{Kd} = \mathbf{f}
\end{aligned} \tag{10.53}$$

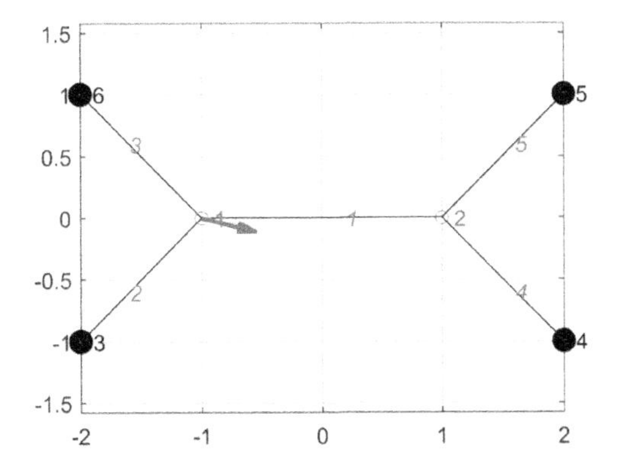

Figure 10.11 A five-bar truss system.

Create a `truss2dMinVolumeComplianceStressConstraint` class that implements Equation (10.53). Check your code against the analytical solutions computed in Exercise 10.8.

Exercise 10.10 Repeat Example 10.19, i.e., find analytical solutions to the buckling constraint problem, but with the additional constraint that the compliance must not exceed the initial compliance.

Exercise 10.11 Consider the problem of minimizing volume subject to compliance and buckling constraints, i.e.,

$$\begin{aligned}
\underset{\{x_1,x_2,\ldots\}}{\text{minimize}} \quad & \frac{\sum x_i A_i^0 l_i}{V^0} \\
\text{s.t.} \quad & \frac{(\mathbf{f}^T\mathbf{d})}{J^0} - 1 \leq 0 \\
& -\frac{4Fl^2}{\pi E A^2} - 1 \leq 0 \\
& x_i > 0 \\
& \mathbf{K}\mathbf{d} = \mathbf{f}
\end{aligned} \tag{10.54}$$

Create a class `truss2dMinVolumeComplianceBucklingConstraint` that implements Equation (10.54). Check your code against the analytical solution computed in Exercise 10.10.

Exercise 10.12 Consider the problem of minimizing volume subject to compliance, stress and buckling constraints, i.e.,

$$\begin{aligned}
\underset{\{x_1,x_2,\ldots\}}{\text{minimize}} \quad & \frac{\sum x_i A_i^0 l_i}{V^0} \\
\text{s.t.} \quad & \frac{(\mathbf{f}^T\mathbf{d})}{J^0} - 1 \leq 0 \\
& \frac{\sigma_i}{\sigma_Y} - 1 \leq 0 \\
& \frac{-\sigma_i}{\sigma_Y} - 1 \leq 0 \\
& -\frac{4Fl^2}{\pi E A^2} - 1 \leq 0 \\
& x_i > 0 \\
& \mathbf{K}\mathbf{d} = \mathbf{f}
\end{aligned} \tag{10.55}$$

Create a `truss2dMinVolumeComplianceStressBucklingConstraint` class that implements Equation (10.55). Check your code against analytical solutions.

Exercise 10.13 Create a `truss2dMinVolumeDisplacementConstraint` class to apply a displacement constraint on specific nodes (as opposed to overall compliance). It must include a function `assignAllowableDisplacement (obj, deltaMax, nodes)`. Implement and test this class; it is recommended that you inherit the `truss2dOptimize` class.

Exercise 10.14 Consider the determinate 3D truss illustrated in Figure 10.12, where the four nodes are located symmetrically as follows:

$$(0,0,0);\ \left(\frac{-1}{2\sqrt{2}},\frac{\sqrt{3}}{2\sqrt{2}},\frac{1}{\sqrt{2}}\right);\ \left(\frac{-1}{2\sqrt{2}},\frac{-\sqrt{3}}{2\sqrt{2}},\frac{1}{\sqrt{2}}\right)$$

Observe that the length of each bar is 1; connectivity is as illustrated. In Chapter 9, we developed analytical solution for the displacement for this truss, given a unit force in the x-direction. Now consider the problem of minimizing volume subject to compliance constraint. Find the optimal cross-sectional areas analytically.

Exercise 10.15 In Chapter 9, we developed a class `truss3d` for analyzing 3D trusses. Develop a `truss3dMinimizeVolumeComplianceConstraint` that minimizes the volume of a generic 3D truss, subject to a compliance constraint. Test your code against the analytical solution from the previous exercise.

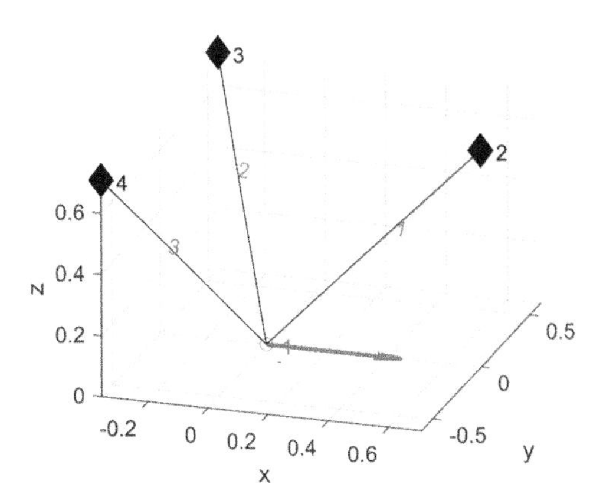

Figure 10.12 Optimizing a determinate 3D truss.

Exercise 10.16 Consider the 3D truss in Figure 10.13 (see Chapter 9 for details). Use the 3D truss optimization class developed in Exercise 10.15 to minimize the volume of the truss.

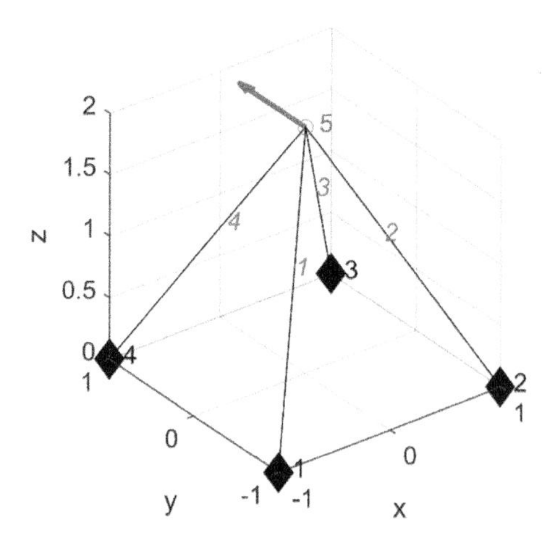

Figure 10.13 A four-bar 3D truss.

11 Gradient Computation

Highlights

1. First-order minimization methods require gradients. By default, optimization algorithms use finite-difference methods such as forward, backward and central difference to compute gradients. Some of the drawbacks of finite-difference methods are discussed.
2. For truss systems, a strategy to compute gradients analytically is described.
3. An alternative strategy based on complex variables to compute gradients is explored.
4. Other approaches to compute gradients, such as automatic differentiation, are also briefly discussed.

In Chapter 10, we focused on various concepts such as numerical scaling to increase the robustness of optimization algorithms. Here, we consider improving the efficiency of these algorithms.

Recall that we relied on MATLAB's `fmincon` to optimize the trusses. Since explicit gradients were not provided, `fmincon` relies on finite difference for computing approximate gradients. There are two main drawbacks with this default strategy: (1) a large number of function calls, and (2) reduced accuracy.

In this chapter, we will study the numerical pitfalls of finite-difference methods. We will then discuss a semi-analytical approach to computing gradients for truss systems, and explore the complex-variable method for computing the gradients.

11.1 Finite Difference in 1D

Engineers often rely on finite-difference methods to numerically compute gradients. Finite difference fundamentally relies on the Taylor series expansion. In 1D, recall the formal Taylor series expansion (see the Appendix, Section A.1):

$$f(x_0 + \Delta x) = f(x_0) + \frac{1}{1!}\frac{df}{dx}\bigg|_{x_0}\Delta x + \frac{1}{2!}\frac{d^2f}{dx^2}\bigg|_{x_0}(\Delta x)^2 + o(\Delta x)^3 \tag{11.1}$$

Retaining only the first derivative,

$$f(x_0 + \Delta x) \approx f(x_0) + \frac{df}{dx}\bigg|_{x_0}\Delta x \tag{11.2}$$

Thus, the first derivative can be approximated via

$$\frac{df}{dx}\bigg|_{x_0} \approx \frac{f(x_0 + \Delta x) - f(x_0)}{\Delta x} \tag{11.3}$$

The approximation introduced in Equation (11.3) leads to a discretization error that increases with increasing step size Δx.

Equation (11.3) is the classic *forward-difference* approximation for computing the first derivative. By replacing Δx with $-\Delta x$ in Equation (11.3), we have

$$\frac{df}{dx}\bigg|_{x_0} \approx \frac{f(x_0 - \Delta x) - f(x_0)}{-\Delta x} \tag{11.4}$$

i.e.,

$$\frac{df}{dx}\bigg|_{x_0} \approx \frac{f(x_0) - f(x_0 - \Delta x)}{\Delta x} \tag{11.5}$$

Equation (11.5) is the *backward-difference* approximation for computing the first derivative.

Finally, consider two special cases of Equation (11.2):

$$f(x_0 + \Delta x/2) \approx f(x_0) + \frac{df}{dx}\bigg|_{x_0}\frac{\Delta x}{2} \tag{11.6}$$

$$f(x_0 - \Delta x/2) \approx f(x_0) - \frac{df}{dx}\bigg|_{x_0}\frac{\Delta x}{2} \tag{11.7}$$

Subtracting Equation (11.7) from Equation (11.6), we have

$$f(x_0 + \Delta x/2) - f(x_0 - \Delta x/2) \approx \frac{df}{dx}\bigg|_{x_0}\Delta x \tag{11.8}$$

This leads to the *central-difference* approximation

$$\frac{df}{dx}\bigg|_{x_0} \approx \frac{f(x_0 + \Delta x/2) - f(x_0 - \Delta x/2)}{\Delta x} \tag{11.9}$$

Observe that all three methods entail subtracting two floating-point values that are typically close to each other, especially when the step size is small. This leads to floating-point (round-off) errors, as illustrated below.

Example 11.1 Find the gradient of the following function at $x = 1$:

$$f(x) = x^2 e^{x-1} \tag{11.10}$$

Use the three finite-difference methods with various step sizes Δx, and compare the accuracy against the analytical derivative.

Solution: Note that the analytical derivative at $x = 1$ is given by

$$\left.\frac{df}{dx}\right|_{1.0} = \left.\left(2xe^{x-1} + x^2 e^{x-1}\right)\right|_{x=1} = 3 \tag{11.11}$$

We now use the three finite-difference methods to compute the numerical gradient. The forward-difference approximation is

$$\left.\frac{df}{dx}\right|_{1.0} \approx \frac{f(1.0 + \Delta x) - f(1.0)}{\Delta x} \tag{11.12}$$

For $x = 1$ the computed gradient is 3.0352, for $\Delta x = 0.001$ it is 3.00350, and so on. One can similarly compute the errors for the backward-difference and central-difference methods.

A MATLAB code, `finiteDifferenceGradient.m`, in the software accompanying this text computes the finite difference for any function (defined via MATLAB) by any of the three methods. A typical example is provided here:

```
>> f = @(x) x^2*exp(x-1);
>> g =  finiteDifferenceGradient(f,1,1e-5,0) % central
g =
   3.000000000030755
>> g =  finiteDifferenceGradient(f,1,1e-5,-1) % backward
g =
   2.999965000205495
>> g =  finiteDifferenceGradient(f,1,1e-5,1) % forward
g =
      3.000035000244594
```

The three errors (absolute values) are illustrated in Figure 11.1 as a function of step size. Observe the following:

- The forward and backward-difference methods behave almost identically. Typically, the central-difference method is more accurate than the other two.
- For all three methods, reducing the step size from, say, 10^{-1} to 10^{-4} reduces the discretization error, and therefore improves the accuracy.

(cont.)

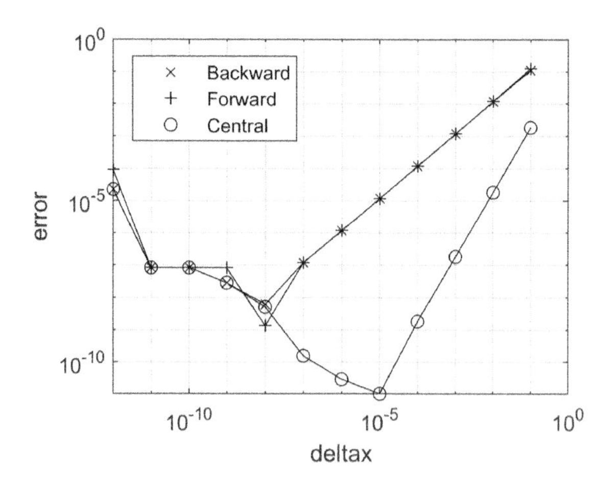

Figure 11.1 Finite-difference errors in computing the gradient for Equation (11.10) at $x = 1$.

- However, small values of the step size, say, 10^{-9}, lead to floating-point (round-off) errors.
- For this problem, the optimal step size is around 10^{-5} for the central-difference method and 10^{-8} for the other two.

We now repeat the experiment for a different function.

Example 11.2 Find the gradient of the following function, via the three finite-difference methods, at $x = -10$:

$$f(x) = e^x - x^2 \tag{11.13}$$

Compare the accuracy against the analytical derivative.

Solution: At $x = -10$, the analytical derivative is given by

$$\left.\frac{df}{dx}\right|_{-10} = 20.000045399929764 \tag{11.14}$$

The finite-difference errors are illustrated in Figure 11.2. Observe that, as before, the central-difference method is the most accurate, and that the errors initially decrease and then start increasing. The optimal value is around 10^{-6} for the central-difference method and 10^{-7} for the other two.

(cont.)

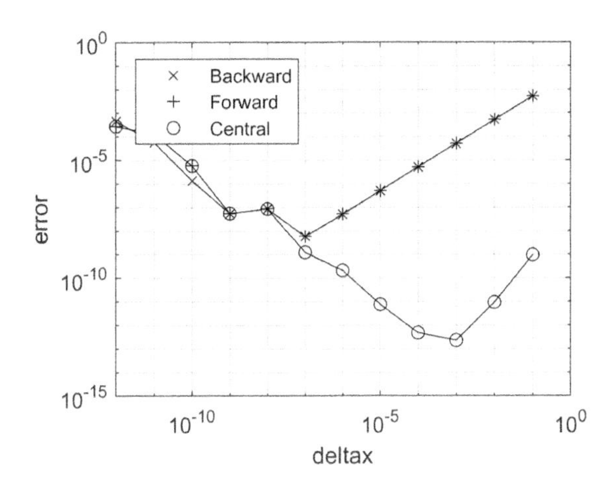

Figure 11.2 Finite-difference errors in computing the gradient for Equation (11.13) at $x = -10$.

In the previous two examples, we assumed that the function could be evaluated to high accuracy. However, numerical noise is unavoidable. In the next example, we consider the impact of noise on finite-difference calculations.

Example 11.3 Consider the following function that exhibits random noise of the order of 10^{-5}:

$$f(x) = e^x - x^2 + 10^{-5}\,\mathrm{rand} \tag{11.15}$$

Find the gradient via the three forward-difference methods at $x = -10$, and compare the accuracy against the analytical derivative (without the noise).

Solution: As before, at $x = -10$ the analytical derivative (without the noise) is given by

$$\frac{df}{dx}\Big|_{-10} = 20.000045399929764 \tag{11.16}$$

The finite-difference errors, after accounting for the noise, are illustrated in Figure 11.3. Observe that there is a significant drop in accuracy. Further, observe that for Δx smaller than the noise level of (10^{-5}), numerical gradients are unreliable.

(cont.)

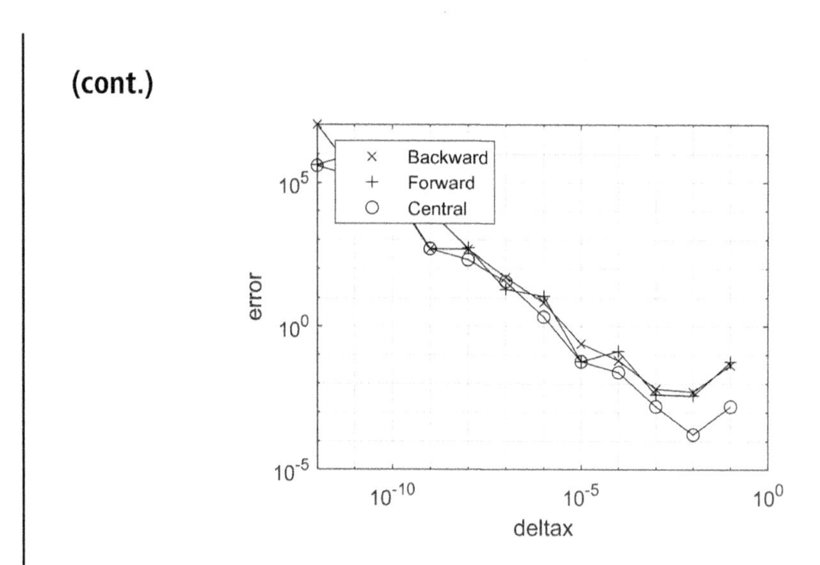

Figure 11.3 Finite-difference errors for the function in Equation (11.15) at $x = -10$.

11.2 Finite Difference in *N* Dimensions

We now consider the generalization of finite-difference methods to higher dimensions. Specifically, consider the N-dimensional function

$$f(x_1, x_2, \ldots, x_N) \tag{11.17}$$

The finite-difference gradients can be computed by generalizing the equations from Section 11.1. For example, the forward-difference gradient with respect to the first variable is given by

$$\frac{\partial f}{\partial x_1} \overset{\text{forward}}{\approx} \frac{f(x_1 + \Delta x_1, x_2, \ldots, x_N) - f(x_1, x_2, \ldots, x_N)}{\Delta x_1} \tag{11.18}$$

Similarly, the backward-difference gradient with respect to the second variable is given by

$$\frac{\partial f}{\partial x_2} \overset{\text{backward}}{\approx} \frac{f(x_1, x_2, \ldots, x_N) - f(x_1, x_2 - \Delta x_2, \ldots, x_N)}{\Delta x_2} \tag{11.19}$$

Observe that the step sizes Δx_1 and Δx_2 need not be equal. Thus, one can define a vector step size

$$\Delta \mathbf{x} = \{\Delta x_1, \Delta x_2, \ldots, \Delta x_N\}^T \tag{11.20}$$

Then the *forward-difference gradient* is formally given by the approximation

$$\frac{\partial f}{\partial x_i} \overset{\text{forward}}{\approx} \frac{f(\mathbf{x} + \mathbf{e}_i \Delta x_i) - f(\mathbf{x})}{\Delta x_i} \tag{11.21}$$

where $\mathbf{e}_i$ is the unit vector in the ith direction. The *backward-difference gradient* is defined as

$$\frac{\partial f}{\partial x_i} \overset{\text{backward}}{\approx} \frac{f(\mathbf{x}) - f(\mathbf{x} - \mathbf{e}_i \Delta x_i)}{\Delta x_i} \tag{11.22}$$

The *central-difference gradient* is defined as

$$\frac{\partial f}{\partial x_i} \overset{\text{central}}{\approx} \frac{f(\mathbf{x} + \mathbf{e}_i \Delta x_i/2) - f(\mathbf{x} - \mathbf{e}_i \Delta x_i/2)}{\Delta x_i} \tag{11.23}$$

Example 11.4 Consider the following function that models the potential energy of a two-spring system:

$$f(u,v) = 0.5 \begin{bmatrix} 100\left(\sqrt{u^2 + (1+v)^2} - 1\right)^2 \\ +50\left(\sqrt{u^2 + (1-v)^2} - 1\right)^2 \end{bmatrix} - (10u + 8v) \tag{11.24}$$

Compute the numerical gradient via all three finite-difference methods at two points, $(1, 10)$ and $(0, 2)$, and compute the numerical errors.

Solution: We shall study the first component of the gradient, and define the error as

$$e_u = \frac{\left| \left.\frac{\partial f}{\partial u}\right|_{\text{exact}} - \left.\frac{\partial f}{\partial u}\right|_{\text{finite difference}} \right|}{\left| \left.\frac{\partial f}{\partial u}\right|_{\text{exact}} \right|} \tag{11.25}$$

The analytical gradient can be computed by exploiting the gradient of the potential energy, derived in Section 5.14. The finite-difference gradient can be computed using the function `finiteDifferenceGradient.m`, discussed in Section 11.1. For example, the central finite difference for the spring potential energy function at (1, 10), with a step size of 1×10^{-5}, can be computed as follows:

```
>> g =  finiteDifferenceGradient(@(x)springFunction(x),[1
10],1e-5,0)

g =

    1.0e+03 *

    0.125424849193223

    1.392716492682666
```

Given the analytical gradient and the numerical gradient, the error can be computed. Figure 11.4 plots the errors for the three methods at (1, 10) for various step sizes. Similarly, Figure 11.5 plots the errors at (0, 2).

(cont.)

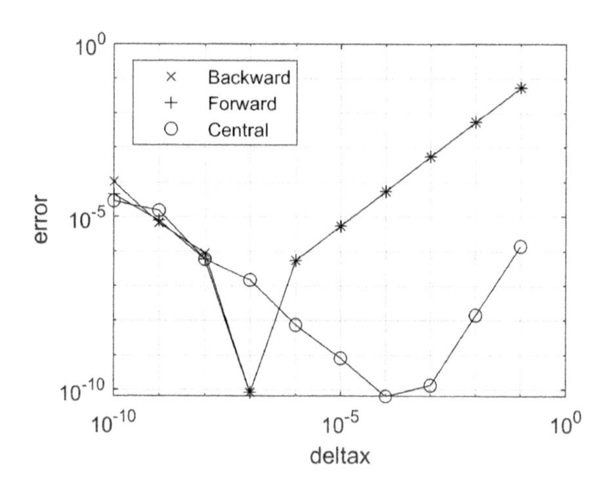

Figure 11.4 Finite-difference errors in gradient of Equation (11.24) at (1, 10) .

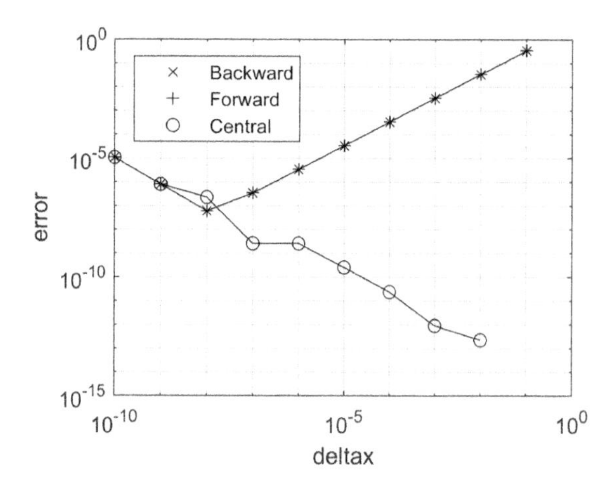

Figure 11.5 Finite-difference errors in gradient of Equation (11.24) at (0, 2).

Observe the unusual behavior of the central-difference method where the error monotonically increases; such behavior is rare but cannot be ruled out. The reader is encouraged to experiment with the finite-difference methods.

The main observations regarding the finite-difference methods are the following:

1. The central-difference method typically fares best, especially for large step size.
2. However, all three methods suffer from both *discretization errors* for large step size, and *floating-point (round-off) errors* for small step size.

3. The optimal choice for step size depends on the function being evaluated and on the point of evaluation. In other words, it is impossible to pick an "optimal" step size a priori; see reference [28] for a thorough discussion. This is a serious drawback of finite-difference methods.
4. In the presence of numerical noise, the step size must be chosen with even greater care.

Despite their drawbacks, finite-difference methods are used extensively in engineering on account of their simplicity. MATLAB's `fmincon` uses forward difference as the default for gradient computation.

Note that finite-difference methods can be easily generalized to extract higher-order derivatives. For example, second-order derivatives, also referred to as the Hessian, are needed in second-order optimization methods such as Newton–Raphson. To derive the finite-difference approximation, consider the 1D Taylor series expansion in Equation (11.1):

$$f(x_0 + \Delta x) \approx f(x_0) + \left.\frac{df}{dx}\right|_{x_0} \Delta x + \frac{1}{2!}\left.\frac{d^2f}{dx^2}\right|_{x_0} (\Delta x)^2 \tag{11.26}$$

By replacing Δx with $-\Delta x$ in Equation (11.26), we have

$$f(x_0 - \Delta x) \approx f(x_0) - \left.\frac{df}{dx}\right|_{x_0} \Delta x + \frac{1}{2!}\left.\frac{d^2f}{dx^2}\right|_{x_0} (\Delta x)^2 \tag{11.27}$$

Adding the two equations,

$$f(x_0 + \Delta x) + f(x_0 - \Delta x) \approx 2f(x_0) + \left.\frac{d^2f}{dx^2}\right|_{x_0} (\Delta x)^2 \tag{11.28}$$

i.e.,

$$\left.\frac{d^2f}{dx^2}\right|_{x_0} \approx \frac{f(x_0 + \Delta x) - 2f(x_0) + f(x_0 - \Delta x)}{(\Delta x)^2} \tag{11.29}$$

Equation (11.29) is the central-difference second-order derivative for 1D functions. Similar expressions can be derived for the forward- and backward-difference methods. For higher-dimensional functions, cross-derivatives must also be computed, given by

$$\frac{\partial^2 f}{\partial x_i \partial x_j} \overset{\text{central}}{\approx} \frac{\begin{pmatrix} f(\mathbf{x} + \mathbf{e}_i\Delta x_i + \mathbf{e}_j\Delta x_j) - f(\mathbf{x} + \mathbf{e}_i\Delta x_i - \mathbf{e}_j\Delta x_j) - \\ f(\mathbf{x} - \mathbf{e}_i\Delta x_i + \mathbf{e}_j\Delta x_j) + f(\mathbf{x} - \mathbf{e}_i\Delta x_i - \mathbf{e}_j\Delta x_j) \end{pmatrix}}{4\Delta x_i \Delta x_j} \tag{11.30}$$

11.3 Analytical Approach

We will now describe a semi-analytical approach for computing the derivative (also referred to as *sensitivity*) of compliance with respect to the cross-sectional areas.

Observe that, at each step of the truss optimization process described in Section 10.2.3, we solved the linear system of equations

$$\mathbf{Ku} = \mathbf{f} \tag{11.31}$$

We then computed the compliance via

$$J = \mathbf{f}^{\mathbf{T}}\mathbf{u} \tag{11.32}$$

Taking the derivative of both equations with respect to any cross-sectional area, we have

$$\mathbf{K}'\mathbf{u} + \mathbf{K}\mathbf{u}' = \mathbf{f}' \tag{11.33}$$

$$J' = (\mathbf{f}^{\mathbf{T}})'\mathbf{u} + \mathbf{f}^{\mathbf{T}}\mathbf{u}' \tag{11.34}$$

where

$$(.)' \equiv \frac{\partial}{\partial A_m} \tag{11.35}$$

In Equation (11.33), $\mathbf{u}'$ is the displacement sensitivity, while $\mathbf{K}'$ is the stiffness matrix sensitivity. Since the external force does not depend on the cross-sectional area, Equations (11.33) and (11.34) reduce to

$$\mathbf{K}'\mathbf{u} + \mathbf{K}\mathbf{u}' = 0 \tag{11.36}$$

i.e.,

$$\mathbf{u}' = -\mathbf{K}^{-1}(\mathbf{K}'\mathbf{u}) \tag{11.37}$$

and

$$J' = \mathbf{f}^{\mathbf{T}}\mathbf{u}' \tag{11.38}$$

The two can be combined as

$$J' = \mathbf{f}^{\mathbf{T}}[-\mathbf{K}^{-1}(\mathbf{K}'\mathbf{u})] \tag{11.39}$$

Taking the transpose on both sides, and noting that the transpose of a scalar is itself, we have

$$J' = -[\mathbf{u}^T(\mathbf{K}')^T\mathbf{K}^{-T}]\mathbf{f} \tag{11.40}$$

Since the stiffness matrix is symmetric,

$$J' = -\mathbf{u}^T\mathbf{K}'\mathbf{K}^{-1}\mathbf{f} \tag{11.41}$$

Further, from Equation (11.31)

$$J' = -\mathbf{u}^T\mathbf{K}'\mathbf{u} \tag{11.42}$$

Equation (11.42) states that, in order to compute the compliance sensitivity, we only need to compute the stiffness sensitivity.

Recall from the analysis of truss systems in Section 9.4.2 that the stiffness matrix is constructed by assembling individual stiffness matrices of the form

$$\mathbf{k}_m = \frac{E_m A_m}{l_m}\begin{bmatrix} \cos^2\alpha_m & \sin\alpha_m \cos\alpha_m \\ \sin\alpha_m \cos\alpha_m & \sin^2\alpha_m \end{bmatrix} \tag{11.43}$$

Therefore,

$$\frac{\partial \mathbf{k}_m}{\partial A_m} = \frac{E_m}{l_m}\begin{bmatrix} \cos^2\alpha_m & \sin\alpha_m \cos\alpha_m \\ \sin\alpha_m \cos\alpha_m & \sin^2\alpha_m \end{bmatrix} \tag{11.44}$$

and

$$\frac{\partial \mathbf{k}_m}{\partial A_l} = \begin{bmatrix} 0 & 0 \\ 0 & 0 \end{bmatrix}; \quad l \neq m \tag{11.45}$$

Thus, to compute the global stiffness matrix sensitivity with respect to any truss member, we only need to assemble the entries for the nodes connected to this member, using Equation (11.44). Combining Equations (11.42) and (11.45), we have

$$\frac{\partial J}{\partial A_m} = -\mathbf{u}_m^T \frac{\partial \mathbf{K}_m}{\partial A_m}\mathbf{u}_m \tag{11.46}$$

where

$$\frac{\partial \mathbf{K}_m}{\partial A_m} = \frac{E_m}{l_m}\begin{bmatrix} c^2 & cs & -c^2 & -cs \\ cs & s^2 & -cs & -s^2 \\ -c^2 & -cs & c^2 & cs \\ -cs & -s^2 & cs & s^2 \end{bmatrix} \tag{11.47}$$

$$c = \cos\alpha_m$$
$$s = \sin\alpha_m$$

and

$$\mathbf{u}_m = \{u_1 \quad v_1 \quad u_2 \quad v_2\} \tag{11.48}$$

corresponds to the displacements at the two nodes associated with that truss bar; see Exercise 11.8.

11.4 Complex-Variable Approach

An alternative and fascinating approach to computing gradients is via the use of complex variables; the basic concept is as follows ([29], [30]).

Consider again the Taylor series

$$f(x_0 + \Delta x) = f(x_0) + \left.\frac{df}{dx}\right|_{x_0} \Delta x + \frac{1}{2!}\left.\frac{d^2 f}{dx^2}\right|_{x_0} (\Delta x)^2 + o(\Delta x)^3 \tag{11.49}$$

Now consider an imaginary step size $i\Delta x$, where i is the complex number $\sqrt{-1}$:

$$f(x_0 + i\Delta x) = f(x_0) + i\left.\frac{df}{dx}\right|_{x_0} \Delta x + \frac{1}{2!}\left.\frac{d^2 f}{dx^2}\right|_{x_0} (i\Delta x)^2 + o(i\Delta x)^3 \tag{11.50}$$

Since $i^2 = -1$, we have

$$f(x_0 + i\Delta x) \approx f(x_0) + i\left.\frac{df}{dx}\right|_{x_0} \Delta x - \frac{1}{2!}\left.\frac{d^2 f}{dx^2}\right|_{x_0} (\Delta x)^2 \tag{11.51}$$

One can now extract the imaginary and real components:

$$\mathrm{Im} f(x_0 + i\Delta x) \approx \left.\frac{df}{dx}\right|_{x_0} \Delta x \tag{11.52}$$

$$\mathrm{Re} f(x_0 + i\Delta x) \approx f(x_0) - \frac{1}{2!}\left.\frac{d^2 f}{dx^2}\right|_{x_0} (\Delta x)^2 \tag{11.53}$$

This leads to the two useful relationships:

$$\left.\frac{df}{dx}\right|_{x_0} \approx \frac{\mathrm{Im} f(x_0 + i\Delta x)}{\Delta x} \tag{11.54}$$

and

$$\left.\frac{d^2 f}{dx^2}\right|_{x_0} \approx \frac{f(x_0) - \mathrm{Re} f(x_0 + i\Delta x)}{0.5\Delta x^2} \tag{11.55}$$

Observe that computing the first derivative in Equation (11.54) does not involve subtracting two close-by floating-point numbers. This essentially eliminates the floating-point errors that we observed for finite-difference methods, and leads to very high accuracy for small step size. Similarly, the second derivative in Equation (11.55) also results in higher accuracy than finite-difference methods.

The downside is that one must be able to compute with complex numbers. Fortunately, computing with complex numbers in MATLAB is straightforward. However, not all programming languages support complex numbers.

Example 11.5 Compute the gradient of the function $f(x) = \sin(2\pi x)$ via complex variables.

Solution: From Equation (11.54),

$$\frac{df}{dx} \approx \frac{\text{Im}[\sin 2\pi(x + i\Delta x)]}{\Delta x} = \frac{\text{Im}[\sin(2\pi x + 2\pi i\Delta x)]}{\Delta x} \tag{11.56}$$

Exploiting standard trigonometric expressions,

$$\frac{df}{dx} \approx \frac{\text{Im}[\sin 2\pi x \cos 2\pi i\Delta x + \cos 2\pi x \sin 2\pi i\Delta x]}{\Delta x} \tag{11.57}$$

For sufficiently small Δx, $\cos 2\pi i\Delta x \approx 1$ and $\sin 2\pi i\Delta x \approx 2\pi i\Delta x$; thus

$$\frac{df}{dx} \approx \frac{\text{Im}[\sin 2\pi x + 2\pi i\Delta x \cos 2\pi x]}{\Delta x} \tag{11.58}$$

Extracting the imaginary component, we have the desired result:

$$\frac{df}{dx} \approx \frac{2\pi\Delta x \cos 2\pi x}{\Delta x} = 2\pi \cos 2\pi x \tag{11.59}$$

This matches the analytical expression.

Now consider a numerical example.

Example 11.6 Use the complex-variable method to find the gradient of the following function at $x = 1$:

$$f(x) = x^2 e^{x-1} \tag{11.60}$$

For various step sizes Δx, compare the accuracy against the analytical derivative.

Solution: Note that the analytical derivative is given by

$$\left.\frac{df}{dx}\right|_{1.0} = \left. 2xe^{x-1} + x^2 e^{x-1}\right|_{x=1} = 3 \tag{11.61}$$

Now, using Equation (11.54),

$$\left.\frac{df}{dx}\right|_{1.0} \approx \frac{\text{Im} f(1.0 + i\Delta x)}{\Delta x} \tag{11.62}$$

The finite-difference and complex-variable errors are illustrated in Figure 11.6. Observe that the complex-variable method remains highly accurate even for very small step size.

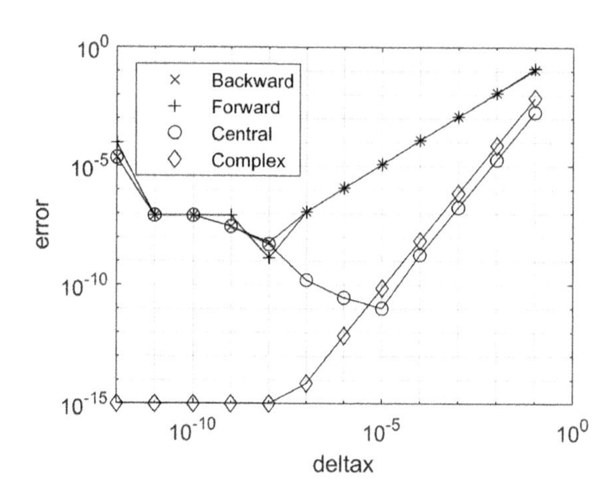

Figure 11.6 Complex-variable versus finite-difference errors for the derivative of the function in Equation (11.60) at $x = 1$.

The complex-variable method can be generalized to higher dimensions. Let $f(\mathbf{x})$ be a real function of N variables. Then the partial derivative with respect to, say, x_1 may be obtained via the following approximation:

$$\frac{\partial f}{\partial x_1} \approx \frac{\operatorname{Im} f(x_1 + i\Delta x_1, x_2, \ldots, x_N)}{\Delta x_1} \tag{11.63}$$

The complex derivative can also be extended to second-order derivatives ([31]).

We provide below a MATLAB implementation for computing the gradient of any function via the complex-variable approach.

```
function [g] = complexVariableGradient(f,x0,h)
% Compute, via complex variable method,
% the gradient of f at x0
N = numel(x0);
g = zeros(N,1); % initialize memory
if (nargin == 2)
    h = 1e-12*ones(N,1);
end
for i = 1:N
    z = x0;
    z(i) = z(i) + sqrt(-1)*h(i);
    fVal = feval(f,z);
    g(i) = imag(fVal)/h(i);
end
```

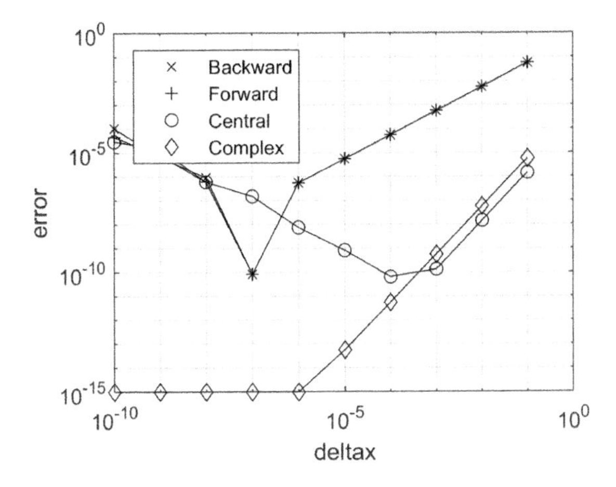

Figure 11.7 Complex-variable versus finite-difference errors for the derivative of Equation (11.64) at (1, 10).

Example 11.7 Consider the following function that models the potential energy of a two-spring system:

$$f(u,v) = 0.5\begin{bmatrix} 100\left(\sqrt{u^2+(1+v)^2}-1\right)^2 \\ +50\left(\sqrt{u^2+(1-v)^2}-1\right)^2 \end{bmatrix} - (10u+8v) \tag{11.64}$$

Use the `complexVariableGradient.m` function to compute the numerical gradient at (1, 10).

Solution: Recall that Equation (11.64) is implemented within the class `twoSpring.m`. We compute the gradient via

```
>> complexVariableGradient(
@(x)twoSpring.potentialEnergy(x),[1;10],1e-10)

ans =

   1.0e+03 *

   0.125424849092006
   1.392716492619549
```

To demonstrate the superiority of the complex-variable method over finite-difference methods, we compute the error in the gradient for the function in Equation (11.64) at (1, 10). This is illustrated, as a function of step size Δx, in Figure 11.7.

As one can observe, the error from the complex-variable method reaches machine precision.

11.5 Gradient of Compliance

We will now see how the complex-variable method can be used for truss optimization. Consider the compliance minimization problem (with appropriate scaling)

$$\begin{aligned}
&\underset{\{x_1,x_2,\ldots\}}{\text{minimize}}\ (\mathbf{f}^T\mathbf{u})/J^0 \\
&\text{s.t} \quad \frac{\sum x_i A_i^0 l_i}{V^0} - 1 \le 0 \\
&\qquad x_i > 0 \\
&\qquad \mathbf{K}\mathbf{u} = \mathbf{f}
\end{aligned} \tag{11.65}$$

Previously, we implemented this problem in MATLAB using `fmincon`, relying entirely on forward difference (default) for gradient computation. Given the drawbacks of the finite-difference method, we will now revisit the truss optimization problems discussed in the previous chapter and attempt to find the gradients using the complex-variable method.

For convenience, let the objective be denoted by

$$\varphi(x_1, x_2, \ldots) = \frac{\mathbf{f}^T\mathbf{u}}{J^0} \tag{11.66}$$

From Equation (11.63), we have, for example,

$$\frac{\partial\varphi}{\partial x_1} = \frac{\text{Im}\ \varphi(x_1 + ih, x_2, \ldots)}{h} \tag{11.67}$$

where h is sufficiently small. Thus, we will evaluate the compliance for a complex value of cross-sectional area! Fortunately, mathematical operations in MATLAB, including matrix inversion, fully support complex variables. The following example illustrates gradient computation for a truss.

Example 11.8 Consider the problem illustrated in Figure 11.8, where a force of (1, −2) N is applied to node 1 and a force of (2, 0) N is applied to node 2. The Young's modulus is 2×10^{11} N/m^2, and the initial cross-sectional area of all bars is 10^{-6} m. Find the gradients in Equation (11.67) at the initial configuration.

Solution: As in Chapter 10, we construct the truss as follows:

```
xy = [0.5 1.5 1.0 0 2.0; 1 1 0 0 0];
```

```
connectivity = [1 2; 1 3; 1 4; 2 3; 2 5; 3 4; 3 5]';
t = truss2dMinComplianceVolumeConstraint(xy,connectivity);
t = t.assignE(2e11);
t = t.assignA(1e-6); % for all members
t = t.fixXofNodes([4 5]);
t = t.fixYofNodes([4 5]);
t = t.applyForce(1,[1;-2]);
t = t.applyForce(2,[2;0]);
```

(cont.)

Before computing the gradients, the truss volume, compliance, etc. are initialized.

```
t = t.initialize();
```

We can now compute the gradient by passing the compliance function as an input parameter (observe that the initial parameters are 1).

```
h = 1e-10; % step size
complexVariableGradient(@(x)t.complianceObjective(x),ones(1,7),h)
ans =
    0.0000
    0.2978
    0.0000
    0.2978
    0.2978
    0.0533
    0.0533
```

Note that the gradient of the volume constraint can be computed analytically. Let

$$g(x_1, x_2, \ldots) = \frac{\sum x_i A_i^0 l_i}{V^0} - 1 \tag{11.68}$$

We have

$$\frac{\partial g}{\partial x_i} = \frac{A_i^0 l_i}{V^0} \tag{11.69}$$

Given these two results, one can extend the compliance minimization class, overriding the default finite-difference method; see Exercise 11.9.

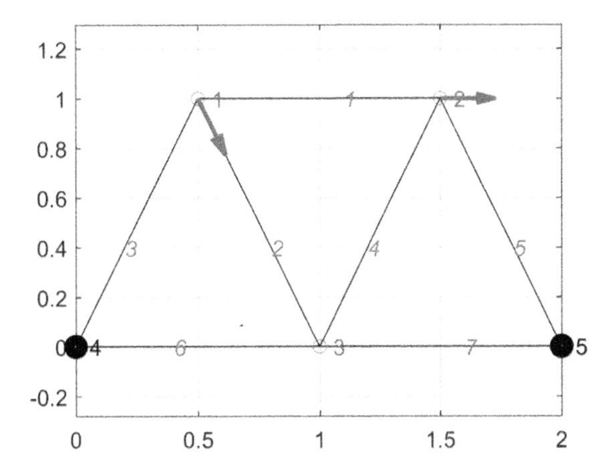

Figure 11.8 An indeterminate truss system.

11.6 Automatic Differentiation

Yet another method that has gained popularity is *automatic differentiation* (AD), which exploits the basic properties of differentiation ([32], [33]). Specifically, consider a function $f_1 = f_2 + f_3$, and suppose the gradients ∇f_2 and ∇f_3 are known. Then the gradient of f may be computed trivially as

$$\nabla f = \nabla f_1 + \nabla f_2 \tag{11.70}$$

On the other hand, suppose $f_1 = f_2 f_3$. Then the chain-rule applies, i.e.,

$$\nabla f_1 = f_2 \nabla f_3 + f_3 \nabla f_2 \tag{11.71}$$

If $f_1 = f_2/f_3$, then

$$\nabla f_1 = \frac{f_3 \nabla f_2 - f_2 \nabla f_3}{(f_3)^2} \tag{11.72}$$

Finally, if $f_1 = f_2(f_3)$, then

$$\nabla f_1 = \nabla f_2 \nabla f_3 \tag{11.73}$$

Thus, the gradient can be computed in a recursive fashion! In other words, we apply Equations (11.70) through (11.73) recursively until we reach a primitive mathematical function such as x^n, $\sin(x)$ or $exp(x)$, whose gradients are known a priori.

The most important aspect of automatic differentiation is that, at a conceptual level, there is *no step-size numerical error*; errors may arise when floating-point numbers are added and subtracted (as is always the case even with function evaluations).

There are two fundamental strategies for implementing automatic differentiation:

1. **Code differentiation** Given a computer implementation of a function $f(\mathbf{x})$ (example: a MATLAB implementation), parse the code, apply the above rules and generate a new piece of code that computes the gradient $\nabla f(\mathbf{x})$. Here, there are two "flavors": forward and backward propagation.
2. **Operator overloading** An alternative approach is to "overload" the definitions of various arithmetic and mathematical functions to directly implement Equations (11.70), (11.71) and (11.72).

Both approaches require that the source code that implements $f(\mathbf{x})$ be readily available. The former approach does not modify this source code, but creates an additional piece of code that must be used for gradient computation. The latter, on the other hand, requires (usually minimal) changes to the source code that implements $f(x)$.

11.7 EXERCISES

Exercise 11.1 Derive an expression for the forward and backward finite-difference equations for the second derivative of a 1D function, analogous to Equation (11.29).

Exercise 11.2 Write a MATLAB function to compute the Hessian of any N-dimensional function via central-, backward- and forward-difference methods. The syntax should be as follows:

```
[H] = finiteDifferenceHessian(f,x0,deltaX,method)
```

Exercise 11.3 Use the `finiteDifferenceHessian` function to find the second derivative of (a) $f(x) = e^x - x^2$ at $x = -10$, and (b) $f(x) = x^2e^{x-1}$ at $x = 1$, using a step size of 0.0001, via all three methods. Compare against the analytical values.

Exercise 11.4 Use the `finiteDifferenceHessian` function to find the Hessian of the two-spring potential energy, using a step size of 0.0001, via all three methods. Compare against analytical results.

Exercise 11.5 Find the slope of the lines in Figure 11.4, prior to round-off errors. What is the order of convergence for the three methods?

Exercise 11.6 The central-difference method is generally more accurate than either the backward- or forward-difference method, but it is not applicable near the boundary limits of a variable, for example when $f(x_0 - h)$ is not defined. Find an expression for the second derivative of a function at $x = x_0$ using $f(x_0)$, $f(x_0 + h)$ and $f(x_0 + 2h)$.

Exercise 11.7 Use the complex-variable method to find the gradient of the following function at $x = 1$:

$$f(x) = x^2e^{x-1}$$

Plot the error with respect to the analytical value for various step sizes.

Exercise 11.8 Write a MATLAB class

```
truss2dMinComplianceVolumeConstraint_Complex << truss2d
```

that exploits the complex gradient of compliance and analytical gradient of the volume constraint (avoiding finite differences). Test your code by optimizing the design in Example 11.8.

Exercise 11.9 Write a MATLAB class

```
truss2dMinVolume Compliance Constraint_Complex<<truss2d
```

that exploits the analytical gradient of the volume objective and complex gradient of compliance constraint (avoiding finite differences). Test your code by optimizing the design in Example 11.8.

12 Finite Element Analysis in 2D

Highlights

1. In this chapter, we address the analysis of 2D elasticity problems using finite element analysis (FEA).
2. While it is beyond the scope of this text to address the theory behind FEA, we will demonstrate its use with a MATLAB implementation.
3. We will pose and solve several elasticity problems using this implementation, and important aspects of FEA will be highlighted.
4. This chapter serves as a foundation for shape optimization, to be discussed in Chapter 13.

12.1 Overview

So far, we have considered the analysis and optimization of trusses. In this chapter we consider the analysis of 2D elastic problems such as the ones in Figure 12.1.

Such 2D problems arise from different situations. For example, consider the 3D problem in Figure 12.2(a). Since the 3D geometry is thin (relative to its overall length and width), the problem may be simplified to the 2D plane-stress problem illustrated in Figure 12.2(b). The simplification from 3D to 2D reduces the computational cost significantly, with minimal loss in accuracy.

To analyze such 2D problems, we will rely on the popular *finite element analysis* (FEA). The fundamental steps in FEA are analogous to truss analysis: creating, assembling and solving a linear system of equations. However, FEA involves an additional step, referred to as *meshing*, to be illustrated in Figure 12.4.

It is beyond the scope of this text to address the theory behind FEA (see, for example, reference [12]). Instead, we will demonstrate the use of FEA through examples. This will serve as a foundation for shape optimization of such problems, to be discussed in Chapter 13.

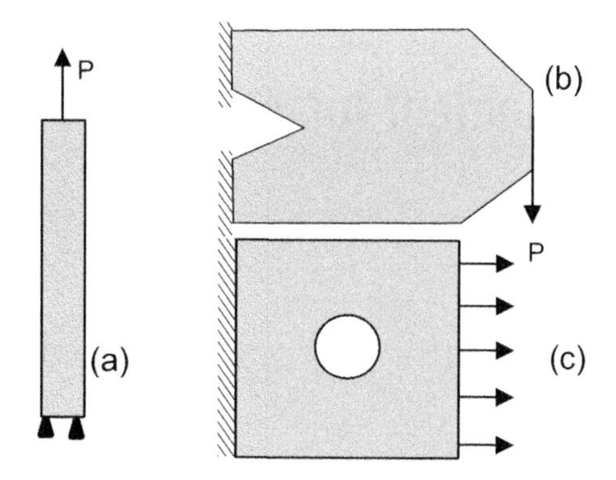

Figure 12.1 Examples of 2D elasticity problems.

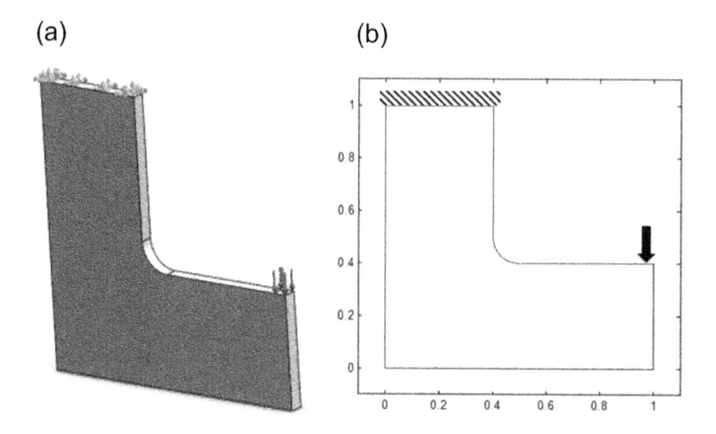

Figure 12.2 (a) A 3D problem over a thin geometry; (b) 2D plane-stress simplification.

12.2 Analysis of a Vertical Bar

The first example that we will consider is the vertical bar in Figure 12.1(a). The first step in FEA is the construction of the underlying geometry. Recall that, in truss analysis, the first step was to create vertices and then define how they are connected to form the truss. Similarly, to create the 2D geometry, we first define the vertices and then connect them to define the boundary.

We will assume that the width of the vertical bar is 1 m and its height is 10 m, as illustrated in Figure 12.3(a); an implicit assumption is that the depth (into the page) is small, say, 0.1 m. Thus its four vertices, on the x-y plane, are located at (0, 0), (0, 1), (1, 10) and (0, 10), as illustrated in Figure 12.3(b). To define the boundary, the four edges are numbered anti-clockwise, as illustrated in Figure 12.3(c), where edge 1 connects vertex 1 to vertex 2, and so on.

Given the above description, we define the vertices and edges via a structure as follows:

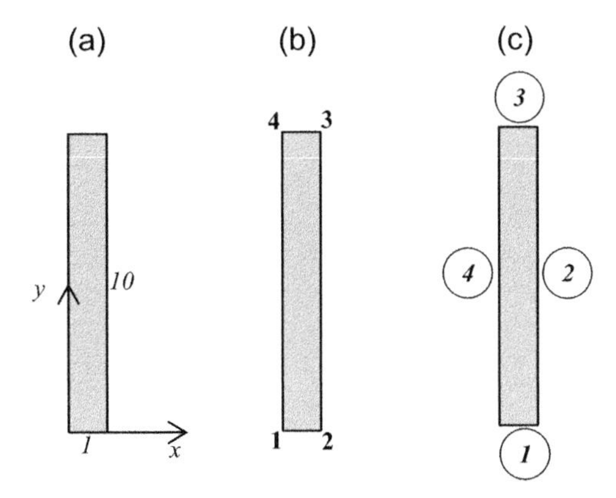

Figure 12.3 Geometric description of the vertical bar.

```
>> verticalBar.vertices = [0 0; 1 0; 1 10; 0 10]'; % in
meters
>> verticalBar.edges = [1 2; 2 3; 3 4; 4 1]';
```

Observe that "vertices" and "edges" are 2×4 matrices; note the apostrophe (') at the end of each of the two matrices.

We now create a finite element object (analogous to a truss object) using the class `TriElasticity.m` (the contents of this class are discussed in Section 12.2.1):

```
fem = TriElasticity(verticalBar);
```

We can perform various geometry operations; for example, we can compute its area via

```
fem.computeArea()
10
```

We can plot the geometry, with and without labels, via

```
fem.plotGeometry();
fem.plotGeometryWithLabels();
```

The reader is encouraged to try these plotting functions.

In creating the `fem` object, we have implicitly created a finite element *mesh*, which will be essential for solving the elasticity problem. To display the mesh, use the function call

```
fem.plotMesh();
```

Figure 12.4 illustrates the mesh, a collection of triangles, with a default number of approximately 500 triangles (in Section 12.2.3, we will study how the number of elements can be changed). These triangles are the basic building blocks for constructing and assembling a stiffness matrix.

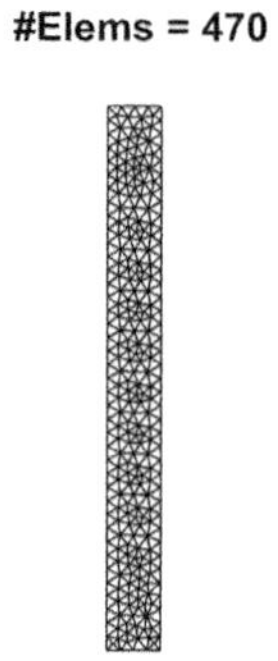

#Elems = 470

Figure 12.4 Finite element mesh of the vertical bar.

Next, we set the material properties. Since lengths are in meters in this example, we will use SI units for material properties. In 2D elasticity, two properties are essential: Young's modulus and the Poisson ratio – here, that of steel:

```
fem = fem.setYoungsModulus(2e11); % in N/m^2
fem = fem.setPoissonsRatio(0.28);
```

12.2.1 Tensile Load

Once the geometry and finite element mesh have been created, and material assigned, we can apply the restraints and loads. Observe from Figure 12.1 and Figure 12.3(c) that edge 1 is restrained from moving along the y-axis; the corresponding MATLAB code is

```
fem = fem.fixYOfEdge(1);
```

In 2D, force is interpreted as force per unit length of thickness. We will assume that a total vertical load of 10 N is applied. Since the depth of the bar is assumed to be 0.1 m, a force per unit length of 100 N/m is applied to edge 3. In MATLAB:

```
fem = fem.applyYForceOnEdge(3,100);
```

To visualize the applied restraints and load, we assemble the underlying stiffness matrix and force vector (analogous to the truss assembly):

```
fem = fem.assemble();
```

We can now plot the mesh:

```
fem.plotMesh();
```

Observe in Figure 12.5 that all nodes on edge 1 have been restrained along the y-axis, while the force has been distributed on all nodes on edge 3.

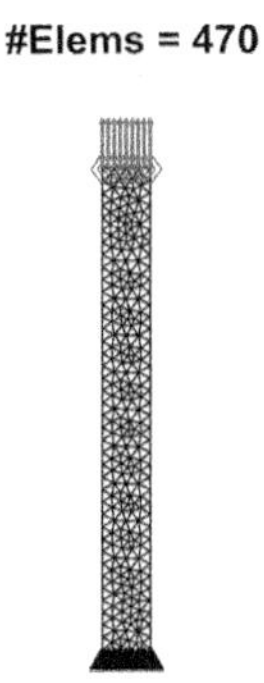

Figure 12.5 Restraint and load on the mesh.

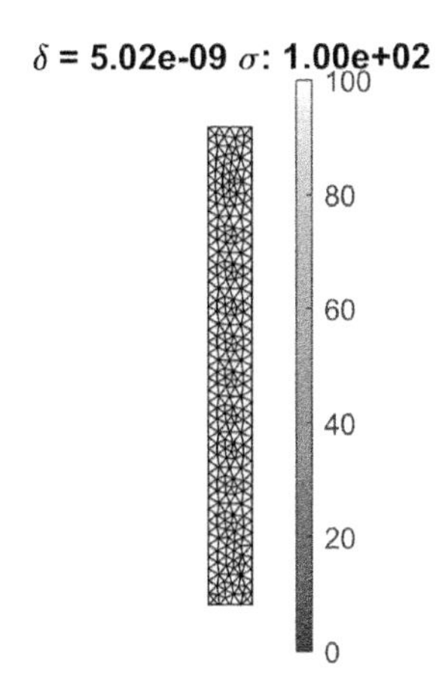

Figure 12.6 Stress plot for a tensile load.

We are now ready to solve the finite element problem and plot the resulting stress.

```
fem = fem.solve();
fem.plotStress();
```

The resulting *von Mises stress plot* (see reference [10]) is illustrated in Figure 12.6. The maximum deformation is 5×10^{-9} m and the maximum stress is 100 Pa.

As in truss analysis, it is important to verify the results in Figure 12.6 against analytical results. For a bar under tension, recall that the maximum displacement and stresses are given by (see reference [10])

$$\delta = \frac{PL}{EA} \tag{12.1}$$

$$\sigma = \frac{P}{A} \tag{12.2}$$

where

$$\begin{aligned} A &: \text{cross-sectional area} \\ L &: \text{length of bar} \\ E &: \text{Young's modulus} \\ P &: \text{applied force} \end{aligned} \tag{12.3}$$

We will simplify these expressions to 2D. If the depth of the bar into the page is d and its width is w (here, 1 m), then the equations reduce to

$$\delta = \frac{PL}{Edw} = \frac{(P/d)L}{Ew} \tag{12.4}$$

$$\sigma = \frac{P}{dw} = \frac{(P/d)}{w} \tag{12.5}$$

Observe that P/d is the force per unit length, 100 N/m. Therefore, the expected displacement is

$$\delta = \frac{(P/d)L}{Ed} = \frac{100 \times 10}{2 \times 10^{11} \times 1} = 5 \times 10^{-9} \text{ m} \tag{12.6}$$

The expected stress is

$$\sigma = \frac{P/d}{w} = \frac{100}{1} = 100 \text{ N/m}^2 \tag{12.7}$$

The FEA results in Figure 12.6 match the analytical results!

12.2.2 Bending Load

We will now apply a bending load instead of a tensile load, while fully restraining edge 1 along both x and y. The restraint is imposed as follows:

```
fem = fem.fixEdge(1);
```

The bending load corresponds to a positive force along the x-axis; the corresponding MATLAB code is as follows (observe that we must reset the force along the y-axis to zero):

```
fem = fem.applyYForceOnEdge(3,0);
fem = fem.applyXForceOnEdge(3,100);
```

We can now assemble and plot the mesh; see Figure 12.7.

```
fem = fem.assemble();
fem.plotMesh();
```

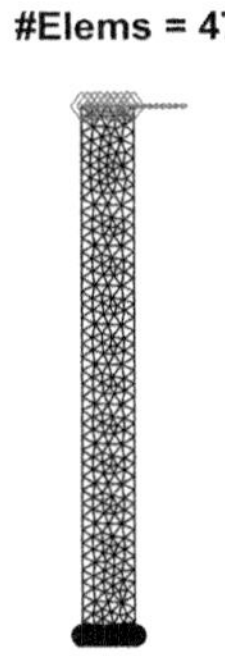

#Elems = 470

Figure 12.7 Restraint and bending load on the mesh.

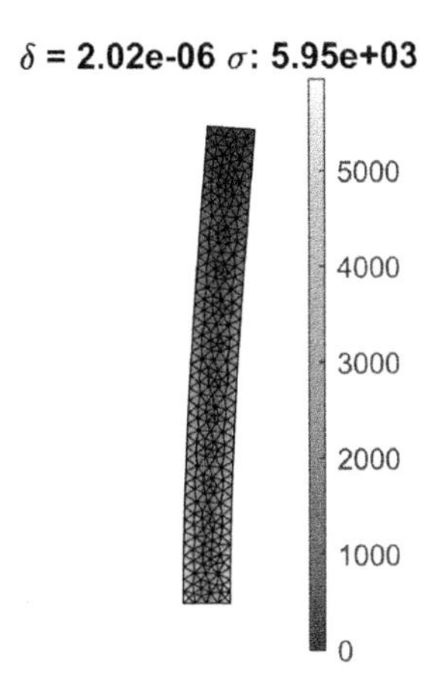

Figure 12.8 Stress plot for a bending load.

We are now ready to solve and plot the resulting stress.

```
fem = fem.solve();
fem.plotStress();
```

The resulting von Mises stress plot is illustrated in Figure 12.8. Note that the deformation has been exaggerated for visual clarity.

For a bending scenario, recall that the expected displacement and stress ([10]) are

$$\delta = \frac{PL^3}{3EI} \tag{12.8}$$

$$\sigma = \frac{PL(w/2)}{I} \tag{12.9}$$

where

$$\begin{aligned} &I: \text{moment of inertia} = (dw^3/12) \\ &w: \text{width of bar} \end{aligned} \tag{12.10}$$

As before, we simplify the equations to 2D:

$$\delta = \frac{(P/d)L^3}{3Ew^3/12} \tag{12.11}$$

$$\sigma = \frac{(P/d)L(w/2)}{w^3/12} \tag{12.12}$$

Therefore, the expected displacement is

$$\delta = \frac{(P/d)L^3}{3Ew^3/12} = \frac{100 \times 10^3}{3 \times 2 \times 10^{11} \times (1)^3/12} = 2 \times 10^{-6}\ \text{m} \tag{12.13}$$

The expected tensile stress is

$$\sigma = \frac{(P/d)L(w/2)}{w^3/12} = \frac{100 \times 10 \times (1/2)}{1^3/12} = 6000\ \text{N/m}^2 \tag{12.14}$$

The computed displacement and stress differ slightly from the analytical results. There are many reasons for this discrepancy:

(a) The Poisson effect is neglected while deriving Equation (12.9).
(b) FEA inevitably suffers from numerical errors.
(c) The stress in Figure 12.8 is the von Mises stress (which is often used for checking against failure), whereas Equation (12.14) represents the tensile stress.

12.2.3 Bending Load with Finer Mesh

An important concept in FEA is the density of the mesh. In the previous two scenarios we used a default of 500 elements. We will now repeat the bending load example with a larger number of elements. In general, the larger the number of elements, the more accurate the result, but at an increased computational cost.

The desired number of elements (here, 1000) is supplied as a second argument to `TriElasticity`:

```
fem = TriElasticity(verticalBar,1000);
fem = fem.setYoungsModulus(2e11);
fem = fem.setPoissonsRatio(0.28);
fem = fem.fixEdge(1);
fem = fem.applyXForceOnEdge(3,100);
fem = fem.assemble();
fem = fem.solve();
```

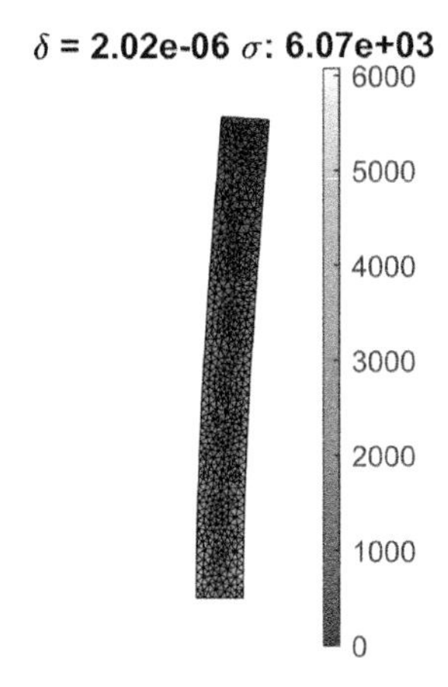

Figure 12.9 Stress plot for the bending load, with 1000 elements.

Figure 12.9 illustrates the deformation and stress plot for a finer mesh.

12.3 MATLAB Implementation

The `TriElasticity` class, employed in the previous section, is built on top of two other classes, `TriMesher` and `Brep2D`, as illustrated in Figure 12.10. There are numerous methods within each class. Instead of describing each of these methods, we will briefly summarize the critical methods and their purpose. The reader is encouraged to study the MATLAB implementation of the three classes.

12.3.1 `Brep2D`

As summarized in Figure 12.10, the `Brep2D` class helps create, manage and plot 2D geometry. The critical methods in this class include:

- `Brep2D`: This is the constructor method that takes as input the geometry structure (such as `verticalBar`).
- `convertBreptoPdeGeom`: The user-supplied geometry is converted into a format that MATLAB in-built functions such as `initmesh` (for creating finite element meshes) can understand.
- `plotGeometryWithLabels`: This method is particularly useful for debugging.
- `computeArea`: This method computes and returns the area of the 2D geometry.

12.3.2 `TriMesher`

The `TriMesher` class helps create, manage and plot triangle finite element meshes. The critical methods in this class include:

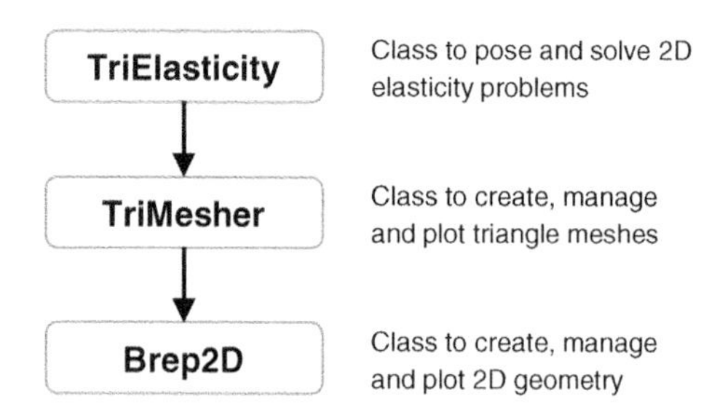

Figure 12.10 The three classes for creating and solving 2D elasticity problems.

- `TriMesher`: This is the constructor method that takes as input (1) the geometry structure (such as `verticalBar`) and (2) optionally, the number of finite elements. It exploits the in-built MATLAB function `initmesh` to create the triangle mesh.
- `plotMesh`: This method is particularly useful for debugging.

There are several utility methods within this class, for example, to evaluate finite element shape functions; these are critical in any FEA implementation. However, discussing the underlying theory is beyond the scope of this text.

12.3.3 `TriElasticity`

The `TriElasticity` class is used for posing and solving 2D elasticity problems. The critical methods in this class include:

- `TriElasticity`: This is the constructor method that takes as input (1) the geometry structure, (2) optionally, the number of finite elements and (3) the type of 2D problem (default is `PlaneStress`).
- `assemble`: Assembles the stiffness matrix and force vector.
- `solve`: Solves the linear system of equations.
- `computeStresses`: Post-processes the displacement results to compute the stresses.
- `plotDeformation`: Plots the computed displacement over the mesh.
- `plotStresses`: Plots the computed von Mises stress over the mesh.

12.4 Analysis of Cantilever Beam

For the next example, consider the cantilever beam in Figure 12.11, where horizontal symmetry is assumed.

To create the geometry, with the origin located at the mid-point of the left wall, the vertices and edges are defined as follows (in this example, units are in inches and pounds).

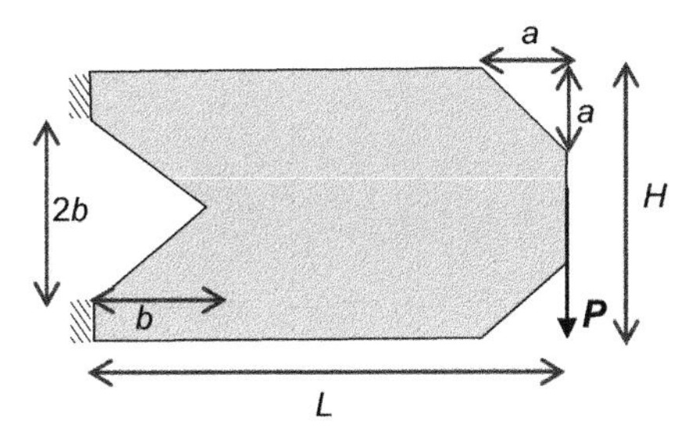

Figure 12.11 The cantilever beam.

```
L = 2.0;%length, unit is inches
H = 1;%height
h = 0.2; % for force application
a = 0.2; % corner cutouts
b = 0.2; % left edge cutout
cantileverBeam.vertices = [b 0;0 -b; 0 -H/2;L-a -H/2;L -H/2+a; L -
h/2; L h/2; L H/2-a; L-a H/2; 0 H/2; 0 b ]';
cantileverBeam.edges = [1 2; 2 3; 3 4; 4 5; 5 6; 6 7; 7 8; 8 9; 9
10; 10 11; 11 1]';
```

Observe the following:

- We have created parameters such as L, H, etc. for clarity (and for shape optimization, to be discussed in Chapter 13).
- We have split the vertical edge on the right into three segments in order to apply the force.
- The "vertices" and "edges" are 2×11 matrices.

We can now create the FEA object, with the default number of elements, set the material to be steel with Young's modulus of 3.05×10^7 psi and Poisson's ratio of 0.28.

```
fem = TriElasticity(cantileverBeam);
fem = fem.setYoungsModulus(3.05e7); % psi
fem = fem.setPoissonsRatio(0.28);
```

One can plot the geometry as follows (see Figure 12.12):

```
fem. plotGeometryWithLabels();grid on;
```

The reader can compute the area and verify it analytically. Next, we apply the restraints and loads. Observe in Figure 12.12 that edge 2 and edge 10 are restrained in all directions; the corresponding MATLAB code is

```
fem = fem.fixEdge([2 10]);
```

We apply a vertical force per unit length of 1000 lb/in on edge 6.

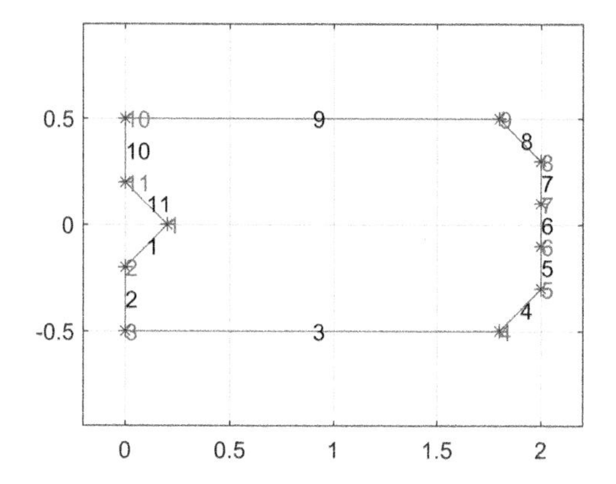

Figure 12.12 Labeling of nodes and edges for the cantilever beam.

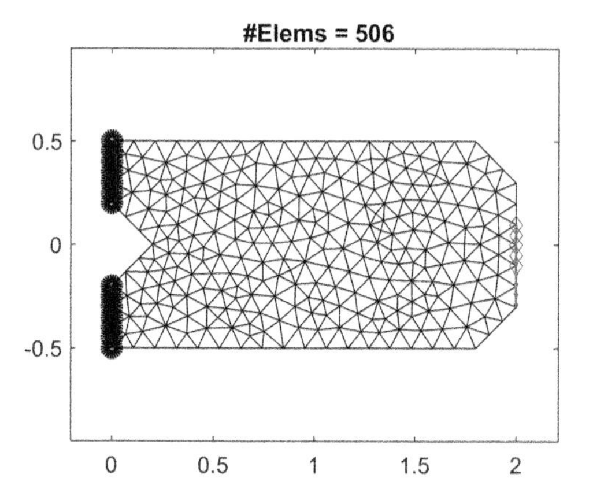

Figure 12.13 Restraint and bending load on the mesh.

```
fem = fem.applyYForceOnEdge(6,-1000);
```

Next we assemble and plot the mesh; see Figure 12.13.

```
fem = fem.assemble();
fem.plotMesh();
```

Finally, we can solve and plot the resulting stress; see Figure 12.14. The maximum displacement is 1.27×10^{-3} in, and the maximum von Mises stress is 14.1 ksi. Unfortunately, there are no analytical solutions to compare against.

```
fem = fem.solve();
fem.plotStress();
```

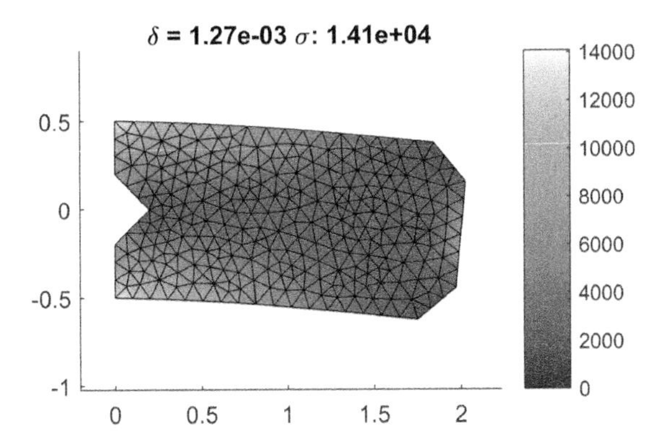

Figure 12.14 Stress results for the cantilever beam.

12.5 Analysis of L-bracket

For the next example, we will consider the problem in Figure 12.15 (length units are in meters).

In order to create the geometry, we will use the node and edge numbering illustrated in Figure 12.16. Note the following:

- The geometry contains an arc; therefore, an additional vertex (vertex 9) is created to facilitate arc construction.
- An additional vertex (vertex 4) and edge (edge 3) are created to facilitate force application.

The corresponding MATLAB code is as follows:

```
L = 0.1;%length in meters
H = 0.04;%height
r = 0.01; % fillet radius
d = 0.005; % for force application
LBracket.vertices = [0 0;L 0; L H; L-d H; H+r H; H H+r; H
L; 0 L; H+r H+r]';
LBracket.edges = [1 2; 2 3; 3 4; 4 5; 5 6; 6 7; 7 8; 8 1];
```

In this code we have not specified that edge 5 is an arc, with vertex 9 as the center. This is specified through an additional array, where the first entry is the edge number and the second entry is the center vertex:

```
LBracket.arcs = [5 9]';
```

Note that (1) `arcs` is a 2×1 matrix (using the transpose operator), and (2) the radius of the arc can be computed from the center and one of the vertices. The

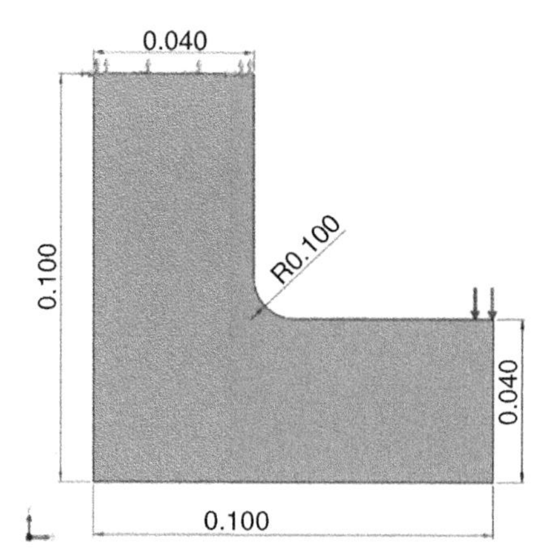

Figure 12.15 The L-bracket problem.

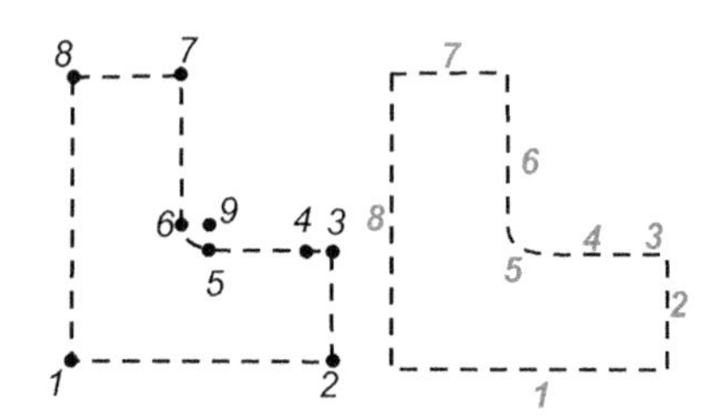

Figure 12.16 Node and edge numbering for the L-bracket.

direction of the arc is determined from the sign of the vertex. If the sign is positive, the arc is assumed to be clockwise; otherwise it is assumed to be anti-clockwise (see Exercise 12.6).

Next, we construct the FEA object, with approximately 1000 elements:

```
fem = TriElasticity(LBracket,1000);
```

We can now plot the geometry via

```
fem.plotGeometryWithLabels();
```

(the reader is encouraged to verify this). Next, we set the material and apply the restraints and loads. Observe in Figure 12.16 that edge 7 is restrained in all directions, while a vertical force per unit length of 100 000 N/m is applied to edge 3; the corresponding MATLAB code is

```
fem = fem.setYoungsModulus(2.1e11);
fem = fem.setPoissonsRatio(0.28);
fem = fem.fixEdge(7);
fem = fem.applyYForceOnEdge(3,-100000);
```

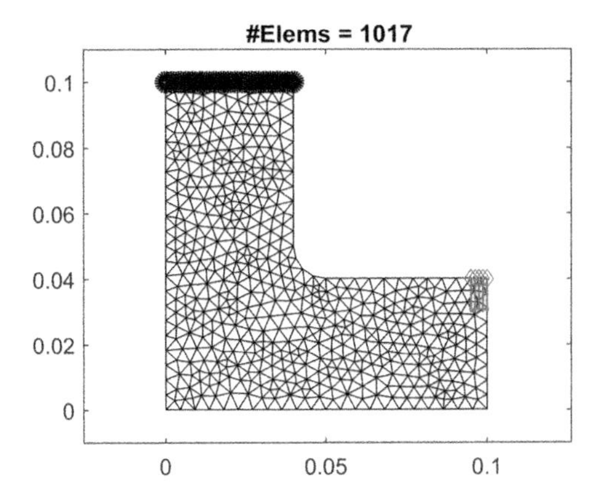

Figure 12.17 Restraint and bending load on the mesh.

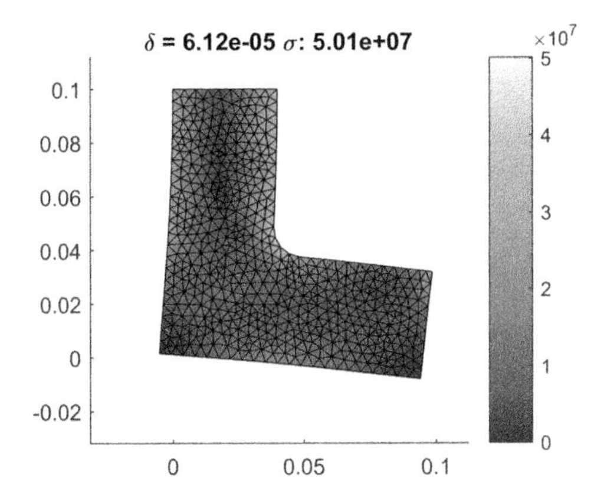

Figure 12.18 Stress results for the L-bracket.

Next, we assemble and plot the mesh:

```
fem = fem.assemble();
fem.plotMesh();
```

Finally, we can solve and plot the resulting stress (see Figure 12.18):

```
fem = fem.solve();
fem.plotStress();
```

12.6 Analysis of a Plate

For the final example, we consider the problem in Figure 12.19; the plate is 0.1 m on each side, and the radius of the hole is 0.01 m. The big difference in this example is the presence of the hole, which makes the definition of the geometry more challenging.

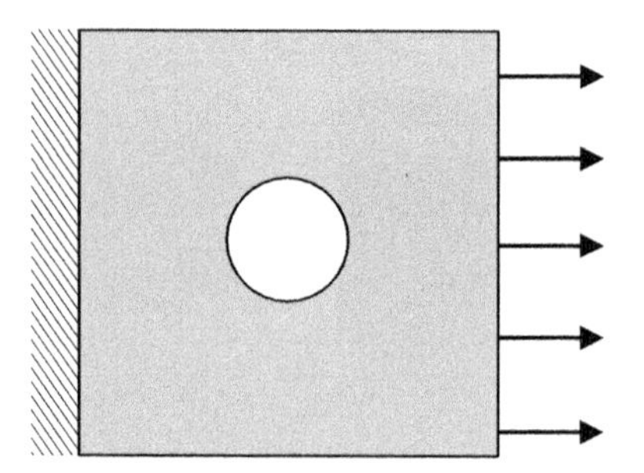

Figure 12.19 The plate with hole problem.

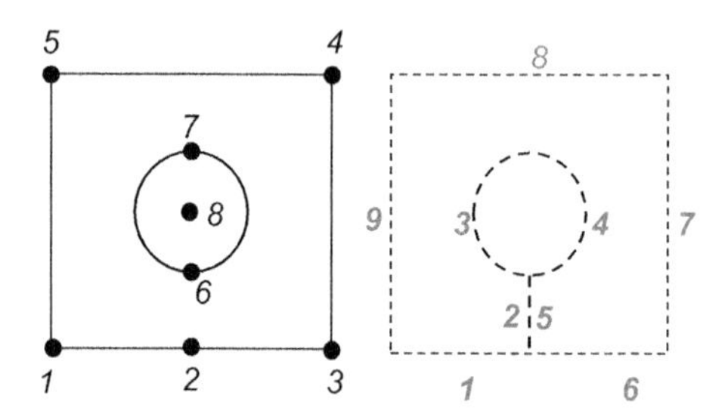

Figure 12.20 Node and edge numbering for the plate with hole.

To create the geometry, we will use the node and edge numbering illustrated in Figure 12.20. Note the following:

- The circle has been split into two arcs for convenience.
- To facilitate description of the boundary by a continuous loop, the bottom edge and the circle are connected through two virtual edges.

The corresponding MATLAB code is as follows:

```
L = 0.1;%length in meters
R = 0.01;%radius
PlateWithHole.vertices = [0 0;L/2 0; L 0; L L; 0 L; L/2
L/2-R; L/2 L/2+R;L/2 L/2]';
PlateWithHole.edges = [1 2; 2 6; 6 7; 7 6; 6 2; 2 3; 3 4; 4
5; 5 1]';
```

Observe that edges 3 and 4 are arcs with vertex 8 as center; therefore,

```
PlateWithHole.arcs = [3 8; 4 8]';
```

Finally, edges 2 and 5 are declared as virtual (not true boundary edges):

```
PlateWithHole.virtual = [2; 5]';
```

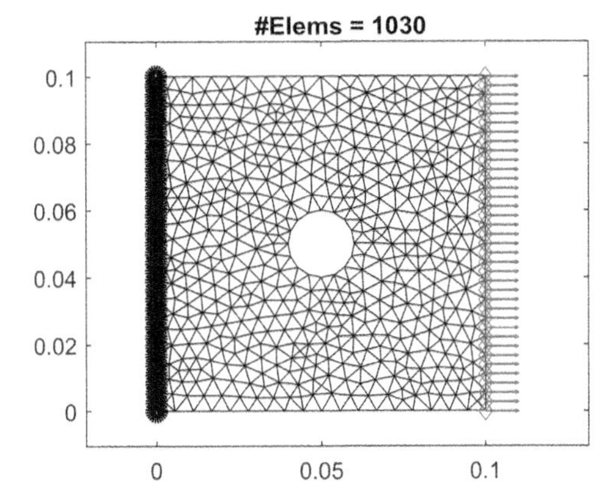

Figure 12.21 Restraint and bending load on the mesh.

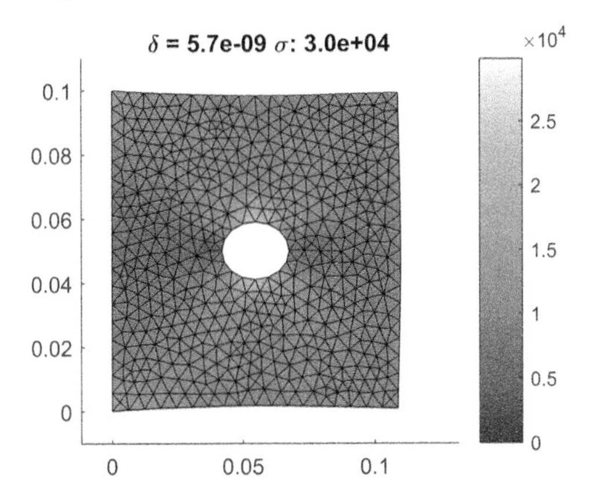

Figure 12.22 Stress results for the plate problem.

Next, we construct the FEA object, with approximately 1000 elements:

```
fem = TriElasticity(PlateWithHole,1000);
```

We can now plot the geometry:

```
fem.plotGeometryWithLabels();
```

(the reader is encouraged to verify this). Next, we set the material, and apply the restraints and loads. Observe in Figure 12.19 that edge 9 is restrained in all directions, while a horizontal force is applied to edge 7; the corresponding MATLAB code is

```
fem = fem.setYoungsModulus(2e11);
fem = fem.setPoissonsRatio(0.28);
fem = fem.fixEdge(9);
fem = fem.applyXForceOnEdge(7,1000);
```

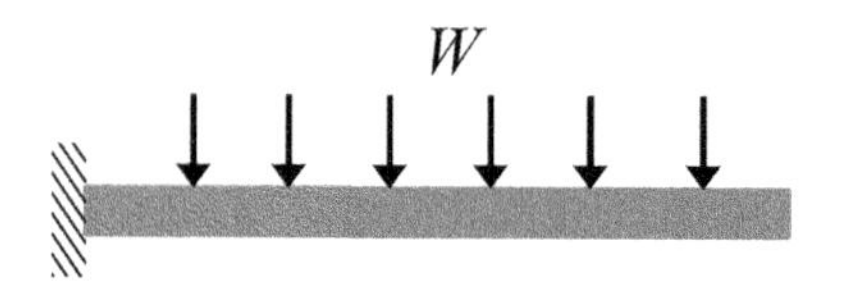

Figure 12.23 A cantilevered uniformly loaded beam.

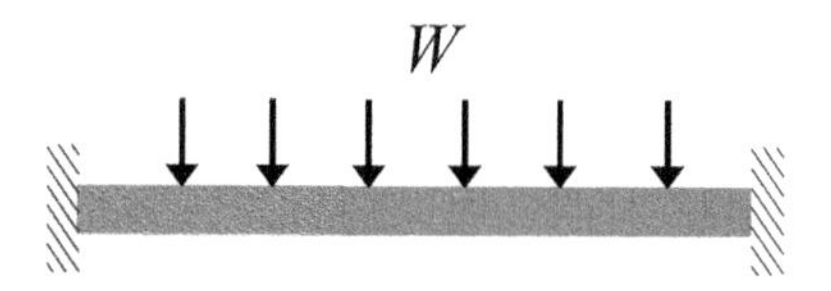

Figure 12.24 A double-cantilevered, uniformly loaded beam.

Next, we assemble and plot the mesh:

```
fem = fem.assemble();
fem.plotMesh();
```

(see Figure 12.21).

Finally, we can solve and plot the resulting stress:

```
fem = fem.solve();
fem.plotStress();
```

(see Figure 12.22).

12.7 EXERCISES

Exercise 12.1 Consider the uniformly loaded beam in Figure 12.23. Recall that the deflection at the tip is given by

$$\delta = \frac{WL^4}{8EI} \tag{12.15}$$

(see reference [10]), where W is the load per unit length. Model the 2D beam of length 1 m and height 0.05 m, with $E = 2 \times 10^{11}$ N/m^2, $\upsilon = 0.28$ and $W = 100$ N/m. Compare the analytical solution against the FEA solution with 1000 elements.

Exercise 12.2 Consider the double-cantilevered, uniformly loaded beam in Figure 12.24. The maximum deflection is given by

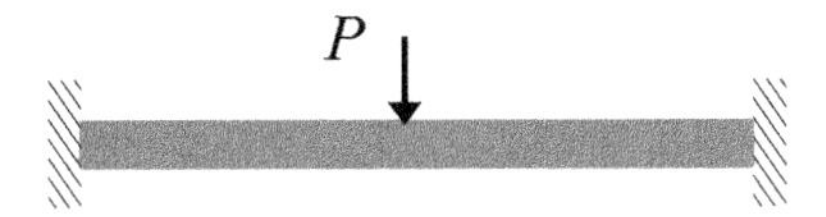

Figure 12.25 A double-cantilevered beam with point load.

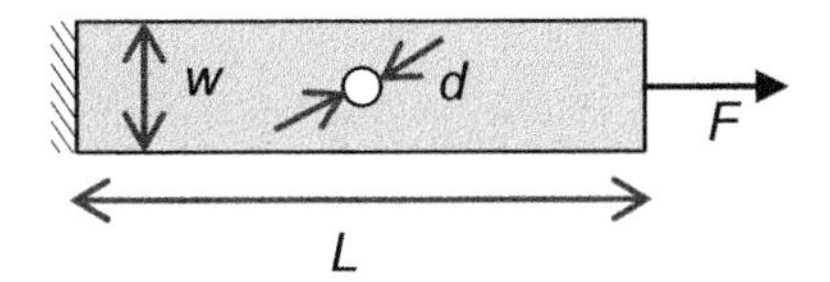

Figure 12.26 Plate with a stress concentration hole.

$$\delta = \frac{WL^4}{384EI} \tag{12.16}$$

(see reference [10]). Now model the 2D beam of length 1 m and height 0.05 m, with $E = 2 \times 10^{11}$ N/m^2, $\upsilon = 0.28$ and $W = 100$ N/m. Compare the analytical solution against the FEA solution with 1000 elements.

Exercise 12.3 Consider the double-cantilevered beam with point load in Figure 12.25. The maximum deflection is given by [10]

$$\delta = \frac{PL^3}{192EI} \tag{12.17}$$

Now model the 2D beam of length 1 m and height 0.05 m, with $E = 2 \times 10^{11}$ N/m^2, $\upsilon = 0.28$ and $P = 100$ N. Compare the analytical solution against the FEA solution with 1000 elements. Note that you have to split the top edge to apply the load.

Exercise 12.4 Consider the plate with a hole in Figure 12.26. For this problem, the *nominal stress* is defined as $\sigma_{\text{nom}} = F/((w - d)t)$, where t is the (virtual) thickness of the plate (into the page). Model the plate with the dimensions $L = 1.0$ m, $w = 0.25$ m and $d = 0.025$ m, with $E = 2 \times 10^{11}$ N/m^2 and $\upsilon = 0.28$. Compute the maximum stress $\sigma_{\max}$ with 2000 elements. Then compute the stress concentration factor K_t, defined as $K_t = \sigma_{\max}/\sigma_{\text{nom}}$. Repeat this experiment for values of d/w from 0.1 to 0.5 in steps of 0.1. Plot the stress concentration values K_t versus d/w.

Exercise 12.5 Create the L-bracket geometry with two cutouts, illustrated in Figure 12.27. Apply a force of 100 000 N/m, as in Figure 12.15. Carry out an FEA with 1000 elements, and plot the stress over the geometry. Next, vary the cutout length from 0.005 m to 0.025 m in steps of 0.005 m, and plot the maximum displacement versus cutout length. Observe that – seemingly

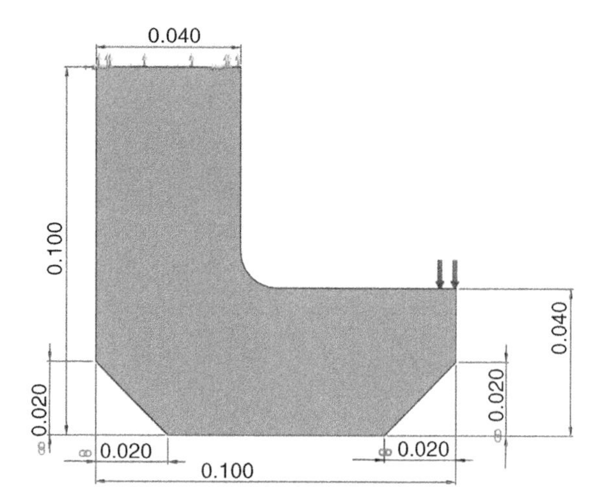

Figure 12.27 L-bracket with cutouts.

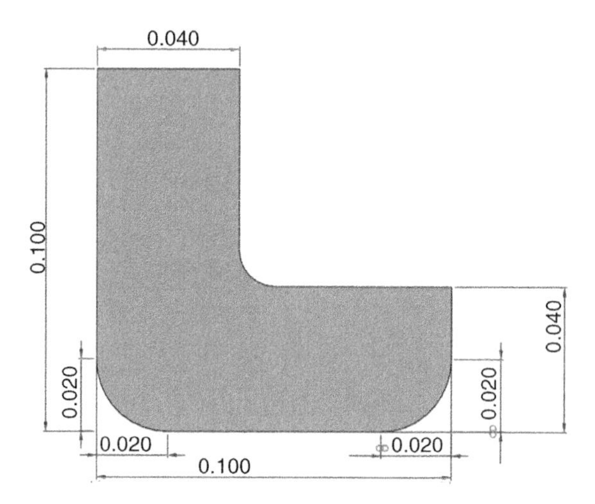

Figure 12.28 L-bracket with rounded cutouts.

contradictorily – the maximum displacement *decreases* with increasing cutout; explain the result.

Exercise 12.6 This exercise is on creating arcs that have an anti-clockwise orientation. Consider the geometry in Figure 12.28 with two rounded cutouts. Observe that, as we traverse around the edges of the geometry (with material to the left), the cutouts have an anti-clockwise orientation, while the fillet has a clockwise orientation. In the MATLAB code provided, arcs can be specified as anti-clockwise by using a negative value for the center vertex in the `arcs` structure. Model the geometry in Figure 12.28, plot with labels and compute its area.

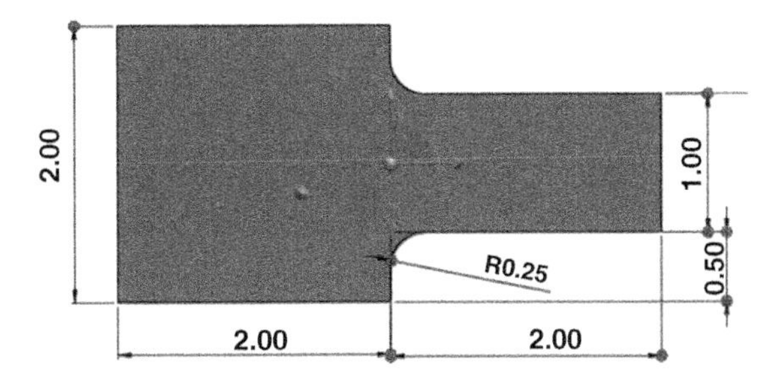

Figure 12.29 A stepped bar with fillets.

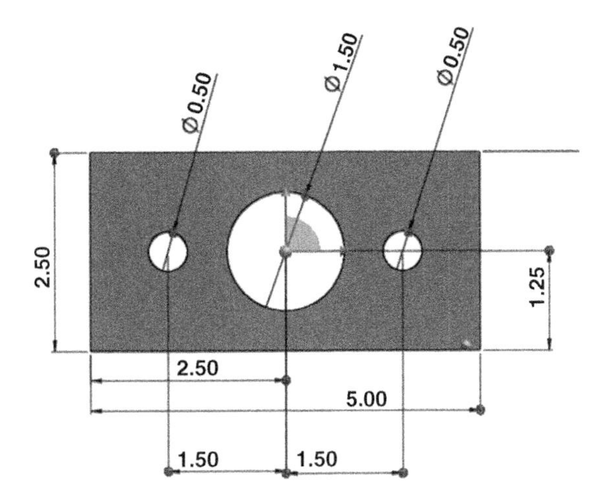

Figure 12.30 Plate with three holes.

Exercise 12.7 Create the geometry illustrated in Figure 12.29. Apply a fixed boundary condition to the left edge, and a positive horizontal force of 100 units on the right edge. Plot the maximum deflection and maximum stress (two different plots) as a function of the number of finite elements, ranging from 500 to 5000 in steps of 500. Observe the smooth behavior of the deflection, versus the non-smooth behavior of the stress.

Exercise 12.8 Model the geometry posed in Figure 12.30. Then, using the `TriElasticity` class, pose and solve the problem illustrated in Figure 12.31. The applied force is 1 unit.

(a) Plot the stress over the geometry.
(b) For a fixed number of mesh elements, plot the maximum stress as a function of the radius of the two smaller holes (with all other parameters remaining constant) in the range 0.25 to 0.65, in steps of 0.05.

For what value of the radius is the stress a minimum?

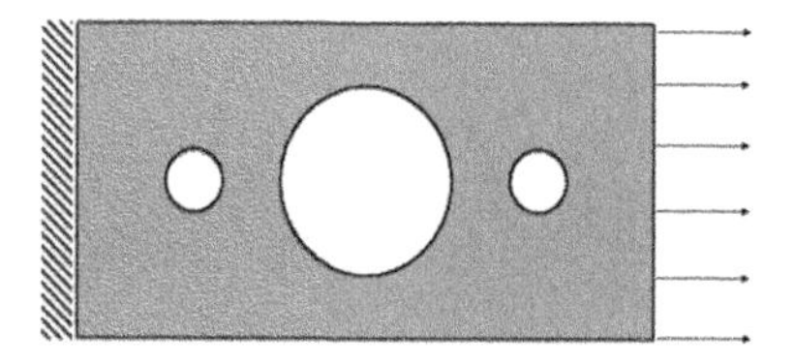

Figure 12.31 A structural problem with the geometry of Figure 12.30.

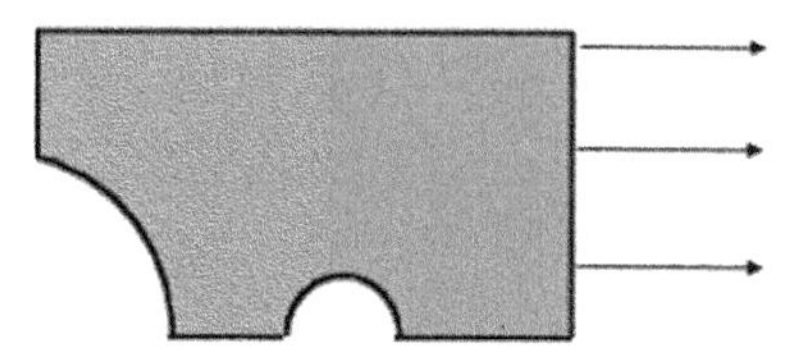

Figure 12.32 Exploiting symmetry.

Exercise 12.9 In this exercise, we will solve the problem in Figure 12.31 by exploiting symmetry. Specifically, instead of modeling the full geometry, model a quarter geometry, as illustrated in Figure 12.32. Then apply symmetry boundary conditions, i.e., restrain the bottom edges along the y-direction, and restrain the left edge along the x-direction; the force is reduced by half. Now solve the problem with 1000 elements; this is equivalent to solving the original problem with 4000 elements, and is the main advantage of exploiting symmetry. (Technically, the boundary conditions in Figure 12.31 are not symmetric, hence quarter-symmetry is an approximation.)

Exercise 12.10 In contrast to the plane-stress problems, if the depth of the 3D geometry is *large* compared to the dimensions of the 2D cross-section, as in Figure 12.33(a), and if the forces do not vary along the length, the 3D problem may be approximated by a 2D plane-strain problem, as illustrated in Figure 12.33(b).

In order to model plane-strain problems, we pass the argument `PlaneStrain` to the `TriElasticity` object:

```
fem = TriElasticity(LBracket,1000,'PlaneStrain');
```

The rest of the code remains unchanged. With this modification, solve the L-bracket problem using the material and load described in Section 12.5. Compare the stress and displacement against the plane-stress problem.

Exercise 12.11 Consider the 3D *axisymmetric* problem posed in Figure 12.34(a), where a uniform pressure is applied at the top, and the bottom face is fixed. The 3D problem may be simplified to a 2D axisymmetric problem over the cross-

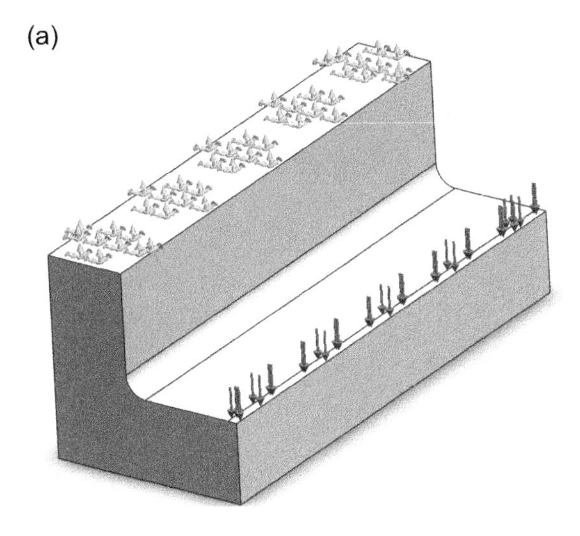

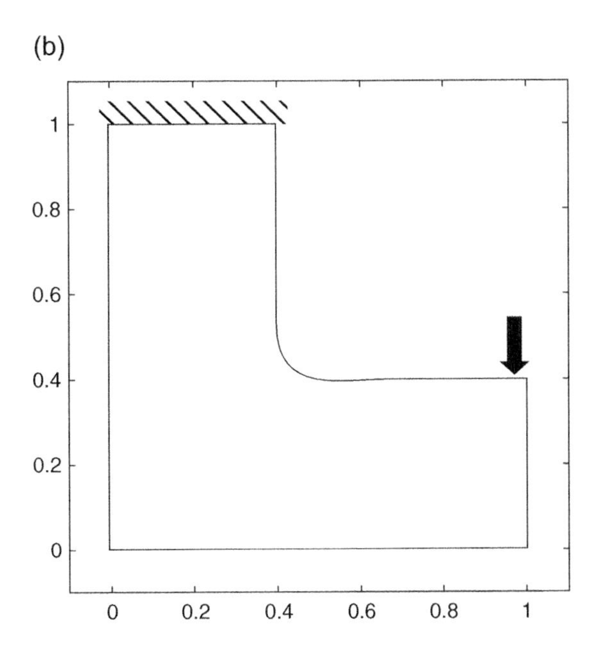

Figure 12.33 (a) A 3D elasticity problem over a long geometry; (b) its 2D plane-strain equivalent.

section illustrated in Figure 12.34(b) (lengths are in meters). It is assumed that the axis of symmetry is the y-axis, passing through the origin.

After creating the 2D geometry structure `cylinder2D` with vertices and edges, construct the axisymmetric model via

```
fem = TriElasticity(cylinder2D,1000,'AxiSymmetric');
```

(a)

(b)

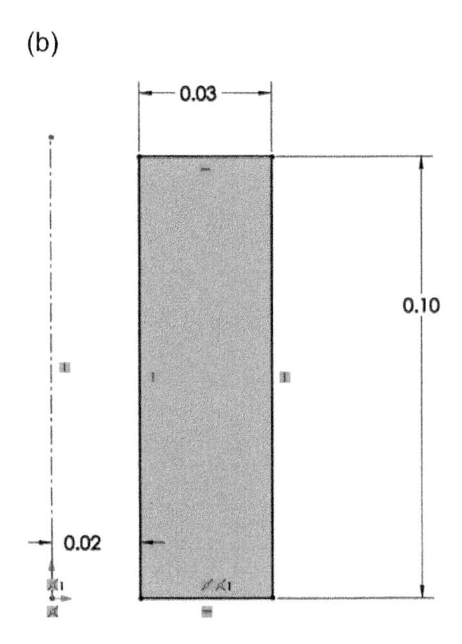

Figure 12.34 (a) A 3D elasticity problem over an axisymmetric cylinder; (b) its corresponding 2D model.

Then restrain the bottom edge, and apply a unit load on the top edge. Solve, and plot the resulting stress.

Exercise 12.12 A barrel problem is shown in Figure 12.35 (lengths are in meters), where the bottom face is fixed and a load is applied at the top. The material is alloy steel. Model this as an axisymmetric problem and find the resulting stress. Note that the outer arc has an anti-clockwise orientation, while the inner arc has a clockwise orientation.

(a)

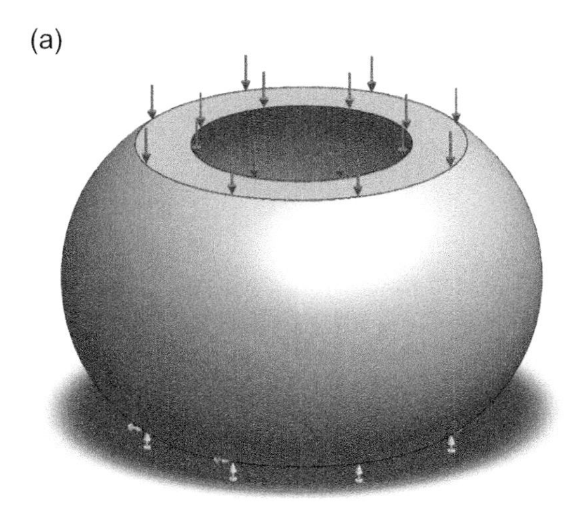

(b)

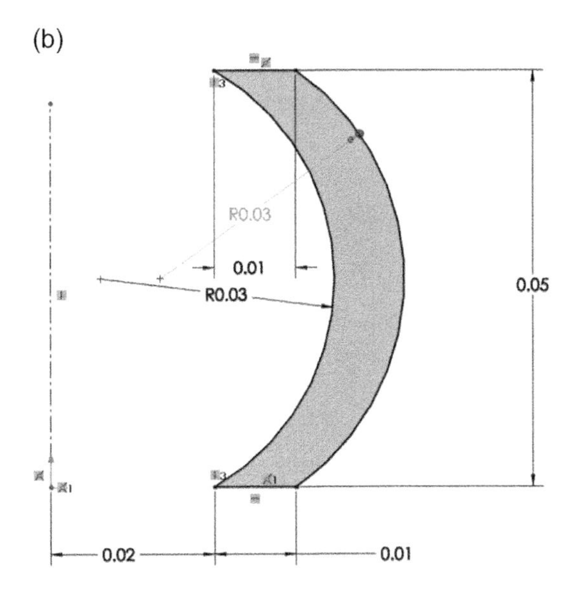

Figure 12.35 (a) A barrel problem; (b) its corresponding 2D model.

13 Shape Optimization in 2D

Highlights

1. In Chapter 12, we analyzed 2D elasticity problems; the focus of this chapter is on shape optimization of such problems.
2. The concept of shape parameters is introduced, followed by a discussion on objectives and constraints in shape optimization.
3. The concept of Pareto-optimal designs is introduced, and the classic compliance minimization problem is discussed within this context.
4. The importance of numerical scaling and gradient computation is revisited. This is followed by a discussion on finite-difference step size and semi-analytic methods of gradient computation.
5. Two MATLAB implementations of shape optimization are presented.

13.1 Shape Optimization and Shape Parameters

Consider the cantilever beam illustrated in Figure 13.1. In the previous chapter, we learned how to use finite element analysis (FEA) to solve this problem.

The objective of this chapter is to explore questions such as whether the *shape* of the cantilever beam can be optimized to minimize the compliance, while keeping the volume constant (and without introducing new features)? Alternatively, can the volume be reduced while keeping the compliance a constant? Such problems are referred to as *shape optimization* problems.

One approach to shape optimization is to directly move the vertices, i.e., to allow the coordinates of the vertices to serve as optimization variables, as in Figure 13.2. However, this can lead to undesirable shapes; for example, loss of symmetry, or even an invalid geometry.

Instead, shape optimization is typically carried out through shape parameters, such as the ones illustrated in Figure 13.3. Recall that these parameters were introduced in the previous chapter to define the shape. We will assume that the parameters L and H are fixed (i.e., do not vary, as indicated by the box around each of them in the figure), while the parameters a and b can be modified to change the shape.

For example, Figure 13.4 illustrates a new shape.

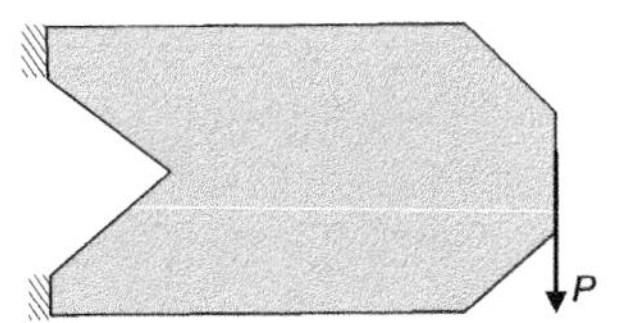

Figure 13.1 A cantilever beam problem.

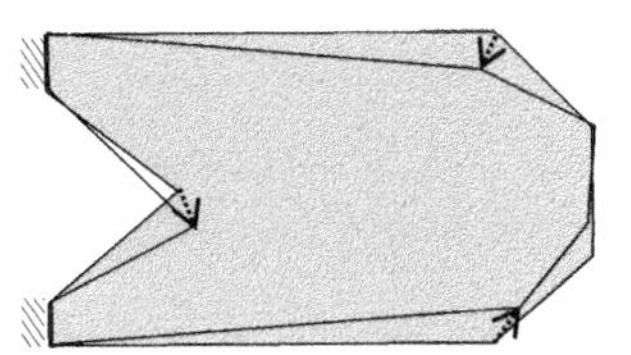

Figure 13.2 Shape control via vertex movement.

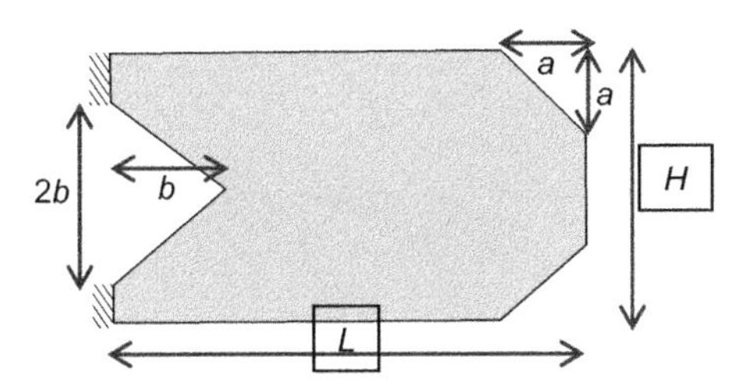

Figure 13.3 Shape parameters that define the geometry.

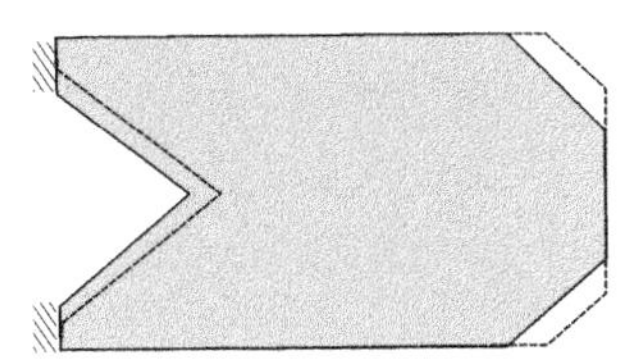

Figure 13.4 Shape control by modification of shape parameters.

13.2 Parametric Studies

We will first carry out a parametric study, where we randomly vary the two shape parameters in Figure 13.3, construct the geometry and compute the volume and compliance (via FEA). Recall that the *compliance* is defined as

$$J = \mathbf{f}\mathbf{T}\mathbf{d} \tag{13.1}$$

(see Chapter 10 on truss optimization). The same definition applies to 2D FEA as well: the lower the compliance, the stiffer the design.

The two shape parameters are varied randomly within the range $a \in [0.05, 0.4]$ and $b \in [0.05, 0.4]$. The range is so chosen as to avoid pathological issues; see the discussion in Section 13.10. The MATLAB implementation for carrying out the parametric study is provided below. Observe that, for each sample, one must construct the geometry, carry out an FEA and evaluate the compliance. Each sample is analyzed using approximately 200 finite elements, and 500 samples are used in this example. The reader is encouraged to study the MATLAB code. Observe that the vertices of the geometry are defined explicitly as functions of a and b.

```
%Parametric Study of compliance versus volume
nSamples = 500;
nElements = 200;
J = zeros(nSamples,1);
V = zeros(nSamples,1);
L = 2.0;%length in inches
H = 1;%height
h = 0.2; % length of force edge
aSamples = 0.05 + (0.4-0.05)*rand(nSamples,1);
bSamples = 0.05 + (0.4-0.05)*rand(nSamples,1);
for i = 1:nSamples
    a = aSamples(i);
    b = bSamples(i);
    cantileverBeam.vertices = [b 0;0 -b; 0 -H/2;L-a -H/2;L
-H/2+a; L -h/2; L h/2; L H/2-a; L-a H/2; 0 H/2; 0 b ]';
    cantileverBeam.edges = [1 2; 2 3; 3 4; 4 5; 5 6; 6 7; 7
8; 8 9; 9 10; 10 11; 11 1];
    fem = TriElasticity(cantileverBeam,nElements);
    fem = fem.setYoungsModulus(3.05e7); % psi
    fem = fem.setPoissonsRatio(0.28);
    fem = fem.fixEdge([2,10]);
    fem = fem.applyYForceOnEdge(6,-1000);
    fem = fem.assemble();
    fem = fem.solve();
    J(i) = fem.computeCompliance();
    V(i) = fem.computeArea();
end
plot(V,J,'.');
xlabel('Volume (V)'); ylabel('Compliance (J)');
```

Figure 13.5 illustrates the resulting volume-versus-compliance scatter plot. Observe in the figure that there is a “barrier” below which there are no designs. In other words, for a given volume, there is a minimum compliance defined by the barrier curve, below which there cannot exist any

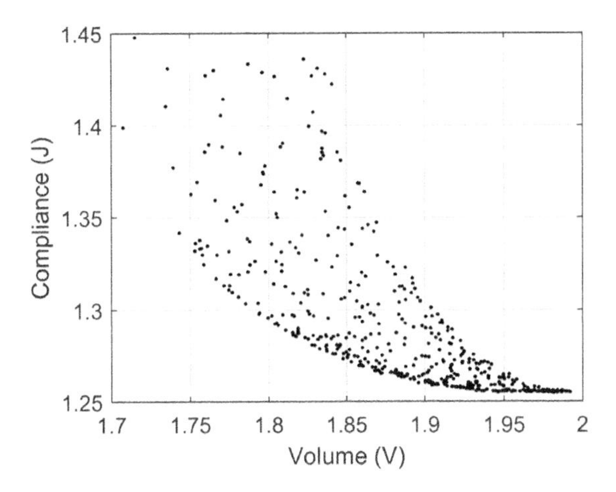

Figure 13.5 A parametric sample of compliance versus volume.

design. Similarly, for a given compliance, there is a minimum volume below which there cannot exist any design. The barrier curve is referred to as the *Pareto-optimal* curve. This curve exists for every structural optimization problem, independent of the type or number of design optimization variables.

To further emphasize this concept, the Pareto-optimal curve is illustrated explicitly in Figure 13.6. Observe the following:

- The region *below* the curve is *infeasible* in that one cannot achieve a combination of compliance and volume in this region, i.e., it is impossible to find a combination of parameters a and b to achieve a design with, say, a volume of 1.75 and compliance of 1.3.
- Designs *far above* the curve are *sub-optimal* and are not of much interest; one can always find better designs with lower volume and/or lower compliance.
- For a given volume, there exists an optimal shape with the lowest compliance; for example, for the design D_0 in Figure 13.6, one can see that there exists a design D_1 with the same volume but with the lowest compliance.
- Similarly, for a given compliance, there exists an optimal shape with the lowest volume; for example, given a design D_0, there exists a design D_2 with the same compliance but with the lowest volume.
- The collection of all designs that lie *on* the Pareto-optimal curve are referred to as *Pareto-optimal designs*. One of the goals of shape optimization is to find such designs.

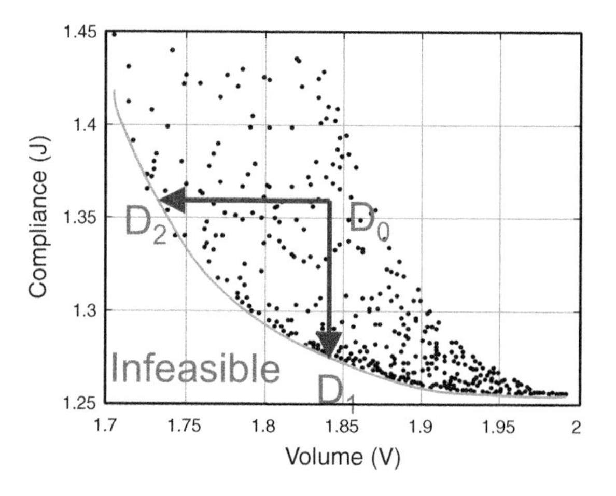

Figure 13.6 Pareto-optimal designs.

13.3 Compliance and Volume Minimization

Given the observations in Section 13.2, one can pose two shape optimization problems:

1. Starting from any design D_0 (for some initial parameter values), with volume V_0 and compliance J_0, find the optimal shape parameters such that the compliance is minimized with an upper bound of V_0 on the volume:

$$\begin{aligned} &\underset{\{a,b\}}{\text{minimize}}\; J \\ &\text{s.t.} \qquad V \leq V_0 \end{aligned} \tag{13.2}$$

2. Starting from any design D_0 (for some initial parameter values), with volume V_0 and compliance J_0, find the optimal shape parameters such that the volume is minimized with an upper bound of J_0 on the compliance:

$$\begin{aligned} &\underset{\{a,b\}}{\text{minimize}}\; V \\ &\text{s.t.} \qquad J \leq J_0 \end{aligned} \tag{13.3}$$

We also impose upper and lower bounds on the parameters to avoid nonsensical results:

$$\begin{aligned} &\underset{\{a,b\}}{\text{minimize}}\; J \\ &\text{s.t.} \qquad V \leq V_0 \\ &\qquad\quad a^{\min} \leq a \leq a^{\max} \\ &\qquad\quad b^{\min} \leq b \leq b^{\max} \end{aligned} \tag{13.4}$$

and

$$\begin{aligned} &\underset{\{a,b\}}{\text{minimize}}\; V \\ &\text{s.t.} \quad J \leq J_0 \\ &\quad\quad a^{\min} \leq a \leq a^{\max} \\ &\quad\quad b^{\min} \leq b \leq b^{\max} \end{aligned} \tag{13.5}$$

The remainder of this chapter will focus on solving such shape optimization problems.

13.4 Scaling for Numerical Robustness

Recall from the discussion on truss optimization in Chapter 10 that numerical scaling is critical for robust numerical optimization. Thus, given an initial set of parameters (a_0, b_0), the design variables are scaled by introducing non-dimensional variables:

$$\begin{aligned} x_1 &= a/a_0 \\ x_2 &= b/b_0 \end{aligned} \tag{13.6}$$

This leads to a set of bounds for the non-dimensional variables:

$$\begin{aligned} x_1^{\min,\max} &= a^{\min,\max}/a_0 \\ x_2^{\min,\max} &= b^{\min,\max}/b_0 \end{aligned} \tag{13.7}$$

Similarly, using the initial values (J_0, V_0) for the compliance and volume, Equations (13.4) and (13.5) are transformed as follows:

$$\begin{aligned} &\underset{\{x_1,x_2\}}{\text{minimize}}\; f = J/J_0 \\ &\text{s.t.} \quad \frac{V}{V_0} - 1 \leq 0 \\ &\quad\quad x_1^{\min} \leq x_1 \leq x_1^{\max} \\ &\quad\quad x_2^{\min} \leq x_2 \leq x_2^{\max} \end{aligned} \tag{13.8}$$

and

$$\begin{aligned} &\underset{\{x_1,x_2\}}{\text{minimize}}\; f = V/V_0 \\ &\text{s.t.} \quad \frac{J}{J_0} - 1 \leq 0 \\ &\quad\quad x_1^{\min} \leq x_1 \leq x_1^{\max} \\ &\quad\quad x_2^{\min} \leq x_2 \leq x_2^{\max} \end{aligned} \tag{13.9}$$

Equations (13.8) and (13.9) are now amenable to numerical optimization.

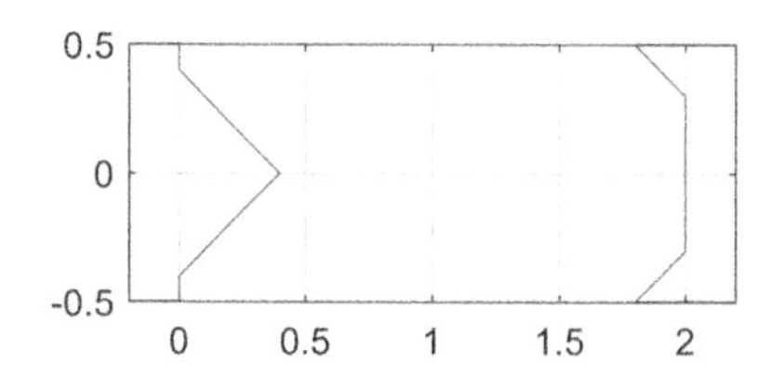

Figure 13.7 Initial design with $a = 0.2$ and $b = 0.4$.

We will first focus on compliance minimization, i.e., on Equation (13.8). Specifically, we start with $a = 0.2$ and $b = 0.4$; the initial design is illustrated in Figure 13.7. The initial compliance is (J_0, V_0), where $J_0 = 1.44$ and $V_0 = 1.8$, i.e., the design lies far above the Pareto-optimal curve; see Figure 13.6. As the figure shows, the expected minimal compliance for a volume of 1.8 is approximately 1.29.

13.5 Numerical Noise from FEA

Unfortunately, a direct implementation of Equation (13.8) with default `fmincon` settings will fail! The fundamental problem is that the objective function, i.e., the compliance computed via FEA, is numerically noisy. This is illustrated in Figure 13.8, where we have computed the compliance for small changes around the initial design (on the order of 0.0001) in the parameter a. Consequently, the default termination criterion of 1×10^{-6} for the objective is unreasonable. The optimization algorithm might never converge, since the numerical error in estimating the compliance far exceeds this value. This is easily remedied by loosening the termination criterion to a relative value in the approximate range 0.01 to 0.05.

The second observation is that numerical errors such as those illustrated in Figure 13.8 will lead to erroneous gradients when the default finite-difference methods are used. To illustrate, consider the forward-difference method for computing the gradient of the compliance:

$$\begin{aligned}\frac{\partial J}{\partial a} &\approx \frac{J(a + \Delta a,\ b) - J(a,\ b)}{\Delta a} \\ \frac{\partial J}{\partial b} &\approx \frac{J(a,\ b + \Delta b) - J(a,\ b)}{\Delta b}\end{aligned} \tag{13.10}$$

Table 13.1 summarizes the estimated derivatives for two different mesh sizes (N is the number of elements) and various step sizes. As one can observe in the table, the estimated gradients are reasonably consistent with respect to parameter b for large step sizes, but are inconsistent (and therefore erroneous) with respect to parameter a. Similar results can be expected for the central- and backward-difference

Table 13.1 Sensitivity of compliance computed via Equation (13.10).

Δa; Δb	$\left.\frac{\partial J}{\partial a}\right\|_{(0.2,0.4)}$		$\left.\frac{\partial J}{\partial b}\right\|_{(0.2,0.4)}$	
	$N = 500$	$N = 1000$	$N = 500$	$N = 1000$
10^{-2}	−0.038	−0.057	0.28	0.29
10^{-3}	−0.64	−0.075	0.261	0.28
10^{-4}	−0.0212	9.1	0.140	0.30
10^{-5}	0.028	9.3	0.255	0.59
10^{-6}	−1058	−57	0.255	3.44
10^{-7}	0.024	1046	0.255	682
10^{-8}	−1040	9108	0.255	318

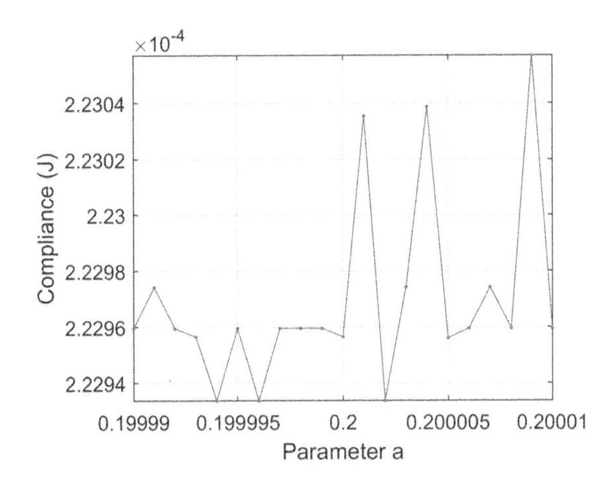

Figure 13.8 Compliance is noisy for small changes in geometric parameter.

methods. These inconsistent gradients may lead to divergence in gradient-based algorithms.

The underlying reason for the numerical noise in FEA is that, as shape parameters are changed, the underlying mesh structure can change dramatically. To illustrate, consider the mesh illustrated in Figure 13.9 for a particular instance of shape parameters: $a = 0.2$ and $b = 0.4$.

Now the parameter a is perturbed slightly to 0.201; the resulting mesh is illustrated in Figure 13.10. Observe that the number of elements in the mesh has changed. This changes the size of the stiffness matrix, resulting in a significant shift in the compliance estimate, and therefore in the consequences.

This problem does not arise in truss systems since the mesh structure is fixed, independent of the change in cross-sectional areas.

So how do we move forward? There are three broad strategies:

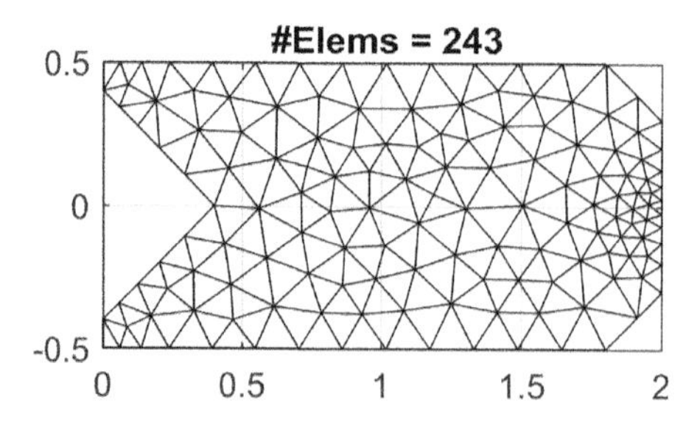

Figure 13.9 Geometry and mesh when and $a = 0.2$ and $b = 0.4$.

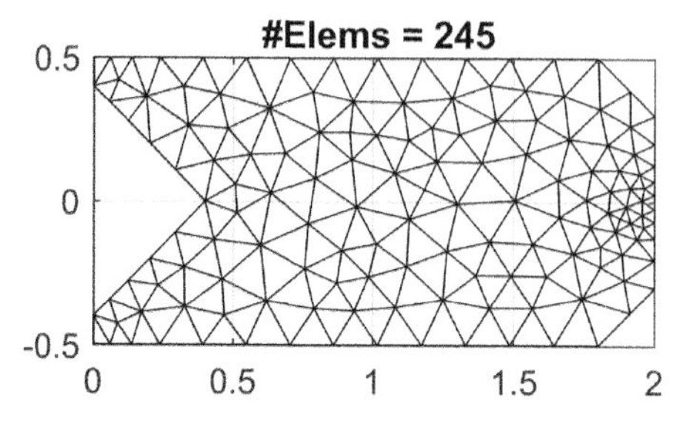

Figure 13.10 Geometry and mesh when $a = 0.201$ and $b = 0.4$.

1. Use large values for the finite-difference step size, and *hope* that the gradients will be sufficiently accurate; an implementation of this is provided below for illustration but is not the recommended choice!
2. Use zeroth-order methods, which don't require gradients (an exercise for the reader). Again, this is not recommended since zeroth-order methods can be inefficient.
3. Develop alternative methods to compute accurate gradients; this is the recommended method and will be discussed.

13.6 Finite-Difference Step Size

As suggested in Table 13.1, one can use a relatively large finite-difference step size, say, in the approximate range 0.001 to 0.01, to partially overcome the numerical noise. This is implemented in the accompanying MATLAB implementation `cantilever2DMinComplianceFD.m`.

The reader is encouraged to study the code, noting the following:

- The class specifically targets the cantilever beam, i.e., it is not a generic shape optimization class. However, it can be suitably modified to optimize other shapes.

- For this reason, we do not define the class as a child class of `TriElasticity`; instead, we use `TriElasticity` within the class.
- To construct the shape optimization object, one must provide the initial values of the two parameters (`a` and `b`), and the desired number of elements to be used in FEA.
- For computing the objective (compliance), one must construct the geometry via `TriElasticity`, and carry out an FEA. However, for computing the constraint (volume), we only need to compute the geometry, i.e., there is no need to carry out an FEA.
- A key part of the implementation that the reader must note is the overriding of the default `fmincon` settings.

```
classdef cantilever2DMinComplianceFD
...
    opt = optimoptions('fmincon',...
          'TolX',obj.myTerminationTolerance,...
          'TolFun',obj.myTerminationTolerance,...
'ConstraintTolerance',obj.myTerminationTolerance,...
'FiniteDifferenceStepSize',obj.myFiniteDifferenceStepSize);
          [xMin,fMin,flag,output] =
       fmincon(@obj.complianceObjective,x0,[],[],[],[],
        LB,UB,@obj.volumeConstraint,opt);
end
```

We now test this class with the default tolerances of 1×10^{-6}. To keep the computational time low, the number of elements is set to 200.

```
nElements = 200;
shapeOpt = shapeOpt.setParamsUpperBound([0.4,0.4]);
shapeOpt = shapeOpt.setParamsLowerBound([0.01,0.01]);
results = shapeOpt.optimize()
```

Unfortunately, the optimization converges to an infeasible point (the volume constraint is not met).

```
Converged to an infeasible point ...
Initial Params:
    0.2000    0.4000
Final Params:
    0.2000    0.2245
Initial/Final Volume:
    1.8000    1.9096
Initial/Final Compliance:
    1.4362    1.2686
#FEAs:
   102
```

We now set the finite-difference tolerance to 0.001 and the termination tolerance to 0.001.

```
nElements = 200;
params0 = [0.2 0.4];% initial shape parameters, a & b
shapeOpt = cantilever2DMinComplianceFD(params0,nElements);
shapeOpt = shapeOpt.setParamsUpperBound([0.4,0.4]);
shapeOpt = shapeOpt.setParamsLowerBound([0.01,0.01]);
shapeOpt = shapeOpt.setTerminationTolerance(0.001);
shapeOpt = shapeOpt.setFiniteDifferenceStepSize(0.001);
results = shapeOpt.optimize()
```

The optimization now converges, and we arrive at the optimal values of $a = 0.38$ and $b = 0.24$, with an optimal of compliance at 1.29 (as expected per Figure 13.5). Observe that an 11 percent decrease in compliance is achieved for a fixed volume.

```
Local minimum found that satisfies the constraints ...
Initial Params:
    0.2000    0.4000
Final Params:
    0.3802    0.2393
Initial/Final Volume:
    1.8000    1.7981
Initial/Final Compliance:
    1.4362    1.2953
#FEAs:
   306
```

The optimal geometry is illustrated in Figure 13.11. The reader is encouraged to "play around" with the parameters, and study the convergence.

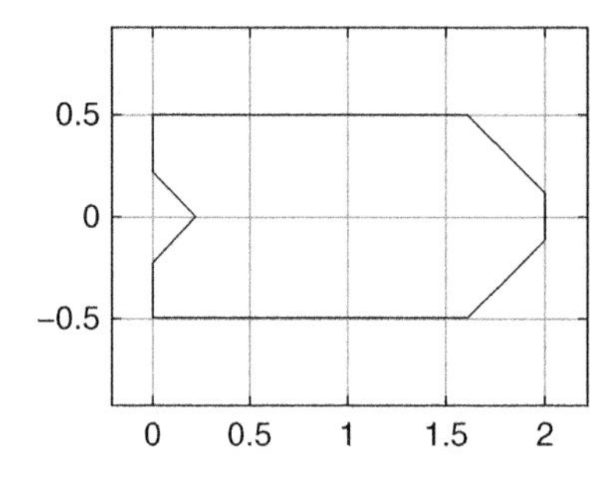

Figure 13.11 Optimized design with $a = 0.38$ and $b = 0.4$.

13.7 Semi-analytic Gradients in FEA

We will now consider an alternative method for evaluating gradients. Recall that, for truss problems, we considered the powerful complex-variable method to compute gradients accurately. Unfortunately, this method does not apply here since the built-in MATLAB mesh generation algorithm cannot handle complex numbers. We must therefore devise an alternative strategy.

Recall from the discussion on truss optimization in Chapter 10 that the sensitivity of compliance with respect to any design variable is given by

$$J' = -\mathbf{d}^T\mathbf{K}'\mathbf{d} \tag{13.11}$$

This expression is valid in shape optimization as well. Thus, the problem reduces to that of computing the sensitivity of the stiffness matrix with respect to a shape parameter.

Now consider the finite-difference estimation

$$K' \equiv \frac{\partial K}{\partial a} \approx \frac{K(a+\Delta a,\, b) - K(a, b)}{\Delta a} \tag{13.12}$$

Evaluating this expression entails the following step. For a particular shape instance the finite element mesh is illustrated in Figure 13.9. We will assume that the stiffness matrix has been computed for this instance. Next, let the shape parameter `a` be increased by a small amount, say, 0.001. Instead of creating a new mesh, the boundary nodes of the original mesh are projected onto the new geometry, thereby retaining the mesh size and stiffness size; see Figure 13.12. The projection method is captured by `projectMeshOntoBrep` in a new class, `cantilever2DMinComplianceSA.m`, to be described at the end of this section.

We can now compute the new stiffness matrix $K(a + \Delta a,\, b)$, and therefore the sensitivity in Equation (13.12).

Thus, an alternative algorithm to compute the compliance sensitivity is as follows:

1. First, compute the stiffness matrix sensitivity K' with respect to each shape parameter, via mesh projection.
2. Then estimate the compliance sensitivity.

Table 13.2 summarizes the computed sensitivities for various mesh sizes and finite-difference step sizes using the projection method. As one can observe, the sensitivities are consistent (and are likely to be accurate). However, it can also be observed that the sensitivity estimate depends on the mesh size, i.e., a large number of elements may be necessary for an accurate estimate. For a full treatment of shape sensitivity techniques, see references [34] and [35].

Table 13.2 Sensitivity of compliance computed via stiffness element sensitivity.

Δa; Δb	$\frac{\partial J}{\partial a}\|_{(0.2,0.2)}$			$\frac{\partial J}{\partial b}\|_{(0.2,0.2)}$		
	$N = 500$	$N = 1000$	$N = 1500$	$N = 500$	$N = 1000$	$N = 1500$
10^{-2}	0.018	0.016	0.016	0.114	0.145	0.160
10^{-3}	0.016	0.014	0.014	0.104	0.134	0.149
10^{-4}	0.015	0.014	0.014	0.104	0.133	0.148
10^{-5}	0.015	0.014	0.014	0.104	0.133	0.148
10^{-6}	0.015	0.014	0.014	0.104	0.133	0.148
10^{-7}	0.015	0.014	0.014	0.104	0.133	0.148
10^{-8}	0.015	0.014	0.014	0.104	0.133	0.148

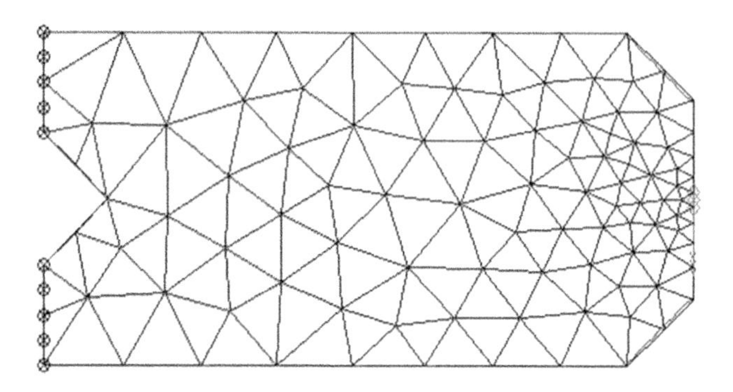

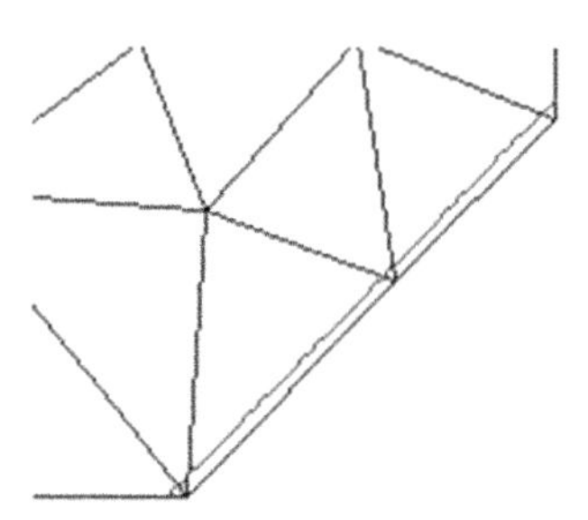

Figure 13.12 Projecting the mesh onto a perturbed geometry.

Also, note that the sensitivity of the non-dimensional objective with respect to non-dimensional parameters is given by

$$\frac{\partial f}{\partial x_1} = \frac{1}{J_0}\frac{\partial J}{\partial x_1} = \frac{1}{J_0}\frac{\partial J}{\partial a}\frac{da}{dx_1} = \frac{a_0}{J_0}\frac{\partial J}{\partial a} \tag{13.13}$$

$$\frac{\partial f}{\partial x_2} = \frac{b_0}{J_0}\frac{\partial J}{\partial b} \tag{13.14}$$

This strategy, based on mesh projection, is implemented in the accompanying code, `cantilever2DMinComplianceSA.m`, which the reader is encouraged to study; the critical method in this class is mesh projection.

```
classdef cantilever2DMinComplianceSA
...
        opt = optimoptions('fmincon',...
            'TolX',obj.myTerminationTolerance,...
            'TolFun',obj.myTerminationTolerance,...
'ConstraintTolerance',obj.myTerminationTolerance,...
            'SpecifyObjectiveGradient',true);
            [xMin,fMin,flag,output] =
fmincon(@obj.complianceObjective,x0,[],[],[],[],
         LB,UB,@obj.volumeConstraint,opt);
...
end
```

We now test this class with a step size of 0.001 for computing Equation (13.12) and a termination tolerance of 0.001.

```
...
shapeOpt = cantilever2DMinComplianceSA(params0,nElements);
...
results = shapeOpt.optimize()
```

We arrive at optimal values of $a = 0.38$ and $b = 0.25$, with the compliance at 1.30 (as expected per Figure 13.5). Observe that only 41 FEA iterations are needed, as opposed to 306 FEA iterations via finite difference.

```
Constraints satisfied ...
Initial Params:
    0.2000    0.4000
Final Params:
    0.3820    0.2521
Initial/Final Volume:
    1.8000    1.7905
Initial/Final Compliance:
    1.4362    1.3000
#FEAs: 41
```

The reader is encouraged to "play around" with the simulation parameters.

13.8 Global Search Method

In Chapter 6, we discussed global optimization methods which avoid (to a large extent) getting trapped by local minima, and therefore are effective in the presence of noise. We will illustrate the use the `GlobalSearch` method for solving the shape optimization problem using the `cantilever2DMinComplianceGS.m` class, which the reader is encouraged to study; observations include the construction of the problem structure.

```
classdef cantilever2DMinComplianceGS
    ...
        function [results] = optimize(obj)
            ...
        problem createOptimProblem('fmincon','objective',
      @obj.complianceObjective, 'x0', x0, 'lb', LB,'ub',
UB, 'nonlcon',@obj.volumeConstraint,'options',opt);
        [xMin,fMin,flag,output] =
run(GlobalSearch,problem);
      ...
      end
```

We now test this class with a step size of 0.001 and a termination tolerance of 0.001:

```
...
shapeOpt = cantilever2DMinComplianceGS(params0,nElements);
...
results = shapeOpt.optimize()
```

We arrive at optimal values of $a = 0.38$, and $b = 0.24$, with the compliance at 1.30 (as expected per Figure 13.5). Observe that 315 FEA iterations were needed, similar to the finite-difference approach.

```
Initial Params:
    0.2000    0.4000
Final Params:
    0.3802    0.2393
Initial/Final Volume:
    1.8000    1.7981
Initial/Final Compliance:
    1.4362    1.2953
#FEAs:
   315
```

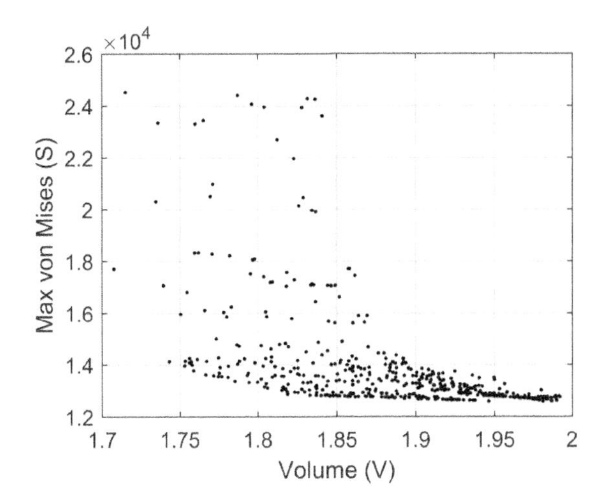

Figure 13.13 A parametric sample of maximum von Mises stress versus volume.

13.9 Stress Scatter Plot

Often, stresses are more critical than compliance in engineering, i.e., stress minimization is often more critical than compliance minimization. Here, we will generate a scatter plot similar to the one in Figure 13.5, with the exception that, instead of compliance, we plot the maximum von Mises stress on the y-axis; this is illustrated in Figure 13.13. The von Mises stress is the most common measure of failure in ductile materials ([12]). Once again, we can observe a barrier curve. Further, for any initial design far from the Pareto-optimal curve, there exists a design with the same volume but lower stress, as well as a design with the same stress but lower volume. However, finding these designs may pose numerical challenges, since stress exhibits significantly more noise than compliance. We leave this as an exercise for the reader.

13.10 Geometric Constraints

In Section 13.7, upper and lower bounds were imposed on the geometric parameters to avoid pathological conditions. In other scenarios, additional geometric constraints may typically be needed. For example, consider the geometry in Figure 13.14, which requires *four* shape parameters, a, b, c and d.

However, geometric constraints must be imposed to avoid the scenario illustrated in Figure 13.15. A coupled geometric constraint $b + d/2 - c \leq 0$ must be imposed to avoid such conditions; this is left as an exercise for the reader.

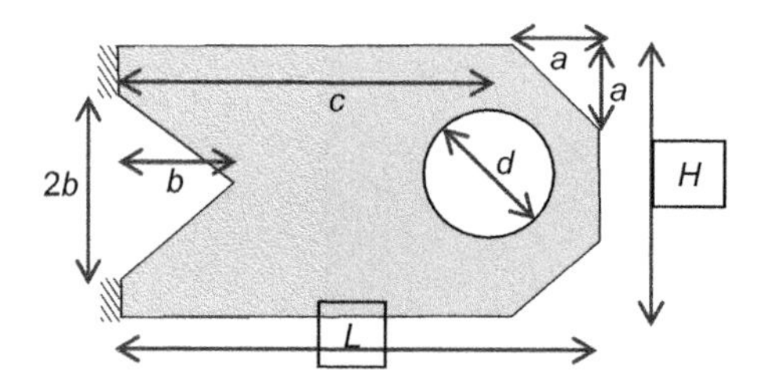

Figure 13.14 Cantilever with four shape parameters.

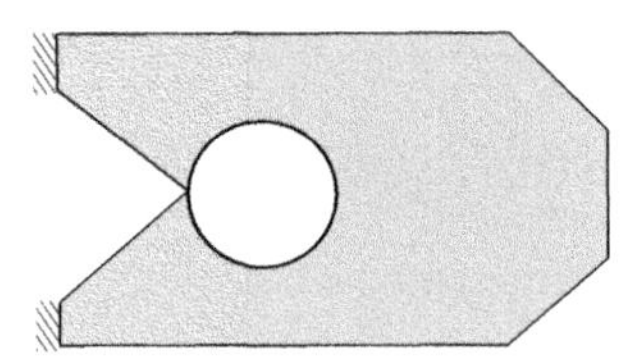

Figure 13.15 Pathological conditions must be avoided through geometric constraints.

13.11 Conclusions

Shape optimization is a rich discipline ([36]), and important in engineering. In this chapter, we have merely "scratched the surface" of shape optimization, emphasizing the critical concepts of shape parameters, Pareto-optimal design, numerical scaling, gradient estimation and geometric constraints. These concepts are further explored in the following exercises, and in Chapters 14 and 15.

13.12 EXERCISES

Exercise 13.1 Modify the MATLAB class `cantilever2DMinComplianceFD.m` to use the default `fmincon` settings. Test the class on the initial design in Figure 13.7; explain the results.

Exercise 13.2 Generate five different initial designs (i.e., five different shape parameter sets) that are far away from the volume-compliance Pareto-optimal curve in Figure 13.5; for example, designs with compliance greater than 1.35 and volume greater than 1.8. Display these designs as points in a volume-versus-compliance plot.

Exercise 13.3 Test the MATLAB class `cantilever2DMinComplianceFD.m` starting with the five designs found in Exercise 13.2. Display the results as five straight lines joining the initial volume and compliance to final volume and compliance.

Exercise 13.4 Change the inequality in `cantilever2DMinComplianceFD.m` to an equality constraint, and repeat Exercise 13.3.

Exercise 13.5 Repeat Exercise 13.3 using `cantilever2DMinComplianceSA.m`.

Exercise 13.6 Repeat Exercise 13.4 using `cantilever2DMinComplianceSA.m`.

Exercise 13.7 Write a MATLAB class `cantilever2DMinVolumeFD.m` that uses the finite-difference approach to solve the volume-minimization problem, subject to compliance (inequality) constraint. Test the class on the five designs found in Exercise 13.2. Display the results as five straight lines joining the initial volume and compliance to final volume and compliance.

Exercise 13.8 Repeat Exercise 13.7 using `cantilever2DMinComplianceSA.m`.

Exercise 13.9 Generate five different initial designs (i.e., five different shape parameter sets) that are far away from the volume-stress Pareto-optimal curve in Figure 13.13; for example, designs with stress greater than 1.6×10^4 and volume greater than 1.8. Display these designs as points in a volume-versus-stress plot.

Exercise 13.10 Write a MATLAB class `cantilever2DMinStressFD.m` that minimizes the maximum stress using the finite-difference method. Test the class for the initial designs from Exercise 13.9 and find the corresponding optimized designs. Where do these lie on the Pareto scatter plot in Figure 13.13?

Exercise 13.11 Create a parametric study for the design illustrated in Figure 13.14, while ensuring that geometric problems such as the one in Figure 13.15 are avoided. Plot the compliance versus volume as in Figure 13.5 for 500 samples, using 200 finite elements.

Exercise 13.12 Write a MATLAB class `cantilever4ParameterMinComplianceFD.m` to solve the four-parameter problem in Figure 13.14, using the finite difference method and imposing additional geometric constraints. Test your class using five different designs that are far from the Pareto-optimal curve created in Exercise 13.11.

Exercise 13.13 Write a MATLAB class `cantilever4ParameterMinComplianceSA.m` to solve the four-parameter problem in Figure 13.14, using the semi-analytic approach and imposing additional geometric constraints. Test your class using five different designs that are far from the Pareto-optimal curve created in Exercise 13.11. Compare the results (accuracy and cost) against those obtained in Exercise 13.12.

Exercise 13.14 Write a MATLAB class `cantilever4ParameterMinComplianceGS.m` to solve the four-parameter problem in Figure 13.14, using the global search approach and imposing additional geometric constraints. Test your class using five different designs that are far from the Pareto-optimal curve created in Exercise 13.11. Compare the results (accuracy and cost) against those obtained in Exercises 13.12 and 13.13.

Exercise 13.15 Write a MATLAB class `cantilever4ParameterMinVolumeFD.m` to solve the four-parameter problem in Figure 13.14, using finite difference and imposing additional geometric constraints. Test your class using five different designs that are far from the Pareto-optimal curve created in Exercise 13.11.

Exercise 13.16 Write a MATLAB class `cantilever4ParameterMinVolumeSA.m` to solve the four-parameter problem in Figure 13.14, using the semi-analytic approach and imposing additional geometric constraints. Test your class using five different designs that are far from the Pareto-optimal curve created in Exercise 13.11. Compare the results (accuracy and cost) against those obtained in Exercise 13.15.

Exercise 13.17 Write a MATLAB class `cantilever4ParameterMinVolumeGS.m` to solve the four-parameter problem in Figure 13.14, using the global search method and imposing additional geometric constraints. Test your class using five different designs that are far from the Pareto-optimal curve created in Exercise 13.11. Compare the results (accuracy and cost) against those obtained in Exercise 13.15 and 13.16.

Exercise 13.18 The cantilever problem in Figure 13.3 may also be parameterized as illustrated in Figure 13.16, using three parameters instead of two. Create a parametric study for the design illustrated in Figure 13.16, while ensuring that geometric problems are avoided. Plot the compliance versus volume for 500 samples, using 200 finite elements.

Exercise 13.19 Write a MATLAB class `cantilever3ParameterMinCompliance FD.m` to solve the three-parameter problem in Figure 13.16, using the finite-difference method and imposing additional geometric constraints. Test your class using five different designs that are far from the Pareto-optimal curve created in Exercise 13.18.

Exercise 13.20 Write a MATLAB class `cantilever3ParameterMinCompliance SA.m` to solve the three-parameter problem in Figure 13.16, using the semi-analytic method and imposing additional geometric constraints. Test your class using five

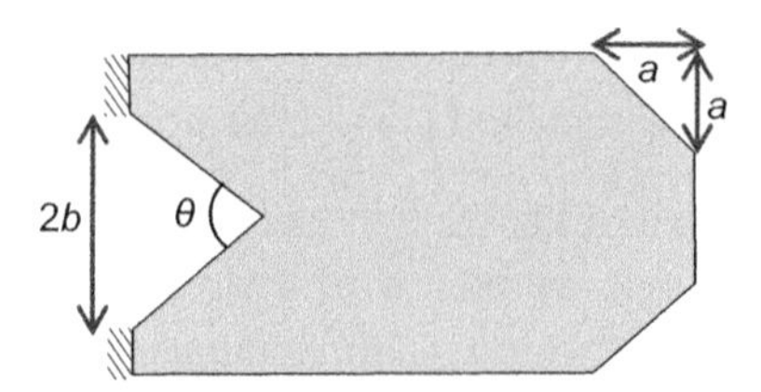

Figure 13.16 A cantilever beam with alternative shape parameters.

different designs that are far from the Pareto-optimal curve created in Exercise 13.18. Compare the results (accuracy and cost) against those obtained in Exercise 13.19.

Exercise 13.21 Write a MATLAB class `cantilever3ParameterMinCompliance GS.m` to solve the three-parameter problem in Figure 13.16, using the global search method and imposing additional geometric constraints. Test your class using five different designs that are far from the Pareto-optimal curve created in Exercise 13.18. Compare the results (accuracy and cost) against those obtained in Exercises 13.19 and 13.20.

Exercise 13.22 Write a MATLAB class `cantilever3ParameterMinVolumeFD.m` to solve the three-parameter problem in Figure 13.16, using the finite-difference method and imposing additional geometric constraints. Test your class using five different designs that are far from the Pareto-optimal curve.

Exercise 13.23 Write a MATLAB class `cantilever3ParameterMinVolume SA.m` to solve the four-parameter problem in Figure 13.16, using the semi-analytic method and imposing additional geometric constraints. Test your class using five different designs that are far from the Pareto-optimal curve created in Exercise 13.18. Compare the results (accuracy and cost) against those obtained in Exercise 13.22.

Exercise 13.24 Write a class `cantilever3ParameterMinVolumeGS.m` to solve the four-parameter problem in Figure 13.16, using the global search method and imposing additional geometric constraints. Test your class using five different designs that are far from the Pareto-optimal curve created in Exercise 13.18. Compare the results (accuracy and cost) against those obtained in Exercises 13.22 and 13.23.

14 Finite Element Analysis in 3D

Highlights

1. In Chapters 12 and 13, we discussed finite element analysis and shape optimization of 2D elasticity problems. In this chapter, we discuss the finite element analysis of 3D elasticity problems, using SOLIDWORKS.
2. We revisit the critical steps in finite element analysis, within the context of SOLIDWORKS. Several sample problems are analyzed and compared against their 2D counterparts.
3. This chapter serves as a foundation for parametric study and shape optimization of 3D problems, to be discussed in Chapters 15 and 16.

14.1 Overview

In this chapter, we illustrate the use of a commercial software, SOLIDWORKS, to carry out finite element analysis (FEA) of 3D problems such as the ones in Figure 14.1.

It is assumed that the reader is familiar with the process of creating 3D models and carrying out basic FEA within SOLIDWORKS. There are several tutorials embedded within SOLIDWORKS for these two tasks; *the reader is urged to study these tutorials before proceeding.* The objective of this chapter is to develop a foundation for parametric study and optimization to be pursued in Chapters 15 and 16.

14.2 Analysis of Cantilever Beam

As the first example, consider the 3D cantilever beam illustrated in Figure 14.2. The cross-sectional dimensions are the same as those of the 2D cantilever beam considered in Chapter 12, and the thickness is 0.1 in. The SOLIDWORKS model is included in the software accompanying this text.

We will use SOLIDWORKS Simulation to set up the static problem illustrated in Figure 14.2. Specifically, as with the 2D cantilever problem, the material is set to alloy steel, with a Young's modulus of 3.05×10^7 psi and Poisson's ratio of

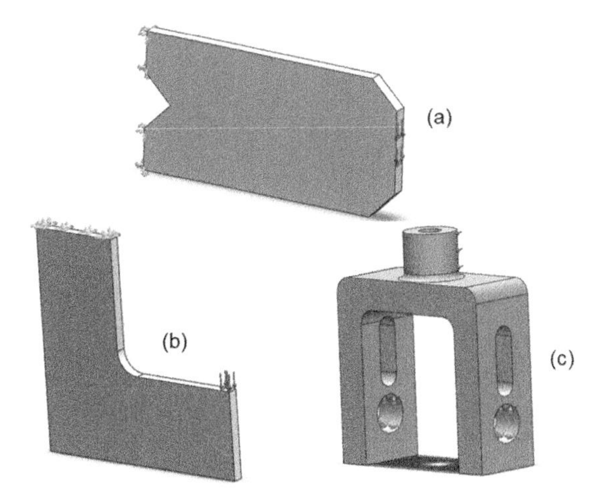

Figure 14.1 Sample 3D structural mechanics problems.

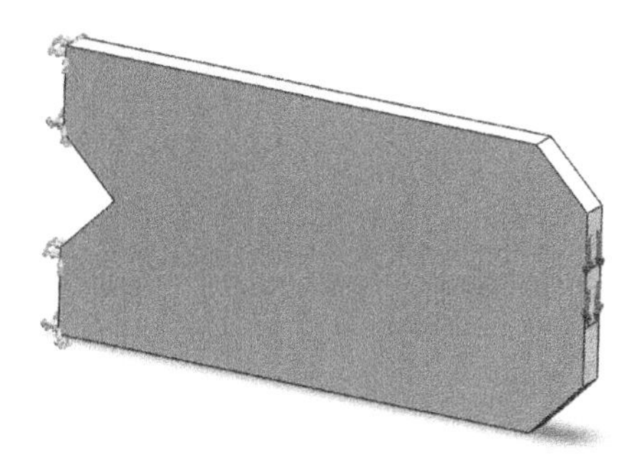

Figure 14.2 3D cantilever beam problem.

0.28. Further, recall that in 2D the loading was 1000 lb per inch of thickness. Here, since the thickness is 0.1 in, we apply a total load of 100 lb.

The default finite element mesh created using SOLIDWORKS is illustrated in Figure 14.3. The mesh consists of about 5000 tetrahedral elements; tetrahedral elements are generalizations of 2D triangles.

FEA simulation is then carried out within SOLIDWORKS (the reader is referred to resources within SOLIDWORKS for FEA tutorials). The resulting displacement plot is illustrated in Figure 14.4. The maximum displacement is 1.27×10^{-3} in, which matches the 2D result of Chapter 12, as the reader can verify.

The von Mises stress is illustrated in Figure 14.5. The maximum stress is 16.5 ksi, which is higher than the 14.1 ksi predicted in 2D. The discrepancy can be attributed to several factors, including assumptions underlying 2D simplification, and FEA errors in 2D and 3D.

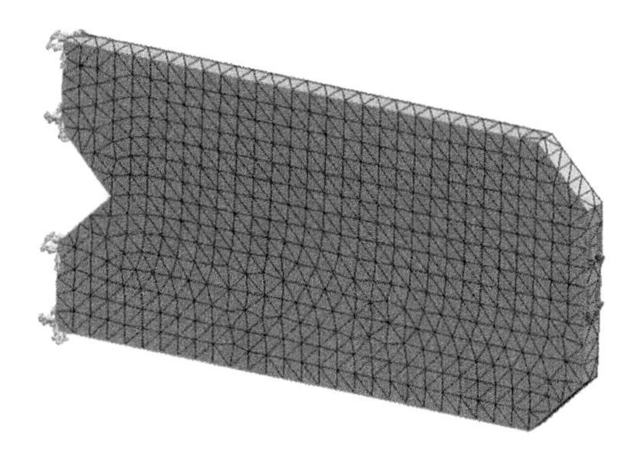

Figure 14.3 Finite element mesh consisting of tetrahedral elements.

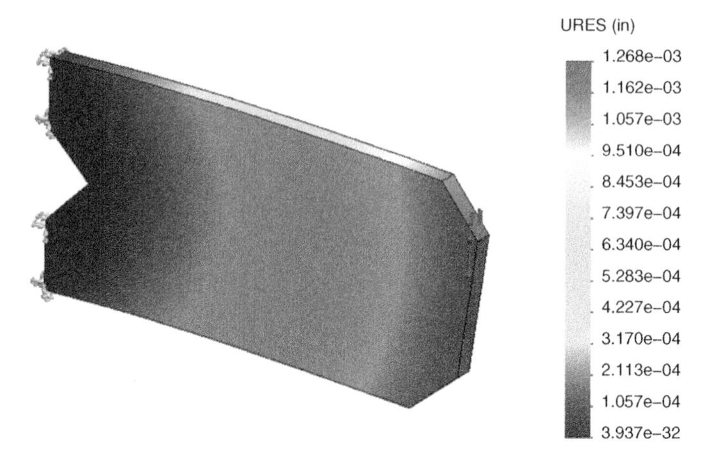

Figure 14.4 Displacement plot for the 3D cantilever beam.

14.3 Analysis of L-bracket

Next consider the 3D L-bracket in Figure 14.6, with a thickness of 0.005 m; the equivalent 2D problem was considered in Chapter 11. The SOLIDWORKS L-bracket model is included with the software accompanying this text.

The material is set to alloy steel, with a Young's modulus of 2.1×10^{11} N/m^2 and Poisson's ratio of 0.28. In 2D, recall that the loading was set to 100 000 N per meter of thickness; since the thickness here is 0.005 m, we apply a total load of 500 N. The default finite element mesh is created using SOLIDWORKS; see Figure 14.7. The mesh consists of about 4000 tetrahedral elements.

FEA simulation is then carried out within SOLIDWORKS. The resulting displacement plot is illustrated in Figure 14.8. The maximum displacement is 6.13×10^{-5} m; this matches the 2D result.

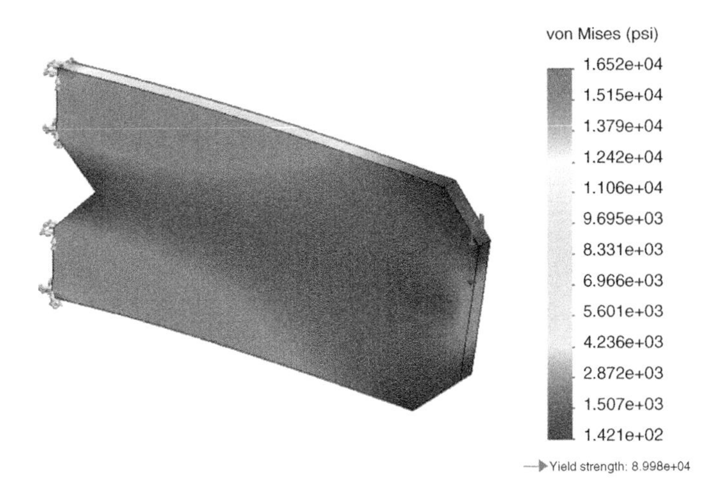

Figure 14.5 Stress plot for the 3D cantilever beam.

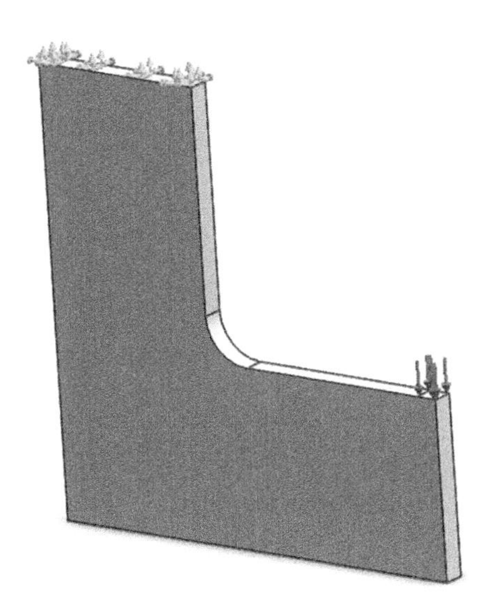

Figure 14.6 3D L-bracket problem.

The von Mises stress is illustrated in Figure 14.9. The maximum stress is 52 MPa, which is slightly higher than the 50 MPa predicted in 2D. The discrepancy once again can be attributed to several factors, including assumptions underlying 2D simplification, and FEA errors in 2D and 3D.

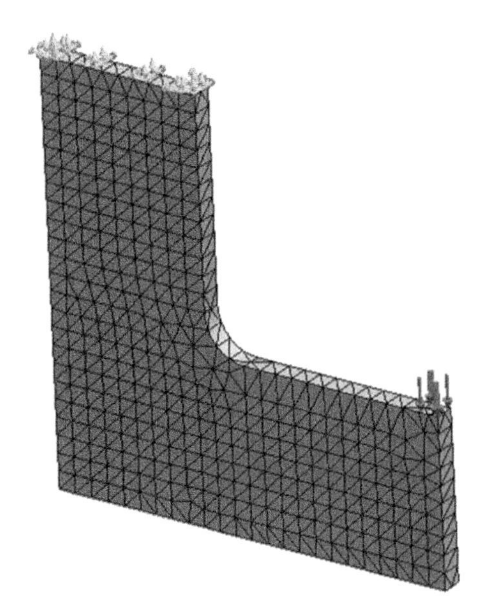

Figure 14.7 Finite element mesh consisting of tetrahedral elements.

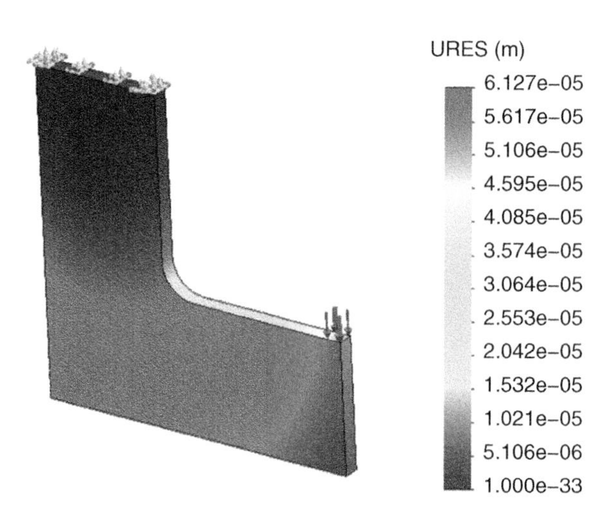

Figure 14.8 Displacement plot for the 3D L-bracket.

14.4 Analysis of Knuckle

As a final example for this chapter, we consider the knuckle problem illustrated in Figure 14.10, where the part is restrained on cylindrical surfaces, and a torque is applied as illustrated. Observe that the problem cannot be reduced to a 2D problem, i.e., it must be solved as a 3D problem. The model is included in the software accompanying this text.

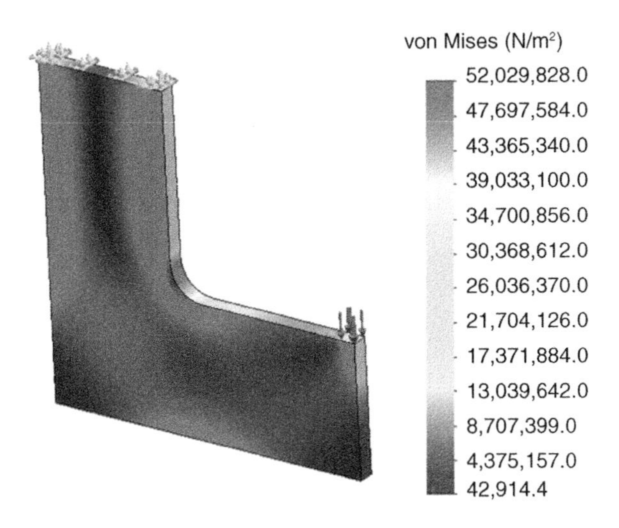

Figure 14.9 von Mises stress plot for the 3D L-bracket.

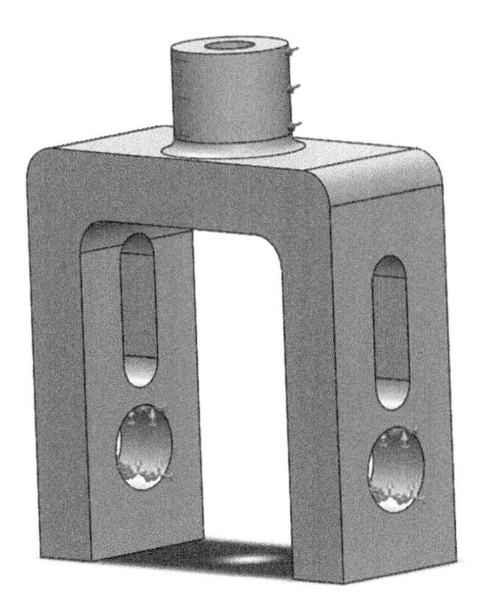

Figure 14.10 Knuckle problem.

The material is set to aluminum 1060, with a Young's modulus of 69×10^{11} N/m^2 and Poisson's ratio of 0.33; the torque is 100 Nm. The default finite element mesh is created using SOLIDWORKS, as illustrated in Figure 14.11. The mesh consists of about 7000 tetrahedral elements.

After performing an FEA, the resulting displacement plot is illustrated in Figure 14.12. The maximum displacement is 2.27×10^{-5} m.

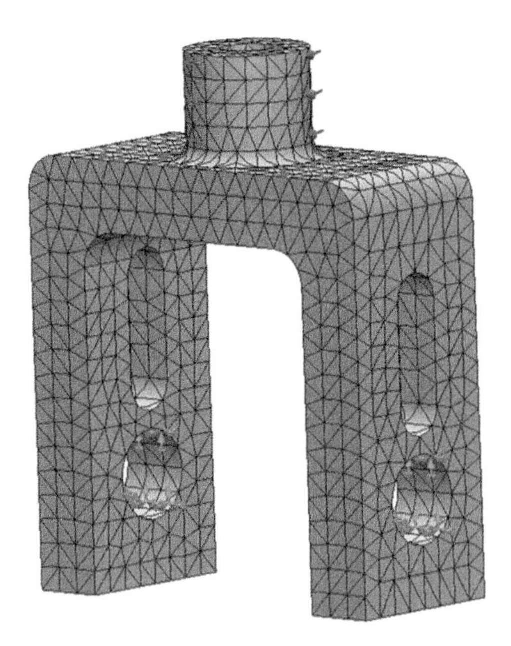

Figure 14.11 Finite element mesh for the knuckle problem.

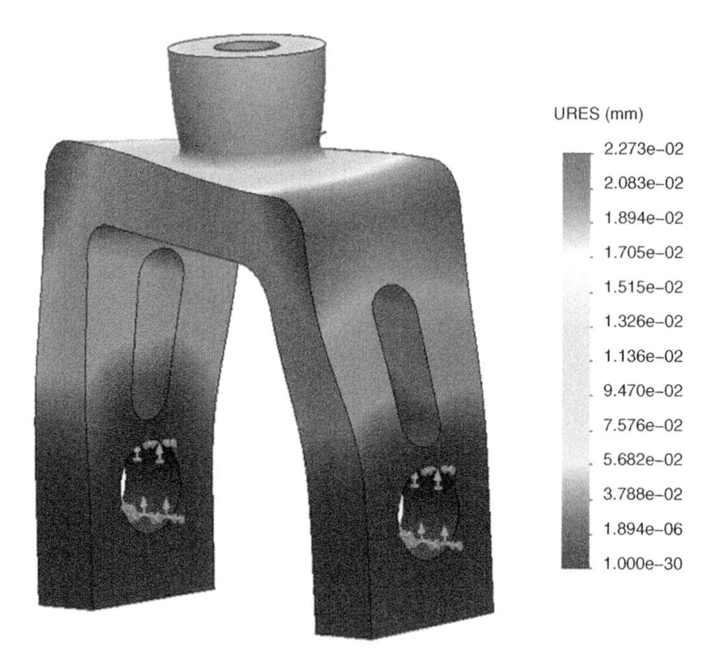

Figure 14.12 Displacement plot for the knuckle problem.

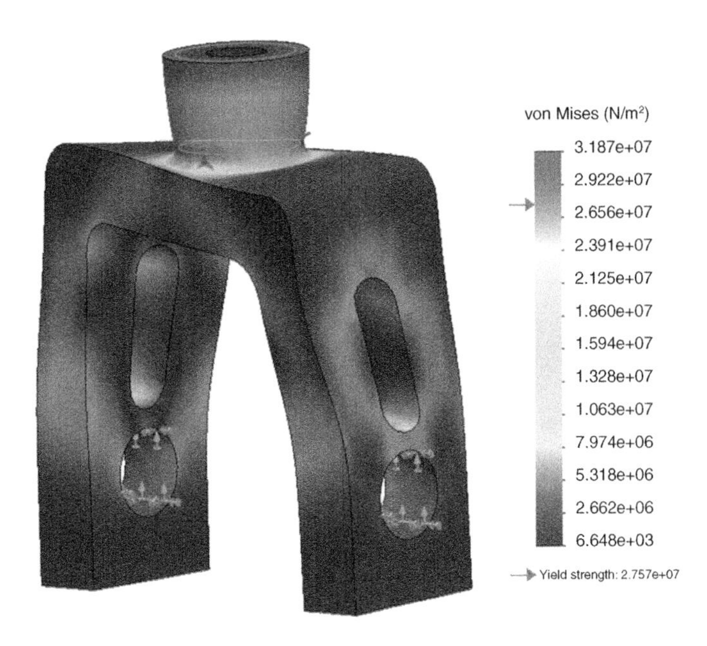

Figure 14.13 von Mises stress plot for the knuckle problem.

The von Mises stress is illustrated in Figure 14.13. The maximum stress is 31 MPa.

In the following exercises, the reader will explore other aspects of FEA, including the impact of mesh refinement.

14.5 EXERCISES

Exercise 14.1 Consider the cantilevered, uniformly loaded beam in Figure 14.14, with length 1 m, width 0.2 m and height 0.05 m, with $E = 2 \times 10^{11}\,\text{N/m}^2$ and $\upsilon = 0.28$. The pressure applied is $1000\,\text{N/m}^2$. Solve the problem using SOLIDWORKS Simulation with default settings, and compare against the analytical solution in Chapter 12.

Exercise 14.2 Consider the double-cantilevered, uniformly loaded beam in Figure 14.15. The properties are as in Exercise 14.1. Solve the problem using SOLIDWORKS Simulation with default settings, and compare against the analytical solution in Chapter 12.

Exercise 14.3 Consider the double-cantilevered beam with a line load, illustrated in Figure 14.16. The dimensions and material are as in Exercise 14.1, and the load

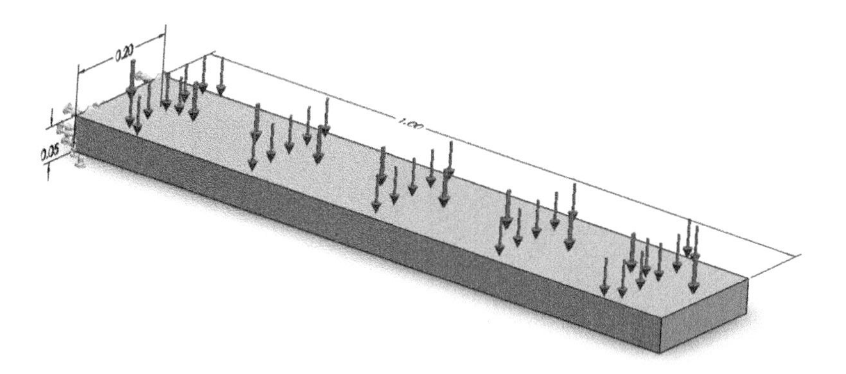

Figure 14.14 A cantilevered, uniformly loaded beam.

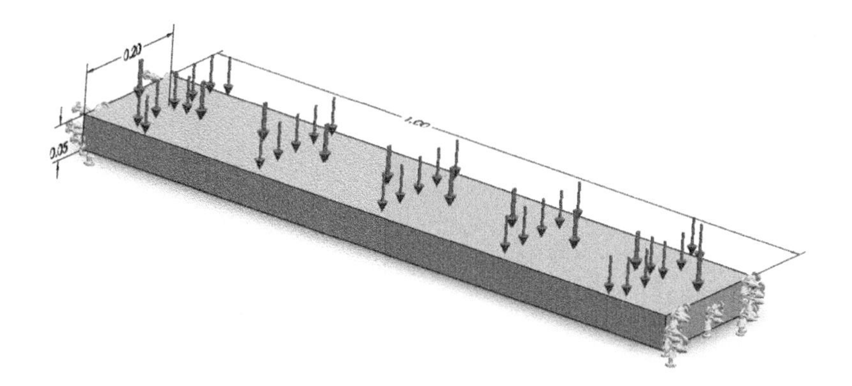

Figure 14.15 A double-cantilevered, uniformly loaded beam.

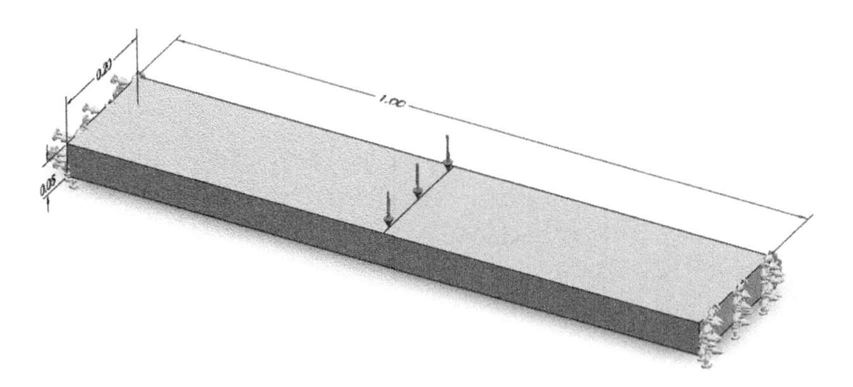

Figure 14.16 A double-cantilevered beam with line load.

applied is 100 N. Solve using SOLIDWORKS Simulation, and compare against the analytical solution in Chapter 12. Note that you have to create an edge on the top face to apply the load.

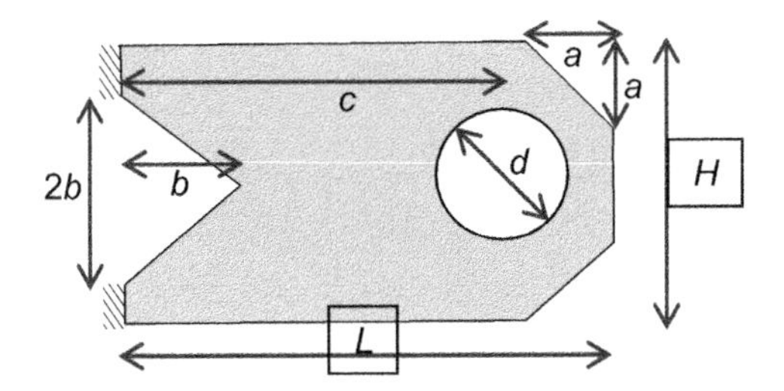

Figure 14.17 A cantilever with four shape parameters.

Exercise 14.4 For the cantilever problem illustrated in Figure 14.2, carry out an FEA for ten different mesh sizes, from 4000 to 20 000 elements. Does the displacement vary significantly? How about the maximum stress?

Exercise 14.5 For the 3D cantilever problem, recall the shape parameters a and b used in 2D. Identify the corresponding dimensions in this model. Carry out an FEA for ten random samples within the range $a \in [0.05, 0.4]$ and $b \in [0.05, 0.4]$, with default mesh parameters. Note down the volume, displacement and maximum stress. Do you observe any trend in displacement versus volume, or stress versus volume?

Exercise 14.6 Construct a 3D cantilever beam with the cross-section illustrated in Figure 14.17, with thickness of 0.1 in, where the parameters are as follows: $L = 2$, $H = 1$, $a = 0.2$, $b = 0.2$, $c = 1.6$, $d = 0.1$. Carry out an FEA for the loading and material as before. Compare the displacement and stress results against those from the original cantilever beam (without the hole).

Exercise 14.7 For the cantilever beam in Figure 14.17, vary the hole location and diameter, keeping other parameters fixed. Carry out an FEA for ten random samples. Note down the volume, displacement and maximum stress. Do you observe any trend in displacement versus volume, or stress versus volume?

Exercise 14.8 For the L-bracket problem illustrated in Figure 14.6, carry out an FEA for ten different mesh sizes, from 4000 to 20 000 elements. Plot the maximum displacement and maximum stress as a function of mesh size.

Exercise 14.9 Modify the L-bracket problem to include the cutouts illustrated in Figure 14.18. With the default mesh size, carry out an FEA. Compare the maximum displacement and maximum stress against those computed previously (without the cutouts). Observe that the cutouts have very little impact on the results.

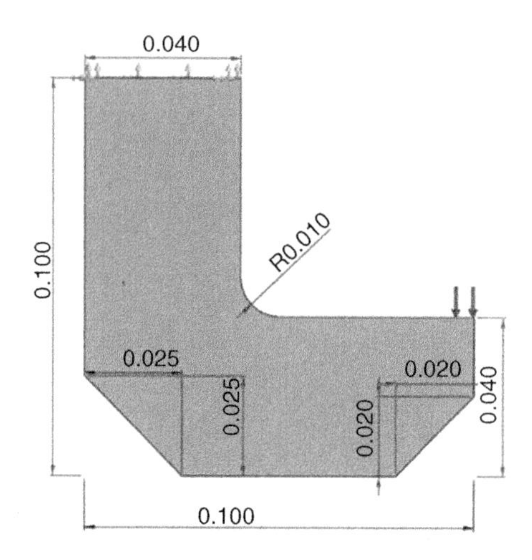

Figure 14.18 L-bracket with cutouts.

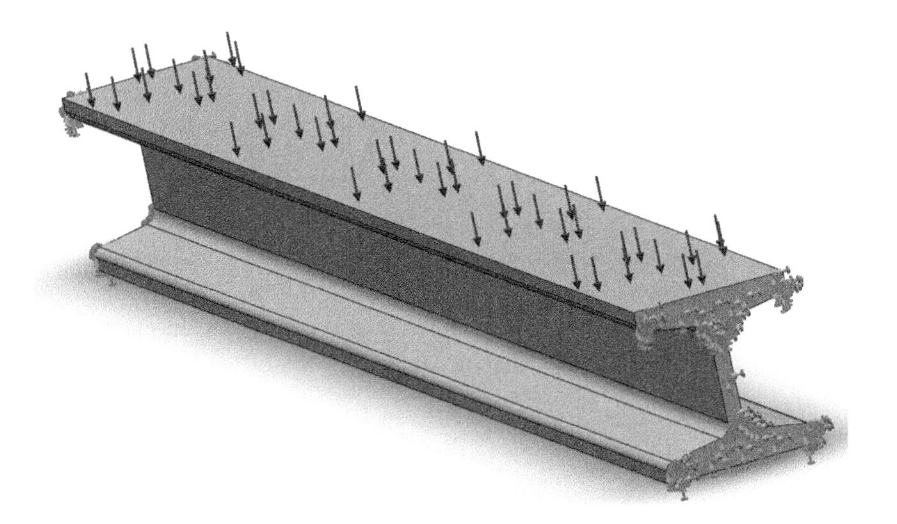

Figure 14.19 I-beam subject to uniform loading.

Exercise 14.10 Consider the I-beam illustrated in Figure 14.19, which is fixed at both ends and subject to a uniform loading of 100 psi. Carry out an FEA on the model (included in the textbook software). Compare the results against idealized beam theory.

15 SOLIDLAB: A SOLIDWORKS–MATLAB Interface

Highlights

1. SOLIDLAB is a collection of MATLAB routines, together with compiled Python code, that serves as an interface between SOLIDWORKS and MATLAB. It was developed at the University of Wisconsin–Madison.
2. Using SOLIDLAB, you can, for example, (1) open a SOLIDWORKS model from within MATLAB, (2) modify part dimensions, (3) query mass properties, (4) carry out finite element analysis, etc.
3. In this chapter, we will demonstrate the use of SOLIDLAB for carrying out an FEA-based parametric study.

15.1 Overview and Installation

SOLIDLAB is a collection of MATLAB routines, together with Windows-compiled Python code, to interact with SOLIDWORKS. It was developed at the Engineering Representations and Simulation Laboratory at the University of Wisconsin–Madison (www.ersl.wisc.edu). Using SOLIDLAB, you can open a SOLIDWORKS model from within MATLAB, modify model dimensions, query mass properties, carry out finite element analysis (FEA), etc.

SOLIDLAB runs on a Windows platform, and has been tested on numerous SOLIDWORKS versions (> 2017) and MATLAB versions (> 2017). SOLIDLAB is included in the software package accompanying this text. When using SOLIDLAB for the first time, you must *install* the software. In Windows, open the SOLIDLAB directory (within the software package accompanying this text), and double-click on `installSOLIDLAB`. You should see a message as illustrated in Figure 15.1.

To *use* SOLIDLAB, a part must be open within SOLIDWORKS. Open the part `solidLabTestPart.sldprt` in SOLIDWORKS (the model is included in the textbook software). Observe in Figure 15.2 that `MainBody` contains a sketch named `MainBodySketch`, and the `SideHoles` feature contains a sketch named `SideHolesSketch`. This will be relevant in Section 15.2.

```
SolidLab installed!

Press any key to continue . . .
```

Figure 15.1 In Windows, double-click `installSOLIDLAB` to install.

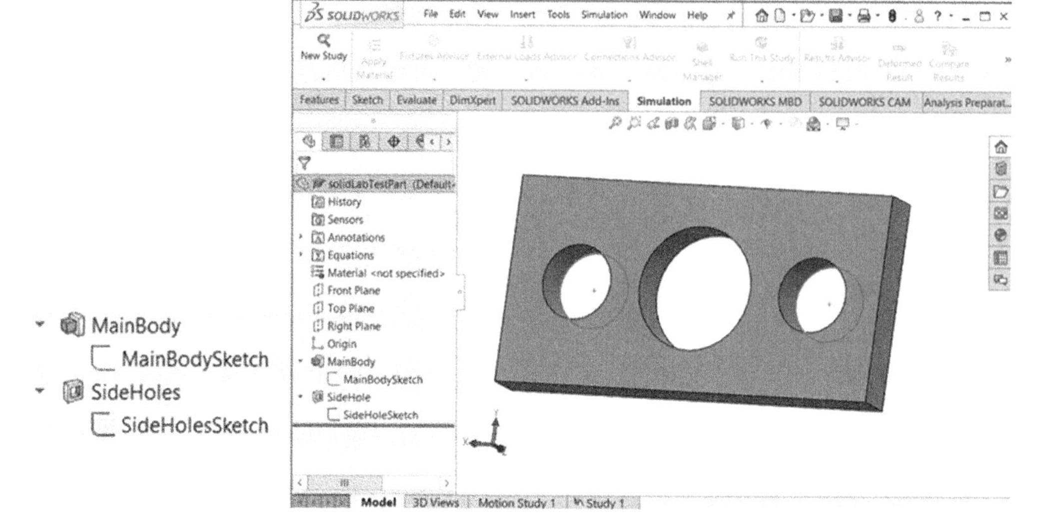

Figure 15.2 `solidLabTestPart.sldprt` opened in SOLIDWORKS.

To test the installation of SOLIDLAB, enter the following command within MATLAB:

```
>> s = solidLab();
```

This creates a SOLIDLAB object. If the connection to SOLIDWORKS is established, you should see a message similar to the following:

```
SOLIDLAB Home: C:\Users\...
SOLIDLAB Version: 2019.02
Connection to SOLIDWORKS established.
```

If the connection cannot be established (typically if SOLIDWORKS is not open), you will see the following message:

```
Connection to SOLIDWORKS could NOT be established.
Make sure SOLIDWORKS is running with a part open.
If you have more than one version of SOLIDWORKS installed, use the
latest version.
```

After successful completion of these tasks, you can proceed to study the examples below.

15.2 Geometric Queries

As the first example, we will obtain the mass properties of the test part via the following command in MATLAB, which returns the mass properties of the active part open in SOLIDWORKS.

```
>> [volume,area,cg,inertia] = s.getMassProperties()
volume =
   5.0230e-05
area =
    0.0153
cg =
    0.0550
    0.0275
    0.0055
inertia =
   1.0e-04 *
    0.1647   -0.0000   -0.0000
   -0.0000    0.5833    0.0000
   -0.0000    0.0000    0.7379
```

Next, we suppress the `SideHoles` feature (see Figure 15.2) via the command

```
>> s.suppress('SideHoles');
```

In the SOLIDWORKS window, you will see that the feature has been suppressed; see Figure 15.3.

Figure 15.3 The `SideHoles` feature suppressed.

Please note the following limitation: *For the current version of SOLIDLAB to work correctly, the feature names in SOLIDWORKS must not contain a space. For example, if the above feature were to be identified as "`Side Holes`", many SOLIDLAB methods might not work as expected.*

One can now re-compute the mass properties:

```
>> [volume,area,cg,inertia] = s.getMassProperties()
volume =
   5.7142e-05
area =
    0.0152
cg =
    0.0550
    0.0275
    0.0055
inertia =
   1.0e-04 *
    0.1671   -0.0000   -0.0000
   -0.0000    0.6704         0
   -0.0000         0    0.8260
```

The reader can compare these values against those computed above. We can revive the `SideHoles` feature via the following command in MATLAB:

```
>> s.unsuppress('SideHoles');
```

This will bring the model back to original state. To save the part, we have

```
>> s.saveSolidworksPart();
```

To illustrate the next SOLIDLAB command, observe in SOLIDWORKS that one can select any of the sketch dimensions, and the name will be displayed in the feature tree menu. For example, in Figure 15.4, the dimension `a@SideHolesSketch@...` is the distance between the center hole and the side holes, with a default value of 35 mm.

We can retrieve all sketch dimensions and their values in SOLIDLAB via the command

```
>> [dimNames,dimValues] = s.getDimensions()
dimNames =
  1×9 cell array
    'D5@MainBodySketch...'     'D3@MainBodySketch...'
'd@SideHolesSketch...'     'D5@SideHolesSketc...'
'D1@MainBody@solid...' 'D4@MainBodySketch...'
'D1@MainBodySketch...'     'D2@MainBodySketch...'
'a@SideHolesSketch...'
dimValues =
   27.5000    55.0000    20.0000    27.5000    11.0000
 33.0000 110.0000    55.0000    35.0000
```

Not all dimensions will be relevant for our study. In particular, consider

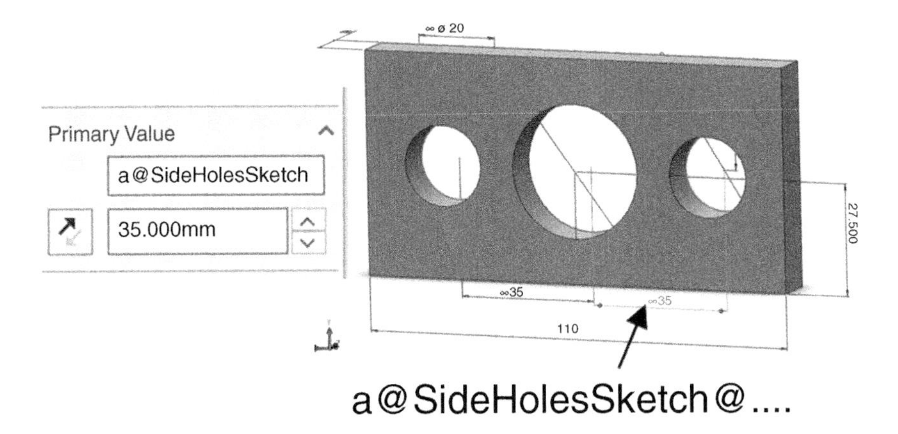

Figure 15.4 Sketch dimensions.

```
>> dimNames{9}
ans =
a@SideHolesSketch@solidLabTestPart.Part
```

and

```
>> dimValues(9)
ans =
   35
```

One can now modify this dimension as follows:

```
>> s.modifyDimensions(dimNames{9},40);
```

This will move the two side holes away from the central hole, as illustrated in Figure 15.5.

Similarly, observe that dimension 3 corresponds to the diameter of the two side holes:

```
>> dimNames{3}
ans =
d@SideHolesSketch@solidLabTestPart.Part
```

Its current value is

```
>> dimValues(3)
ans =
   20
```

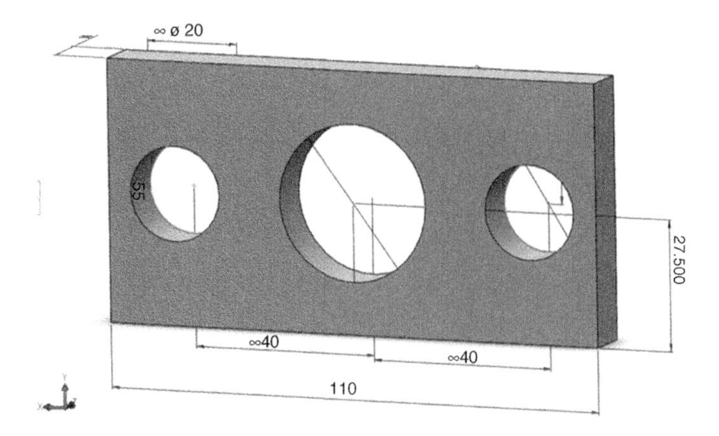

Figure 15.5 Modifying the distance between the central hole and side holes.

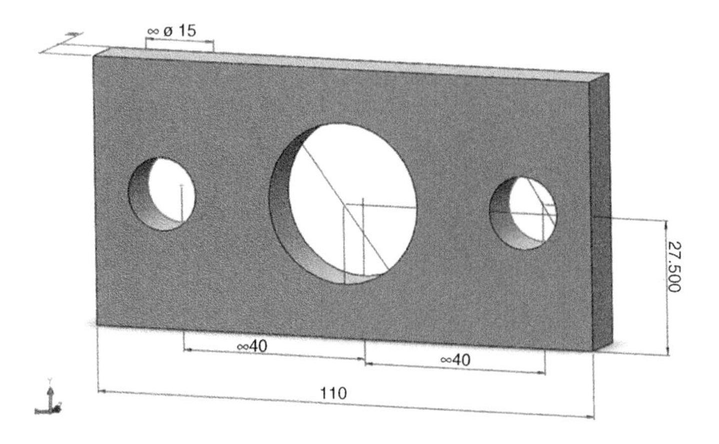

Figure 15.6 Modifying the diameter of the side holes.

We can reduce the diameter to 15 mm via

```
>> s.modifyDimensions(dimNames{3},15);
```

This will reduce the diameter of the two side holes, as illustrated in Figure 15.6.
One can modify multiple dimension values simultaneously:

```
>> s.modifyDimensions(dimNames([3,9]),[20,35]);
```

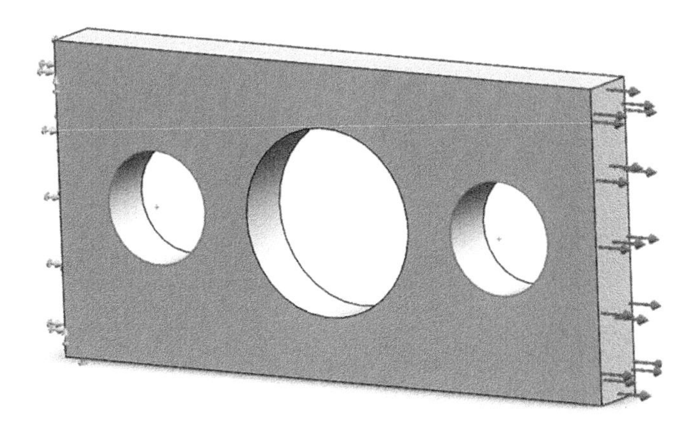

Figure 15.7 A finite element study.

The part returns to its original state, and we can now save the part.

```
>> s.saveSolidworksPart();
```

15.3 FEA Queries

We will now use SOLIDLAB to carry out FEA-related operations. *We will assume that an FEA study has been created within SOLIDWORKS.* For example, for `solidLabTestPart`, a static FEA has been created; see Figure 15.7 (this static FEA is called "Study 1" in SOLIDWORKS).

In order to use FEA-related commands in SOLIDLAB, you must be in the Simulation mode in SOLIDWORKS, i.e., you must carry out an FEA in SOLIDWORKS before proceeding.

We will create a finite element mesh for this study using SOLIDLAB, as follows:

```
>> s.createMesh();
```

The resulting mesh is displayed in Figure 15.8 (you may need to explicitly display the mesh in SOLIDWORKS).

One can execute the study (i.e., run FEA) using SOLIDLAB, as follows (see Figure 15.9).

```
>> s.solveStudy();
```

One can also extract the maximum stress and displacement via

```
>> [maxVonMises,maxDisp] = s.getResults()
Receiving FEA results ...
maxVonMises =

   494803136

maxDisp =

   1.0636e-04
```

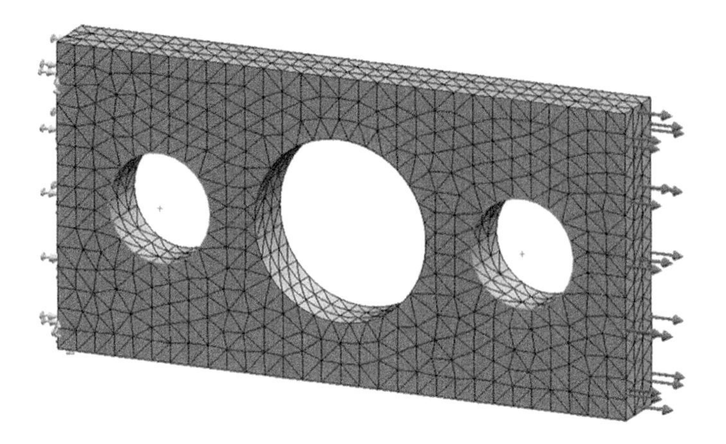

Figure 15.8 Finite element mesh.

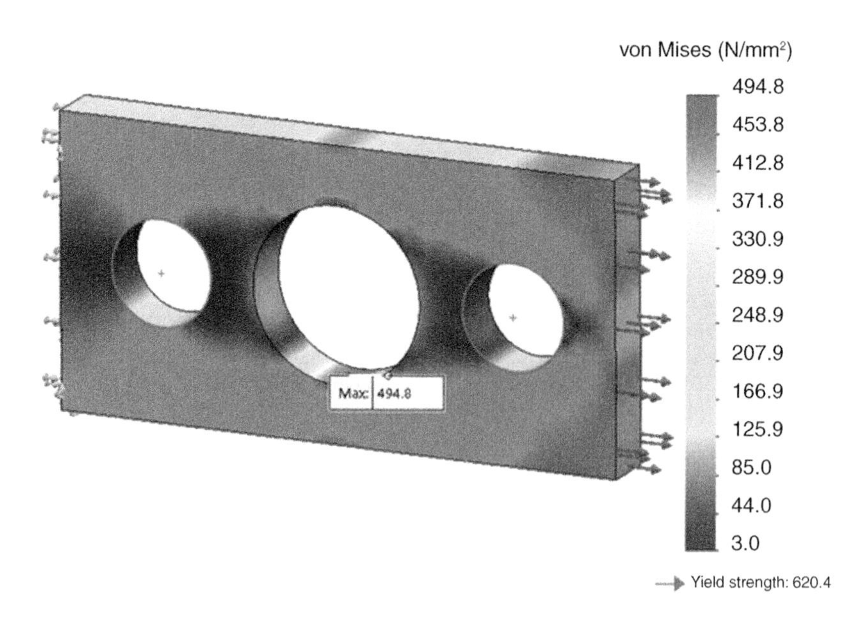

Figure 15.9 Finite element stress plot.

15.4 Displacement Scatter Plot

With these basic tools, we can now carry out a parametric study by varying the shape parameters and studying the FEA results. Specifically, we will (1) vary the diameter of the two holes between 15 mm and 25 mm, and (2) vary the hole offset between 32 mm and 38 mm. The range of dimensions will ensure that the part is always valid. For each set of parameters, we will compute the volume, carry out an FEA and extract the maximum displacement and stress. We will then plot these results.

The MATLAB code `cantilever3DParametricStudy.m` for carrying out a parametric study is given below; the reader is encouraged to compare this code against one developed in Section 13.2 for the parametric study of the 2D cantilever beam. Important differences here are that (1) we are relying on SOLIDWORKS for geometry update and FEA, and (2) we extract the displacement from SOLIDWORKS, as opposed to the compliance (since the latter is challenging to extract in SOLIDWORKS). Before executing the code, ensure that the test part is open in SOLIDWORKS and that you are in the Simulation mode within SOLIDWORKS.

```
%Create solidLab object
s = solidLab();
% get all dimension names and values
[dimNames,dimValues] = s.getDimensions();
nSamples = 200;
Displacement = zeros(nSamples,1);
Stress = zeros(nSamples,1);
Volume = zeros(nSamples,1);
dSamples = 15 + (25-15)*rand(nSamples,1); % diameter
aSamples = 32 + (38-32)*rand(nSamples,1); % distance
for i = 1:nSamples
    i/nSamples % to monitor progress
    d = dSamples(i);
    a = aSamples(i);
    s.modifyDimensions(dimNames([3,9]),[d,a]);
    s.saveSOLIDWORKSPart();
    [volume,area,cg,inertia] = s.getMassProperties();
    s.createMesh();
    s.solveStudy();
    [maxVonMises,maxDisp] = s.getResults();
    Displacement(i) = maxDisp;
    Stress(i) = maxVonMises;
    Volume(i) = volume;
end
% reset to original values
s.modifyDimensions(dimNames([3,9]),[20,35]);
s.saveSOLIDWORKSPart();
plot(Volume, Displacement,'*');
```

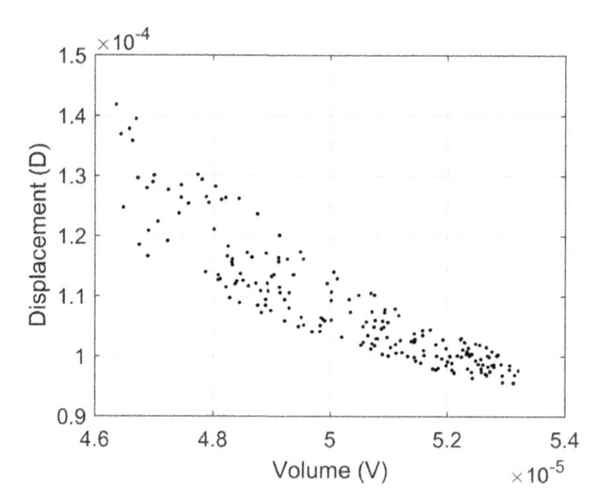

Figure 15.10 Displacement versus volume.

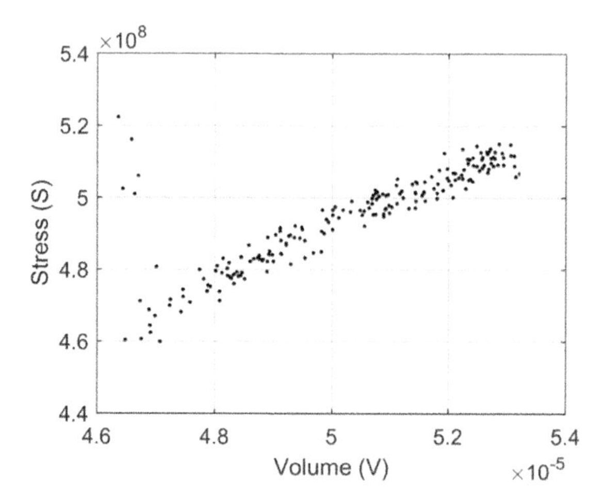

Figure 15.11 Maximum von Mises stress versus volume.

The displacement-versus-volume scatter plot is illustrated in Figure 15.10. One can visualize a barrier Pareto-optimal curve that is monotonic, i.e., the displacement of all Pareto-optimal designs increases with decreasing volume.

15.5 Stress Scatter Plot

A plot of maximum von Mises stress versus volume scatter is presented in Figure 15.11. Observe that, unlike the displacement curve, the stress Pareto-optimal curve does not exhibit a monotonic behavior! (We will explore the

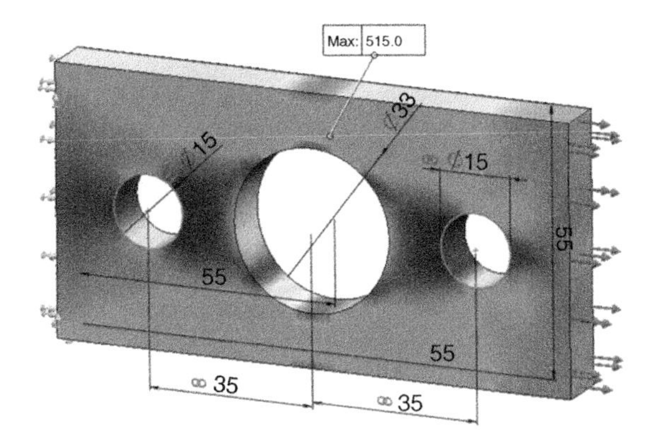

Figure 15.12 Maximum stress for $d = 15\,\text{mm}$ occurs at the larger hole.

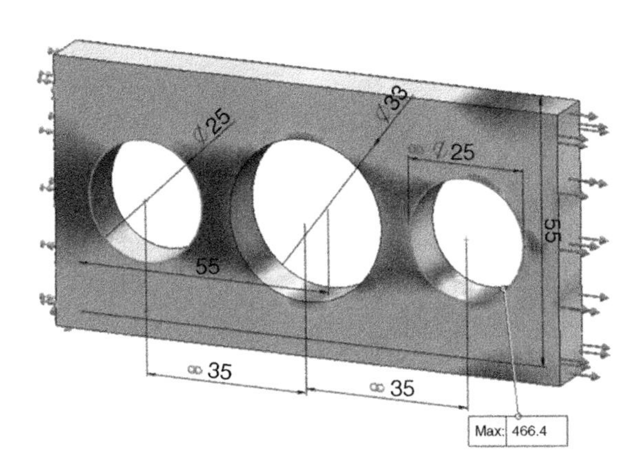

Figure 15.13 Maximum stress for $d = 25\,\text{mm}$ occurs at the smaller hole.

underlying reason below.) This implies that it is possible to reduce volume and reduce maximum stress simultaneously – a win-win situation in engineering. However, the maximum stress plot is not smooth, and this will pose a problem during shape optimization.

Recall that the diameter of the two holes was varied between 15 mm and 25 mm. In particular, the stress plot when the diameter is 15 mm is illustrated in Figure 15.12. The maximum stress is around 515 MPa and it occurs at the circumference of the larger hole.

On the other hand, consider the stress plot when the diameter is 25 mm, illustrated in Figure 15.13. The maximum stress is around 466 MPa, and occurs on the circumference of the smaller hole! This explains the reduction in stress in Figure 15.11. As the hole diameter is increased further, at some point the stress will start increasing again, leading to the non-monotonic behavior.

15.6 Conclusions

In engineering optimization, one must often work with multiple software packages; for example, a package for computer-aided design (CAD) modeling and another for optimization. A script-driven interface between such software packages is often essential. In this chapter, we have demonstrated one such interface, namely SOLIDLAB, which bridges SOLIDWORKS and MATLAB. The interface functions implemented in SOLIDLAB are only a handful (and will be expanded in the future), but they are sufficient for finite-element-driven design optimization.

15.7 EXERCISES

Exercise 15.1 Consider the cross-link illustrated in Figure 15.14 (the SOLIDWORKS model is part of the software accompanying this text). Using SOLIDLAB, find (1) the volume of the part, and (2) the values of the two dimensions, `D3@layout@crosslink.Part` and `D2@layout@crosslink.Part`.

Exercise 15.2 For the cross-link example in Figure 15.14, use SOLIDLAB to vary the value of `D2@layout@crosslink.Part` between 65 mm and 75 mm, in steps of 1 mm, and plot the surface area and volume of the part as a function of this dimension (two separate plots). Fit straight lines $V = aD_2 + b$ and $A = cD_2 + d$.

Exercise 15.3 Continuing with Exercise 15.2, set the dimension of `D2@layout@crosslink.Part` to its original value. Use SOLIDLAB to vary the value of `D3@layout@crosslink.Part` between 10 mm and 20 mm, in steps of 1 mm, and plot the surface area and volume of the part as a function of this dimension (two separate plots). Fit straight lines $V = eD_3 + f$ and $A = cD_3 + h$.

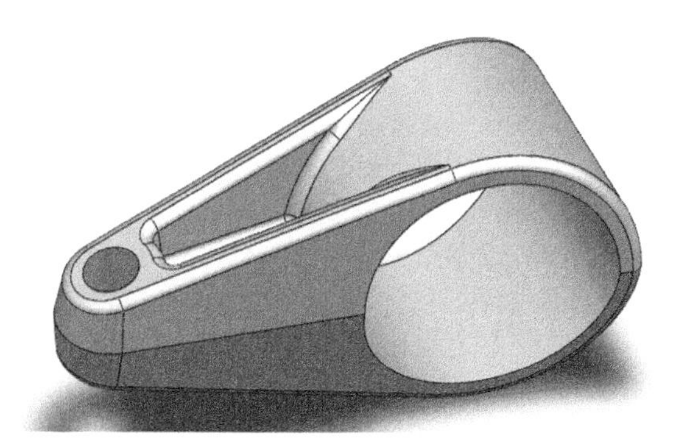

Figure 15.14 Cross-link.

Exercise 15.4 Recall that you can create a mesh in SOLIDLAB via the command

```
s.createMesh();
```

One can also pass an optional argument to control relative mesh size, for example, using the command

```
s.createMesh(0.9);
```

This will create a mesh whose element edge length is approximately 90 percent of the default length, i.e., it will create a denser mesh. Using this command, create a mesh with 0.9 relative size, and carry out an FEA on `solidLabTestPart`. Do the displacement results change significantly? How about the stress results?

Exercise 15.5 One can get access to the mesh via the command

```
[nodes,elems] = s.getMesh();
```

This will return two arrays, where `nodes` is a [`numNodes`, 3] array that specifies coordinates of the mesh, while `elems` is a [`numElems`, 4] array that specifies, for each tetrahedral element, the four nodes that make up that element. Carry out a parametric study by varying the relative mesh size between 0.5 and 1.5, in steps of 0.1, on `solidLabTestPart`. Then, for each mesh size, carry out an FEA. Plot the maximum displacement versus number of nodes. Plot the maximum (von Mises) stress versus number of nodes. What are your observations on the smoothness of the two results?

Exercise 15.6 Recall the 3D cantilever problem discussed in Section 14.2 (see Figure 14.2). Further, recall the shape parameters a and b used in 2D. Identify the corresponding dimensions in SOLIDWORKS for this model. Now vary the two parameters within the range $a \in [0.05, 0.45]$ and $b \in [0.05, 0.45]$. Carry out an FEA for each parameter set (with default mesh parameters). Plot displacement versus volume and stress versus volume, for 200 samples. Compare the stress plot against the one in Figure 15.11; what are your observations?

Exercise 15.7 Recall the knuckle problem discussed in Section 14.4 (see Figure 14.10). The slots are controlled by three critical parameters (length, width and location with respect to the restrained holes). Pick a reasonable range for each of these parameters, that ensures a valid design. Carry out an FEA for each parameter set (with default mesh parameters). Plot displacement versus volume and stress versus volume, for 200 samples.

Exercise 15.8 Recall the L-bracket problem posed in Exercise 14.9 (see Figure 14.18). Further, recall the shape parameters (a and b) used in 2D. Identify the corresponding dimensions in this model. Now vary the two parameters randomly within the range $a \in [0.05, 0.4]$ and $b \in [0.05, 0.4]$. Carry out an FEA for each parameter set (with default mesh parameters). Plot (1) displacement versus volume and (2) maximum von Mises stress versus volume.

Exercise 15.9 Recall the 3D cantilever-with-hole problem posed in Exercise 14.7 (see Figure 14.17). There are now four shape parameters: a, b, c and d. Vary these parameters within a reasonable range. Assume a thickness of 0.01. Carry out an FEA for each parameter set (with default mesh parameters). Plot displacement versus volume and stress versus volume.

16 Shape Optimization Using SOLIDLAB

Highlights

1. In this chapter, we illustrate shape optimization of 3D elasticity problems, using SOLIDWORKS and MATLAB, with SOLIDLAB as the interface.
2. We revisit the critical concepts of shape optimization, and illustrate them using benchmark problems.
3. Unlike in 2D shape optimization, we will be limited to finite-difference-based gradient estimation.

16.1 Overview

In Chapter 14, we considered the 3D finite element analysis (FEA) of the cantilever illustrated in Figure 16.1. In this chapter, we will discuss 3D shape optimization.

Conceptually, this is similar to shape optimization of the 2D cantilever considered in Chapter 13, with shape parameters a and b (see Figure 16.2). The main differences are that (1) we will rely on SOLIDWORKS for FEA, and (2) we will minimize the displacement instead of minimizing compliance, since there is no direct way of extracting compliance from SOLIDWORKS FEA. Further, we will only consider the finite-difference approach to shape optimization, since the mesh-projection method used in 2D is hard to implement using SOLIDWORKS.

16.2 Displacement Minimization

We pose the shape optimization problem as follows:

$$\begin{aligned} &\underset{\{a,b\}}{\text{minimize}} \ \delta_{max} \\ &\text{s.t.} \quad V \le V_0 \\ &\qquad 0.01 \le a \le 0.4 \\ &\qquad 0.01 \le b \le 0.4 \end{aligned} \tag{16.1}$$

The initial values for the parameters are $a_0 = 0.2$ and $b_0 = 0.4$; the initial volume at these parameters is $V_0 = 0.18$ in^3. A load of 100 lb is applied, as discussed in

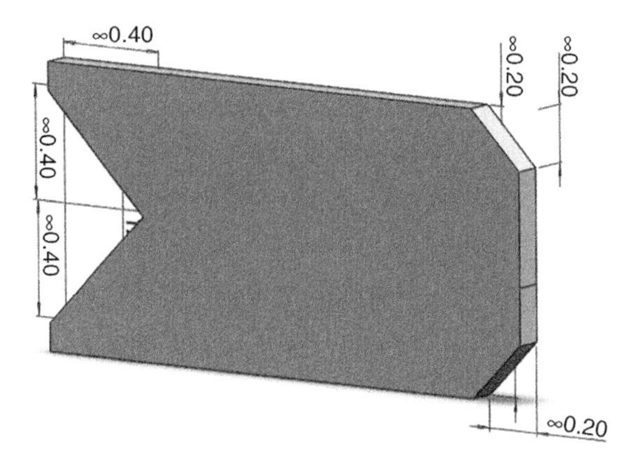

Figure 16.1 The 3D cantilever beam.

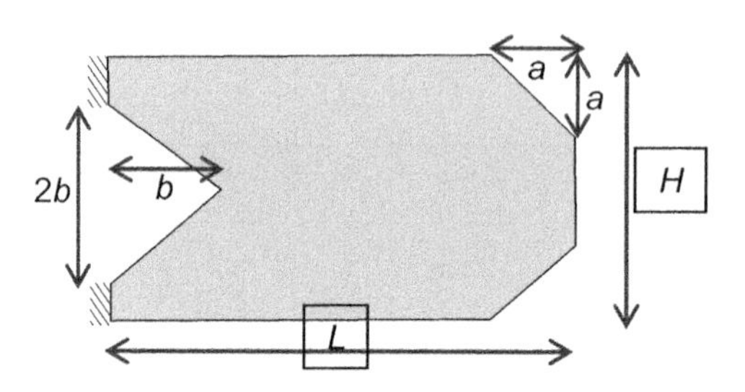

Figure 16.2 Shape parameters that define the geometry.

Chapter 14. Further, to keep the computational cost low, a coarse mesh with about 5000 elements is used in SOLIDWORKS for FEA. The initial maximum displacement $\delta_0 = 1.455 \times 10^{-3}$ in. As before, we introduce the necessary scaling via the non-dimensional variables

$$\begin{aligned} x_1 &= a/a_0 \\ x_2 &= b/b_0 \end{aligned} \tag{16.2}$$

leading to

$$\begin{aligned} &\underset{\{x_1,x_2\}}{\text{minimize}} f = \delta_{\max}/\delta_0 \\ &\text{s.t.} \quad \frac{V}{V_0} - 1 \leq 0 \\ &\qquad x_1^{\min} \leq x_1 \leq x_1^{\max} \\ &\qquad x_2^{\min} \leq x_2 \leq x_2^{\max} \end{aligned} \tag{16.3}$$

The MATLAB class `cantilever3DMinimizeDeflectionFD`, summarized below, implements Equation (16.3) using the finite-difference method for gradient computation. The reader is encouraged to study the implementation, paying careful attention to the following:

- This class specifically targets the 3D cantilever, but can be suitably modified to optimize other shapes.
- Critical functions are `getMaxDisplacementAtParams` and `getVolumeAtParams`. These rely on SOLIDWORKS for computing the objective (displacement) and the constraint (volume). To compute the objective, one must rebuild the geometry and carry out an FEA. However, to compute the constraint (volume) it is sufficient to rebuild the geometry.
- The optimizer uses the finite-difference approach for sensitivity calculations. The default `fmincon` finite-difference step sizes are replaced with user-defined values. The sequential quadratic programming (SQP) algorithm was found to be the most effective, and is used here.
- One of the private variables is the `solidLab` structure; this is initialized once by the constructor and used later by other member functions. This eliminates the overhead of re-creating this structure during optimization.
- We make use of the optimization parameter `PlotFcns` to continuously display the relative objective value.

```
classdef cantilever3DMinimizeDeflectionFD
    properties(GetAccess = 'public', SetAccess =
'private')
        mySOLIDLAB;
        myInitialShapeParams; % used for scaling
        ...
    end
    methods
        function obj =
cantilever3DMinimizeDeflectionFD(params0)
            obj.myInitialShapeParams = params0;
            obj.mySOLIDLAB = solidLab(); %initialization
            ...
        end
        function maxDisp =
getMaxDisplacementAtParams(obj,params)
            obj.mySOLIDLAB.modifyDimensions({
'a@Sketch1@Cantilever.Part',
'b@Sketch1@Cantilever.Part'},params);
            obj.mySOLIDLAB.createMesh();
            obj.mySOLIDLAB.solveStudy();
            [~,maxDisp] = obj.mySOLIDLAB.getResults();
        end
        function volume = getVolumeAtParams(obj,params)
            obj.mySOLIDLAB.modifyDimensions({
'a@Sketch1@Cantilever.Part',
```

```
'b@Sketch1@Cantilever.Part'},params);
            volume = obj.mySOLIDLAB.getMassProperties();
        end
        ...
        function [results] = minimizeMaxDisplacement(obj)
            ...
            [xMin,fMin,flag,output] =
fmincon(@obj.displacementObjective,x0,[],[],[],[],LB,UB,@o
bj.volumeConstraint,opt);
        ...
        end
    end
end
```

We now test this class with a relative objective tolerance of 0.025, relative parameter tolerance of 0.025 and finite-difference step size of 0.01.

```
params0 = [0.2 0.4];% initial shape parameters, a & b
shapeOpt = cantilever3DMinimizeDeflectionFD(params0);
shapeOpt = shapeOpt.setParamsUpperBound([0.4,0.4]);
shapeOpt = shapeOpt.setParamsLowerBound([0.01,0.01]);
shapeOpt = shapeOpt.setObjectiveTolerance(0.025);
shapeOpt = shapeOpt.setParamTolerance(0.025);
shapeOpt = shapeOpt.setFiniteDifferenceStepSize(0.01);
results = shapeOpt.minimizeMaxDisplacement();
```

The optimizer converges to the design in Figure 16.3, with $a = 0.39$ and $b = 0.23$. The final volume is the same as the initial volume (because of the constraint), while the maximum displacement is reduced by 10 percent.

```
Initial Params:
    0.2000    0.4000
Final Params:
    0.3855    0.2286
   1.0e-05 *
    0.2950    0.2948
Initial/Final Displacement:
   1.0e-04 *
    0.4087    0.3674
#FEAs:
    54
```

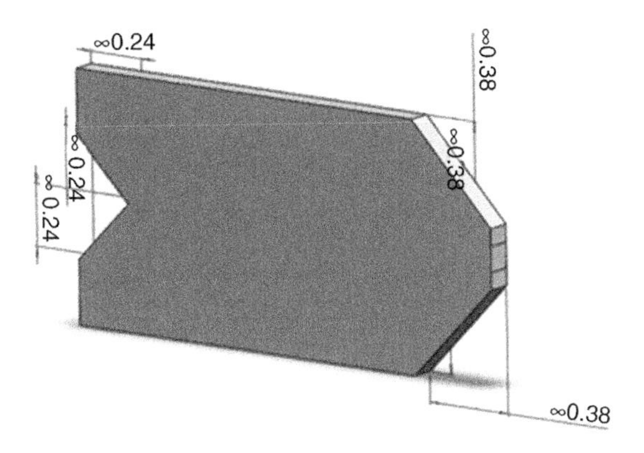

Figure 16.3 The optimized 3D cantilever beam.

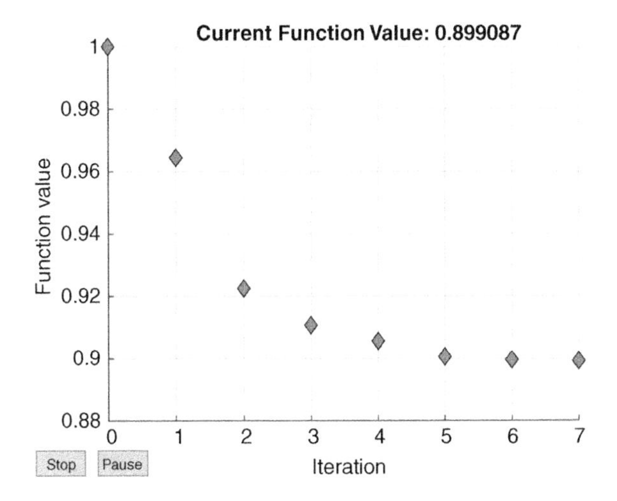

Figure 16.4 Convergence for compliance minimization: relative objective versus iteration.

Recall that, in 2D, the optimal parameters were $a = 0.36$ and $b = 0.26$, and an 11 percent reduction in compliance was achieved.

Figure 16.4 illustrates the convergence of the relative objective.

16.3 Displacement Minimization Using Global Search

Recall that we previously used global search techniques to address minimization of objective functions in the presence of noise. We now demonstrate the use of the `GlobalSearch` method to solve the displacement minimization problem.

The MATLAB class `cantilever3DMinimizeDeflectionGS`, summarized below, implements Equation (16.3) using the global search method, which in turn uses `fmincon` with finite difference for local optimization. The critical part of the code is summarized below:

```
classdef cantilever3DMinimizeDeflectionGS
    ...
        function [results] = minimizeMaxDisplacement(obj)
            ...
            opt = optimoptions('fmincon',...
            'Algorithm','SQP',...
            'TolX',obj.myParamTolerance,...
            'TolFun',obj.myObjectiveTolerance,...
            'PlotFcns',@optimplotfval,...
'FiniteDifferenceStepSize',obj.myFiniteDifferenceStepSize)
;
            problem = createOptimProblem
('fmincon','objective', @obj.displacementObjective, ...
                'x0', x0, 'lb', LB,'ub', UB,
'nonlcon',@obj.volumeConstraint,'options',opt);

            GS = GlobalSearch('MaxTime',600);
        [xMin,fMin,flag,output] = run(GS,problem);
        ...
        end
    end
end
```

The results are summarized below.

```
Initial Params:
    0.2000    0.4000
Final Params:
    0.3855    0.2286
Initial/Final Volume:
   1.0e-05 *
    0.2950    0.2948
Initial/Final Displacement:
   1.0e-04 *
    0.4087    0.3674
#FEAs:
    54
```

16.4 Stress Minimization

We will next consider stress minimization using the design problem illustrated in Figure 16.5. As discussed in Section 15.5, the two design parameters are (1) the distance between the side holes and the center hole and (2) the diameters of the two side holes.

Recall that the maximum von Mises stress function is noisy (see Section 15.5). We therefore use the zeroth-order method `fminsearchcon`, discussed in Section 7.6, to solve the following shape optimization problem:

$$\begin{aligned} &\underset{\{a,d\}}{\text{minimize}} \ \sigma_{\max} \\ &\text{s.t.} \qquad V \leq V_0 \\ &\qquad\qquad 33 \leq a \leq 37 \\ &\qquad\qquad 15 \leq d \leq 30 \end{aligned} \tag{16.4}$$

The initial values for the parameters are $a_0 = 35$ and $d_0 = 20$; the initial volume at these parameters is $V_0 = 50\,230\ \text{mm}^3$. The initial maximum von Mises stress is 494 MPa. As before, we introduce the necessary scaling via the non-dimensional variables

$$\begin{aligned} x_1 &= a/a_0 \\ x_2 &= b/b_0 \end{aligned} \tag{16.5}$$

leading to

$$\begin{aligned} &\underset{\{x_1,x_2\}}{\text{minimize}} \ f = \sigma_{\max}/\sigma_0 \\ &\text{s.t.} \qquad \frac{V}{V_0} - 1 \leq 0 \\ &\qquad\qquad x_1^{\min} \leq x_1 \leq x_1^{\max} \\ &\qquad\qquad x_2^{\min} \leq x_2 \leq x_2^{\max} \end{aligned} \tag{16.6}$$

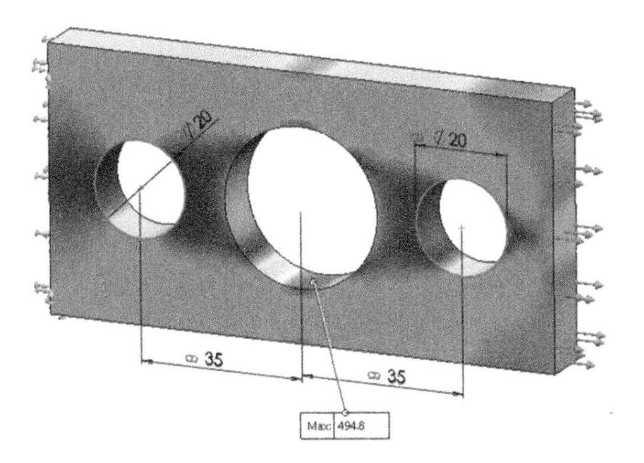

Figure 16.5 The three-hole optimization problem: initial design with $a = 35$, $d = 20$ and maximum stress of 494 MPa.

The MATLAB class included below solves this problem using `fminsearchcon`. The reader is encouraged to study this implementation, paying careful attention to the following:

- This class specifically targets the test part problem but can be suitably modified to optimize other shapes.
- The critical functions are `getMaxStressAtParams` and `getVolumeAtParams`. These rely on SOLIDWORKS for computing the objective (maximum von Mises stress) and the constraint (volume). To compute the objective, one must rebuild the geometry and carry out an FEA. However, to compute the constraint, it is sufficient to rebuild the geometry.
- The private variable `solidLab` is initialized once by the constructor and used later by other member functions.

```
classdef solidLabTestPart3DMinimizeStress
    properties(GetAccess = 'public', SetAccess =
'private')
        mySOLIDLAB;
        myInitialShapeParams; % used for scaling
        ...
    end
    methods
        function obj =
solidLabTestPart3DMinimizeStress(params0)
            obj.myInitialShapeParams = params0;
            obj.mySOLIDLAB = solidLab(); % initialization
...
         end
         function maxStress =
getMaxStressAtParams(obj,params)
obj.mySOLIDLAB.modifyDimensions({'a@SideHolesSketch@solidL
abTestPart.Part',...
'd@SideHolesSketch@solidLabTestPart.Part'},params);
            obj.mySOLIDLAB.createMesh();
            obj.mySOLIDLAB.solveStudy();
            [maxStress] = obj.mySOLIDLAB.getResults();
        end
        function volume = getVolumeAtParams(obj,params)
```

```
obj.mySOLIDLAB.modifyDimensions({'a@Sketch1@Cantilever.Par
t',
'b@Sketch1@Cantilever.Part'},params);
            volume = obj.mySOLIDLAB.getMassProperties();
        end
        ...
        function [results] = minimizeMaxStress(obj)
            ....
            [xMin,~,flag,output] = fminsearchcon(@(x)
obj.stressObjective(x),x0,LB,UB,[],[],@(x)
obj.volumeConstraint(x),opt);
            ...
        end
    end
```

We now use this class with a relative objective tolerance of 0.05 and a relative parameter tolerance of 0.25, and a finite-difference step size of 0.01.

```
params0 = [0.2 0.4];% initial shape parameters, a & b
params0 = [35 20];% initial shape parameters, a & d
shapeOpt = solidLabTestPart3DMinimizeStress(params0);
shapeOpt = shapeOpt.setParamsUpperBound([37,30]);
shapeOpt = shapeOpt.setParamsLowerBound([33,15]);
shapeOpt = shapeOpt.setParamTolerance(0.05); % relative
shapeOpt = shapeOpt.setObjectiveTolerance(0.05); %
relative
results = shapeOpt.minimizeMaxStress()
```

The optimizer converges to the design in Figure 16.6 with $a = 33.46$ and $d = 25.09$.

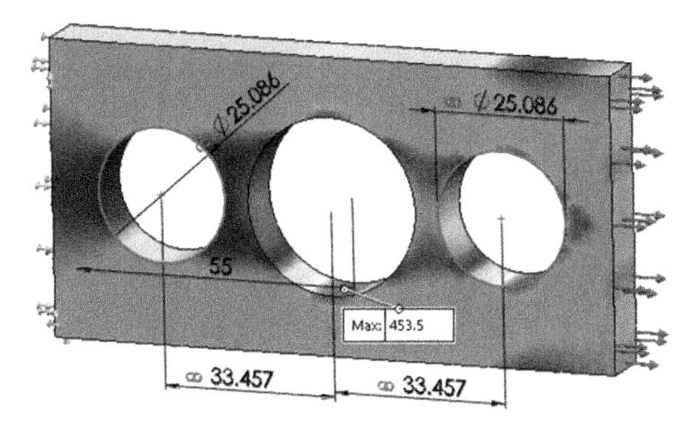

Figure 16.6 The three-hole optimization problem: final design with $a = 33.46$, $d = 25.09$ and maximum stress of 453MPa.

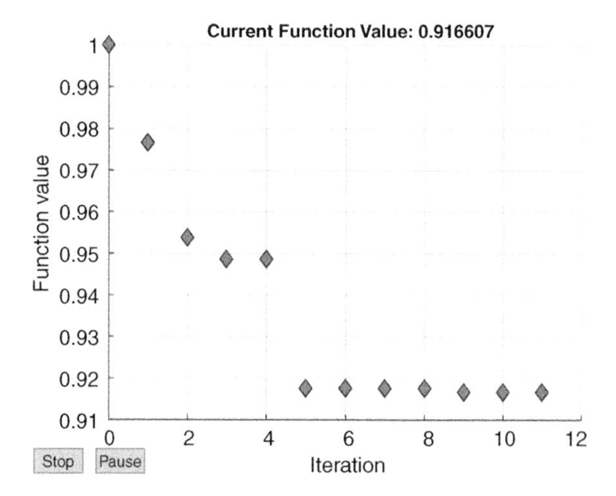

Figure 16.7 Convergence history for stress minimization: relative objective versus iteration number.

The final volume is the same as the initial volume (because of the constraint), while the maximum stress is reduced by about 9 percent.

Figure 16.7 illustrates the convergence of the relative objective.

The reader is encouraged to use the global search technique to solve the stress minimization problem.

16.5 Conclusions

The main conclusions are the following:

- The general principles of optimization (for example, numerical scaling) applicable to simple truss optimization are equally application to the more demanding 3D shape optimization.
- However, 3D shape optimization poses additional challenges ([37]):
  - It is computationally more demanding.
  - It often involves the coupling of two more software packages.
  - Special precautions need to be taken to avoid failures (example: geometry reconstruction failure).
  - The presence of noise can hinder convergence.

16.6 EXERCISES

Exercise 16.1 Solve the displacement minimization problem in Equation (16.3) using the zeroth-order method `fminsearchcon`. Compare your results against those obtained using `fmincon` and global search.

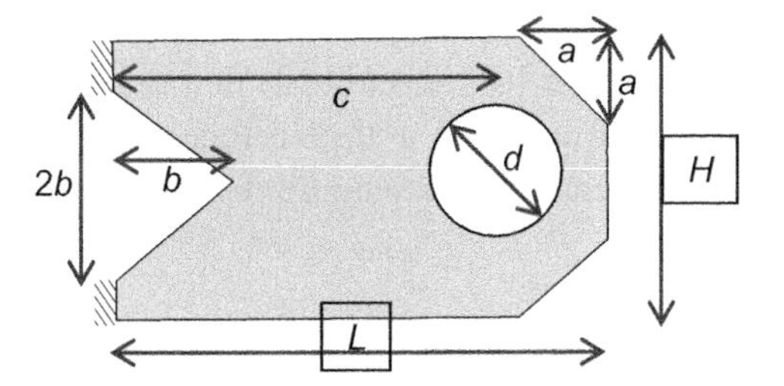

Figure 16.8 Cantilever problem.

Exercise 16.2 Solve the stress minimization problem using `fmincon` (with finite difference) and the global search method. Compare your results against those obtained using `fminsearchcon`.

Exercise 16.3 Modify the displacement minimization problem in Equation (16.3) into a volume-minimization problem, subject to displacement constraint. Solve using the three methods (`fmincon`, global search and `fminsearchcon`), and compare the results.

Exercise 16.4 Consider the 3D cantilever-with-holes, whose 2D sketch is illustrated in Figure 16.8 (assume a thickness of 0.01). There are now four shape parameters, a, b, c and d. Pose and solve the displacement minimization problem, subject to volume constraint, using three methods (`fmincon`, global search and `fminsearchcon`). Assume the following initial values for the parameters: $a = 0.2$, $b = 0.4$, $c = 1.2$, $d = 0.2$. Note that additional constraints are needed to avoid geometry reconstruction issues. Are the results consistent with the displacement-volume scatter plot (see Exercise 15.9)?

Exercise 16.5 Modify the problem in Exercise 16.4 to a volume-minimization problem, subject to displacement constraint. Solve using the three methods. Are the results consistent with the displacement-volume scatter plot (see Exercise 15.9)?

Exercise 16.6 Consider the knuckle problem discussed in Exercise 15.7, and illustrated in Figure 16.9. The slots are controlled by three critical parameters (length, width and location with respect to the restrained holes). Pose and solve a displacement minimization problem, subject to volume constraints, using the three methods. Pick a reasonable initial value for each of these parameters.

Exercise 16.7 Consider the L-bracket illustrated in Figure 16.10. With the shape parameters a and b, pose and solve a displacement minimization problem, subject to volume constraint, using the three methods. Pick a reasonable initial value for the parameters.

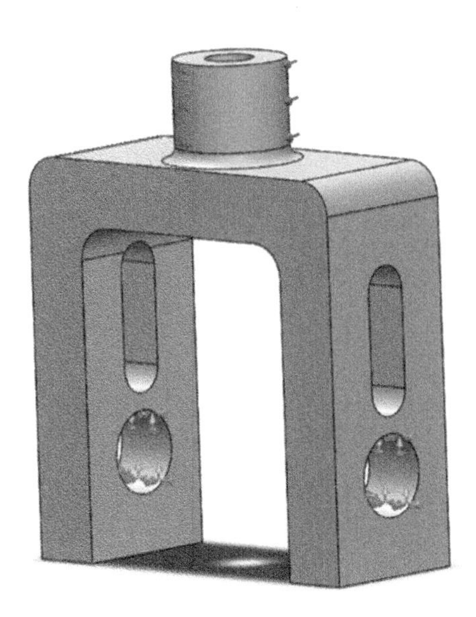

Figure 16.9 Knuckle problem.

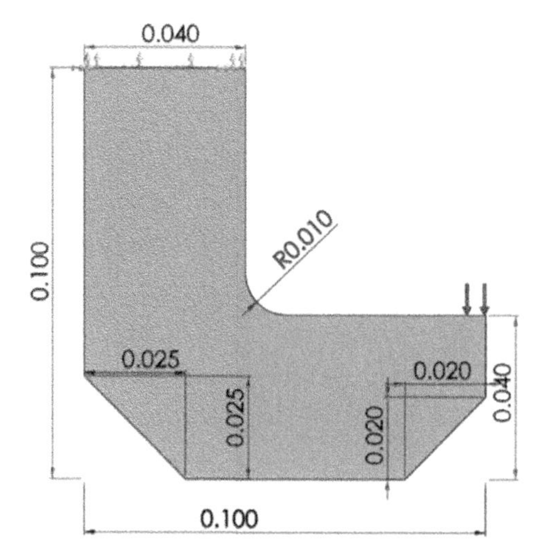

Figure 16.10 L-bracket problem.

Appendix

This appendix covers some mathematical aspects of optimization that were not covered in the main text.

A.1 Taylor Series

For infinitely differentiable functions, the notion of Taylor series expansion plays an important role in optimization. In 1D, recall the formal Taylor series expansion

$$f(x+\Delta x) = f(x) + \frac{1}{1!}\frac{df}{dx}\Big|_x \Delta x + \frac{1}{2!}\frac{d^2 f}{dx^2}\Big|_x (\Delta x)^2 + o(\Delta x)^3 \tag{A.1}$$

To extend this to 2D, the concept of derivative is generalized to the gradient of a 2D function $f(x, y)$, as follows:

$$\nabla f(\mathbf{x}) \equiv \begin{Bmatrix} \dfrac{\partial f}{\partial x} \\ \dfrac{\partial f}{\partial y} \end{Bmatrix} \tag{A.2}$$

Similarly, the second derivative is generalized to the concept of a Hessian:

$$\mathbf{H}_f(\mathbf{x}) = \begin{bmatrix} \dfrac{\partial^2 f}{\partial x^2} & \dfrac{\partial^2 f}{\partial x \partial y} \\ \dfrac{\partial^2 f}{\partial x \partial y} & \dfrac{\partial^2 f}{\partial y^2} \end{bmatrix} \tag{A.3}$$

The Taylor series expansion of a 2D function is defined as follows:

$$f(\mathbf{x}+\Delta\mathbf{x}) = f(\mathbf{x}) + \frac{1}{1!}\Delta\mathbf{x}^T \nabla f(\mathbf{x}) + \frac{1}{2!}\Delta\mathbf{x}^T \mathbf{H}_f(\mathbf{x}) \Delta\mathbf{x} + o(\|\Delta\mathbf{x}\|)^3 \tag{A.4}$$

Example A.1 Find the Taylor series expansion of $f(x, y) = x^2 - 2xy + 4y^2 - 2x - 4y - 5$ about (a) (0, 0), and (b) (2, 1).

Solution: Observe that the gradient of the function is

$$\nabla f = \begin{Bmatrix} 2x - 2y - 2 \\ -2x + 8y - 4 \end{Bmatrix}$$

and the Hessian is

$$\mathbf{H}_f(\mathbf{x}) = \begin{bmatrix} 2 & -2 \\ -2 & 8 \end{bmatrix}$$

Observe that the Hessian happens to be a constant. Furthermore, third- and higher-order derivatives vanish for this function. Thus, its Taylor series about any point consists of *exactly* three terms:

$$f(x_0 + \Delta x,\, y_0 + \Delta y) = f(x_0, y_0) + \{\Delta x\, \Delta y\}\nabla f(x_0, 0) + \frac{1}{2}\{\Delta x\, \Delta y\}\mathbf{H}_f(x_0, y_0)\begin{Bmatrix} \Delta x \\ \Delta y \end{Bmatrix}$$

The Taylor series about the origin is given by

$$f(0 + \Delta x,\, 0 + \Delta y) = f(0,0) + \{\Delta x \quad \Delta y\}\nabla f(0,0) + \frac{1}{2}\{\Delta x \quad \Delta y\}\mathbf{H}_f(0,0)\begin{Bmatrix} \Delta x \\ \Delta y \end{Bmatrix}$$

i.e.,

$$f(\Delta x, \Delta y) = -5 + \{\Delta x \quad \Delta y\}\begin{Bmatrix} -2 \\ -4 \end{Bmatrix} + \frac{1}{2}\{\Delta x \quad \Delta y\}\begin{bmatrix} 2 & -2 \\ -2 & 8 \end{bmatrix}\begin{Bmatrix} \Delta x \\ \Delta y \end{Bmatrix}$$

The Taylor series of the same function about the point (2, 1) is given by

$$f(2 + \Delta x,\, 1 + \Delta y) = -9 + \{\Delta x \quad \Delta y\}\begin{Bmatrix} 0 \\ 0 \end{Bmatrix} + \frac{1}{2}\{\Delta x \quad \Delta y\}\begin{bmatrix} 2 & -2 \\ -2 & 8 \end{bmatrix}\begin{Bmatrix} \Delta x \\ \Delta y \end{Bmatrix}$$

i.e.,

$$f(2 + \Delta x,\, 1 + \Delta y) = -9 + \frac{1}{2}\{\Delta x \quad \Delta y\}\begin{bmatrix} 2 & -2 \\ -2 & 8 \end{bmatrix}\begin{Bmatrix} \Delta x \\ \Delta y \end{Bmatrix}$$

In higher dimensions, the Taylor series of the function $f(x)$ is given by

$$f(\mathbf{x} + \Delta\mathbf{x}) = f(\mathbf{x}) + \Delta\mathbf{x}^T\nabla f(\mathbf{x}) + \frac{1}{2}\Delta\mathbf{x}^T\mathbf{H}_f(\mathbf{x})\Delta\mathbf{x} + o(\|\Delta x\|)^3 \tag{A.5}$$

where the gradient is defined as

$$\nabla f|_{\mathbf{x}^*} = \begin{Bmatrix} \dfrac{\partial f}{\partial x_1} \\ \dfrac{\partial f}{\partial x_2} \\ \vdots \\ \dfrac{\partial f}{\partial x_N} \end{Bmatrix}_{\mathbf{x}^*} \tag{A.6}$$

and the Hessian is defined as the symmetrix matrix

$$\mathbf{H}_f(\mathbf{x}) = \begin{bmatrix} \dfrac{\partial^2 f}{\partial x_1^2} & \dfrac{\partial^2 f}{\partial x_1 \partial x_2} & \cdots & \dfrac{\partial^2 f}{\partial x_1 \partial x_N} \\ & \dfrac{\partial^2 f}{\partial x_2^2} & \cdots & \dfrac{\partial^2 f}{\partial x_2 \partial x_2} \\ & & \cdots & \\ Sym & & & \dfrac{\partial^2 f}{\partial x_N^2} \end{bmatrix} \tag{A.7}$$

A.2 Optimality Theorems

With the concepts of Taylor series expansion in mind, we can establish that the gradient of a multi-dimensional function vanishes at all local minima. This is the necessary condition for a local minimum.

Theorem A.1 If $\mathbf{x}^*$ *is a local minimum of a differentiable function* f(x), *then the gradient of the function vanishes at that point, i.e.,*

$$\nabla f|_{\mathbf{x}^*} = \left\{ \begin{array}{c} \frac{\partial f}{\partial x_1} \\ \frac{\partial f}{\partial x_2} \\ \vdots \\ \frac{\partial f}{\partial x_N} \end{array} \right\}_{\mathbf{x}^*} = \left\{ \begin{array}{c} 0 \\ 0 \\ \vdots \\ 0 \end{array} \right\} \tag{A.8}$$

Proof: Let $\mathbf{x}^* = \{x_1^* \quad x_2^* \quad \ldots \quad x_N^*\}^T$ be a local minimum; the Taylor series about $\mathbf{x}^*$ is given by

$$f(\mathbf{x}^* + \Delta\mathbf{x}) = f(\mathbf{x}^*) + \Delta\mathbf{x}^T \nabla f|_{\mathbf{x}^*} + o(||\Delta\mathbf{x}||)^2 \tag{A.9}$$

where $\Delta\mathbf{x} = \{\Delta x_1 \ \Delta x_2 \ \ldots \ \Delta x_N\}^T$ is a small increment, and the gradient is defined in Equation (A.6). By definition, $f(\mathbf{x}^*)$ must be less than both $f(\mathbf{x}^* + \Delta\mathbf{x})$ and $f(\mathbf{x}^* - \Delta\mathbf{x})$ for sufficiently small $\Delta\mathbf{x}$, i.e.,

$$f(\mathbf{x}^*) \le f(\mathbf{x}^* \pm \Delta\mathbf{x}) = f(\mathbf{x}^*) \pm \Delta\mathbf{x}^T \nabla f(\mathbf{x}^*) + o(||\Delta\mathbf{x}||^2)$$
$$\Rightarrow \quad 0 \le \pm \Delta\mathbf{x}^T \nabla f(\mathbf{x}^*) + o(||\Delta\mathbf{x}||^2)$$

Since this must be true for all $\Delta\mathbf{x}$, let $\Delta\mathbf{x} = \{\Delta x_1 \ 0 \ \ldots \ 0\}^T$, i.e.,

$$0 \le \pm \Delta x_1 \frac{\partial f}{\partial x_1}|_{\mathbf{x}^*} + o(\Delta x_1{}^2)$$

In the limit $\Delta x_1 \to 0$ we have

$$0 \le \pm \frac{\partial f}{\partial x_1}|_{\mathbf{x}^*} \Rightarrow \quad 0 \le \frac{\partial f}{\partial x_1}|_{\mathbf{x}^*} \le 0$$

i.e.,

$$\frac{\partial f}{\partial x_1}|_{\mathbf{x}^*} = 0$$

One can repeat this argument by choosing $\Delta\mathbf{x} = \{0 \ \ \Delta x_2 \ \ldots \ 0\}^T$, resulting in

$$\frac{\partial f}{\partial x_2}|_{\mathbf{x}^*} = 0$$

Thus, all components of the gradient must vanish at a local minimum, i.e.,

$$\left\{\begin{array}{c}\frac{\partial f}{\partial x_1}\\ \frac{\partial f}{\partial x_2}\\ \vdots\\ \frac{\partial f}{\partial x_N}\end{array}\right\}_{\mathbf{x}^*} = \left\{\begin{array}{c}0\\0\\ \vdots\\ 0\end{array}\right\} \tag{A.10}$$

i.e., $\nabla f(\mathbf{x}^*) = 0$.

Theorem A.2 *Let $\mathbf{x}^*$ be a stationary point of $f(\mathbf{x})$. Further, let $f(\mathbf{x})$ be twice differentiable at $\mathbf{x}^*$, and let $\mathbf{H}_f(\mathbf{x}^*)$ be the Hessian matrix of the function at that point, defined per Equation (A.7). Then:*

(a) If all the eigenvalues of $\mathbf{H}_f(\mathbf{x}^*)$ are positive, then $\mathbf{x}^*$ is a local minimum.
(b) If all the eigenvalues of $\mathbf{H}_f(\mathbf{x}^*)$ are negative, then $\mathbf{x}^*$ is a local maximum.
(c) If some the eigenvalues of $\mathbf{H}_f(\mathbf{x}^*)$ are positive while others are negative, then $\mathbf{x}^*$ is a saddle point of $f(\mathbf{x})$, i.e., neither a local minimum nor a local maximum.
(d) If some of the eigenvalues of $\mathbf{H}_f(\mathbf{x}^*)$ are zero (while the rest of the eigenvalues are all of the same sign), then the test is inconclusive.

Proof: Recall the Taylor series expansion about a stationary point:

$$f(x^* \pm \Delta x) = f(x^*) \pm \left.\frac{df}{dx}\right|_{x^*} \Delta x + \frac{1}{2}\left.\frac{d^2 f}{dx^2}\right|_{x^*} \Delta x^2 \pm o(\Delta x)^3 \tag{A.11}$$

At higher order, since $\mathbf{x}^*$ is a stationary point, the gradient vanishes, i.e.,

$$f(\mathbf{x}^* + \Delta\mathbf{x}) = f(\mathbf{x}^*) + \frac{1}{2}\Delta\mathbf{x}^T\mathbf{H}^*\Delta\mathbf{x} + o(\|\Delta x\|)^3 \tag{A.12}$$

We now establish that the second term is always positive if all the eigenvalues of $\mathbf{H}^*$ are positive.

Observe that, by definition, the matrix $\mathbf{H}^*$ is symmetric. Therefore, all its eigenvalues are real and mutually orthogonal. Further, suppose $\{\rho_i, \mathbf{x}^*{}_i\}_{i=1,\dots,N}$ are the eigenpairs of $\mathbf{H}^*$, i.e.,

$$\mathbf{H}^*\mathbf{v}_i = \rho_i\mathbf{v}_i \tag{A.13}$$

Further, any vector $\Delta\mathbf{x}$ can be expressed as a linear combination of $\mathbf{v}_i$, i.e.,

$$\Delta\mathbf{x} = \sum_i a_i\mathbf{v}_i$$

This leads to

$$\mathbf{H}^*\Delta\mathbf{x} = \mathbf{H}^*\sum_i a_i\mathbf{v}_i = \sum_i a_i\mathbf{H}^*\mathbf{v}_i = \sum_i a_i\rho_i\mathbf{v}_i$$

Thus, the second-order term in the Taylor series reduces to

$$\frac{1}{2}\mathbf{x}^T\mathbf{H}^*\Delta\mathbf{x} = \frac{1}{2}\left(\sum_i a_i\mathbf{v}_i\right)^T\mathbf{H}^*\Delta\mathbf{x} = \frac{1}{2}\left(\sum_i a_i\mathbf{v}_i\right)^T\left(\sum_i a_i\rho_i\mathbf{v}_i\right)$$

Further, since the eigenvectors are orthonormal, i.e., $\mathbf{v}_j^T\mathbf{v}_i = 0,\ i \neq j$, and $\mathbf{v}_i^T\mathbf{v}_i = 1$,

$$\frac{1}{2}\Delta\mathbf{x}^T\mathbf{H}^*\Delta\mathbf{x} = \frac{1}{2}\sum_i a_i^2\rho_i$$

leading to

$$f(\mathbf{x}^* \pm \Delta\mathbf{x}) = f(\mathbf{x}^*) + \frac{1}{2}\sum_i a_i^2\rho_i + o(||\Delta\mathbf{x}||^2)$$

From this follows the proof.

Theorem A.3 *Consider the equality-constrained optimization problem*

$$\begin{aligned} &\text{minimize}_{\mathbf{x}}\ f(\mathbf{x}) \\ &\text{s.t.} \qquad h(\mathbf{x}) = 0 \end{aligned} \tag{A.14}$$

A stationary point $\mathbf{x}^*$ *satisfies*

$$\begin{aligned} &\nabla f(\mathbf{x}^*) + \mu^*\nabla h(\mathbf{x}^*) = 0 \\ &h(\mathbf{x}^*) = 0 \end{aligned} \tag{A.15}$$

for some scalar μ^*. *Further, the stationary point is a local minimum if the matrix*

$$\mathbf{H}_f(\mathbf{x}^*) + \mu^*\mathbf{H}_h(\mathbf{x}^*) \tag{A.16}$$

is positive definite.

Proof: We provide here a sketch of the proof; see reference [8] for further details.

Consider the first two terms of the Taylor series approximation of $f(\mathbf{x})$ about a point $\mathbf{x}^*$:

$$f(\mathbf{x}^* + \Delta\mathbf{x}) \approx f(\mathbf{x}^*) + \Delta\mathbf{x}^T\nabla f(\mathbf{x}^*) \tag{A.17}$$

Observe that the increment $\Delta\mathbf{x}$ cannot be arbitrary; it must satisfy the constraint. In other words, the increment must be such that

$$h(\mathbf{x}^* + \Delta\mathbf{x}) \approx 0$$

i.e.,

$$h(\mathbf{x}^*) + \Delta\mathbf{x}^T\nabla h(\mathbf{x}^*) \approx 0 \tag{A.18}$$

But $h(\mathbf{x}^*) = 0$; thus the increment $\Delta\mathbf{x}$ must be such that

$$\Delta\mathbf{x}^T \nabla h(\mathbf{x}^*) = 0 \tag{A.19}$$

In other words, if $\mathbf{x}^*$ is a local minimum, it must satisfy

$$\begin{aligned} \Delta\mathbf{x}^T \nabla f(\mathbf{x}^*) &= 0 \\ \Delta\mathbf{x}^T \nabla h(\mathbf{x}^*) &= 0 \end{aligned} \tag{A.20}$$

This is equivalent to

$$\nabla f(\mathbf{x}^*) + \mu^* \nabla h(\mathbf{x}^*) = 0 \tag{A.21}$$

for some real value (positive or negative) of μ^*. In addition, the stationary point must satisfy the constraint

$$h(\mathbf{x}^*) = 0 \tag{A.22}$$

To establish Equation (A.16), consider the first three terms of the Taylor series approximation of $f(\mathbf{x})$ and $h(\mathbf{x})$ about a stationary point $\mathbf{x}^*$:

$$f(\mathbf{x}^* + \Delta\mathbf{x}) \approx f(\mathbf{x}^*) + \Delta\mathbf{x}^T \nabla f(\mathbf{x}^*) + \frac{1}{2}\Delta\mathbf{x}^T \mathbf{H}_f(\mathbf{x}^*)\Delta\mathbf{x} \tag{A.23}$$

where $\Delta\mathbf{x}$ is such that

$$h(\mathbf{x}^* + \Delta\mathbf{x}) = h(\mathbf{x}^*) + \Delta\mathbf{x}^T \nabla h(\mathbf{x}^*) + \frac{1}{2}\Delta\mathbf{x}^T \mathbf{H}_h(\mathbf{x}^*)\Delta\mathbf{x} = 0 \tag{A.24}$$

Since $h(\mathbf{x}^*) = 0$ and $h(\mathbf{x}^* + \Delta\mathbf{x}) = 0$, we have

$$\Delta\mathbf{x}^T \nabla h(\mathbf{x}^*) + \frac{1}{2}\Delta\mathbf{x}^T \mathbf{H}_h(\mathbf{x}^*)\Delta\mathbf{x} = 0 \tag{A.25}$$

Now multiply Equation (A.24) by μ^* and add to Equation (A.23):

$$f(\mathbf{x}^* + \Delta\mathbf{x}) \approx \left\{ \begin{aligned} & f(\mathbf{x}^*) \\ & + \Delta\mathbf{x}^T [\nabla f(\mathbf{x}^*) + \mu^* \nabla h(\mathbf{x}^*)] \\ & + \frac{1}{2}\Delta\mathbf{x}^T [\mathbf{H}_f(\mathbf{x}^*) + \mu^* \mathbf{H}_h(\mathbf{x}^*)]\Delta\mathbf{x} \end{aligned} \right\} \tag{A.26}$$

From the definition of the stationary point, the second term vanishes, resulting in

$$f(\mathbf{x}^* + \Delta\mathbf{x}) = f(\mathbf{x}^*) + \frac{1}{2}\Delta\mathbf{x}^T [\mathbf{H}_f(\mathbf{x}^*) + \mu^* \mathbf{H}_h(\mathbf{x}^*)]\Delta\mathbf{x} \tag{A.27}$$

For $\mathbf{x}^*$ to be a minimum, this quadratic term must be positive for all $\Delta\mathbf{x}$ that are tangent to $h(\mathbf{x}^*)$. A sufficient (but not necessary) condition is that the matrix $\mathbf{H}_f(\mathbf{x}^*) + \mu^* \mathbf{H}_h(\mathbf{x}^*)$ be positive definite.

References

[1] SolidWorks Corporation, SOLIDWORKS. www.solidworks.com, 2005.

[2] The Mathworks, Inc., MATLAB. www.mathworks.com, 2005.

[3] J. S. Arora, *Introduction to Optimum Design*. Elsevier, 2004.

[4] G. N. Vanderplaats, *Numerical Optimization Techniques for Engineering Design*, 3rd edn. Vanderplaats Research and Development, Inc., 2001.

[5] S. S. Rao, *Engineering Optimization: Theory and Practice*. John Wiley & Sons, 2019.

[6] A. Messac, *Optimization in Practice with MATLAB® for Engineering Students and Professionals*. Cambridge University Press, 2015.

[7] P. Y. Papalambros and D. J. Wilde, *Principles of Optimal Design: Modeling and Computation*. Cambridge University Press, 2000.

[8] J. Nocedal and S. J. Wright, *Numerical Optimization*, 2nd edn. Springer Science + Business Media, 2006.

[9] D. P. Bertsekas, *Constrained Optimization and Lagrange Multiplier Methods*. Academic Press, 2014.

[10] F. Beer, E. R. Johnston, Jr., J. DeWolf, and D. Mazurek, *Mechanics of Materials*. McGraw-Hill Education, 2011.

[11] R. C. Juvinall and K. M. Marshek, *Fundamentals of Machine Component Design*, 5th edn. John Wiley & Sons, 2011.

[12] I. H. Shames and C. L. Dym, *Energy and Finite Element Methods in Structural Mechanics*. Hemisphere Publishing, 1985.

[13] K. Sigmon and T. A. Davis, *MATLAB Primer*, 7th edn. CRC Press, 2004.

[14] P. Venkataraman, *Applied Optimization with MATLAB Programming*. John Wiley & Sons, 2009.

[15] G. Strang, *Introduction to Applied Mathematics*. Wellesley-Cambridge Press, 1986.

[16] A. R. Conn, K. Scheinberg, and L. N. Vicente, *Introduction to Derivative-Free Optimization*. SIAM, 2009.

[17] J.-F. Bonnans, J. C. Gilbert, C. Lemaréchal, and C. A. Sagastizábal, *Numerical Optimization: Theoretical and Practical Aspects*. Springer Science + Business Media, 2006.

[18] J. A. Nelder and R. Mead, "A simplex method for function minimization," *The Computer Journal*, vol. 7, no. 4, pp. 308–313, 1965.

[19] R. Horst and P. M. Pardalos, *Handbook of Global Optimization*. Nonconvex Optimization and Its Applications, vol. 2. Springer Science + Business Media, 1995.

[20] A. Törn and A. Žilinskas, *Global Optimization*. Lecture Notes in Computer Science, vol. 350. Springer, 1989.

[21] P. J. Van Laarhoven and E. H. L. Aarts, "Simulated

annealing," in *Simulated Annealing: Theory and Applications*, pp. 7–15. Springer, 1987.

[22] J. K. Karlof, *Integer Programming: Theory and Practice*. CRC Press, 2006.

[23] P. C. Hansen, V. Pereyra, and G. Scherer, *Least Squares Data Fitting with Applications*. JHU Press, 2013.

[24] R. T. Marler and J. S. Arora, "Survey of multi-objective optimization methods for engineering," *Structural and Multidisciplinary Optimization*, vol. 26, pp. 369–395, 2004.

[25] W. McGuire, R. H. Gallagher, and R. D. Ziemian, *Matrix Structural Analysis*. John Wiley & Sons, 2000.

[26] U. Kirsch, *Structural Optimization: Fundamentals and Applications*. Springer, 1993.

[27] K. Svanberg, "The method of moving asymptotes: A new method for structural optimization," *International Journal for Numerical Methods in Engineering*, vol. 24, no. 2, pp. 359–373, 1987.

[28] J. Iott, R. T. Haftka, and H. M. Adelman, *Selecting Step Sizes in Sensitivity Analysis by Finite Differences*, NASA Technical Memorandum 8632. NASA, 1985.

[29] J. R. Martins, P. Sturdza, and J. J. Alonso, "The complex-step derivative approximation," *ACM Transactions on Mathematical Software (TOMS)*, vol. 29, no. 3, pp. 245–262, 2003.

[30] W. Squire and G. Trapp, "Using complex variables to estimate derivatives of real functions," *SIAM Review*, vol. 40, no. 1, 1998.

[31] G. Lantoine, R. P. Russell, and T. Dargent, "Using multicomplex variables for automatic computation of high-order derivatives," *ACM Transactions on Mathematical Software (TOMS)*, vol. 38, no. 3, 2012.

[32] A. Griewank, "On automatic differentiation." In M. Iri and K. Tanabe, eds., *Mathematical Programming: Recent Developments and Applications*. Mathematics and Its Applications, vol. 6, no. 6, pp. 83–107. Springer Science + Business Media, 1989.

[33] J. Su and J. E. Renaud, "Automatic differentiation in robust optimization," *AIAA Journal*, vol. 35, no. 6, pp. 1072–1079, 1997.

[34] K. K. Choi and N.-M. Kim, *Structural Sensitivity Analysis and Optimization 1*. Springer, 2005.

[35] D. A. Tortorelli and W. Zixian, "A systematic approach to shape sensitivity analysis," *International Journal of Solids and Structures*, vol. 30, no. 9, pp. 1181–1212, 1993.

[36] M. P. Bendsoe, "Optimal shape design as a material distribution problem," *Structural Optimization*, vol. 1, pp. 193–202, 1989.

[37] M. H. Imam, "Three-dimensional shape optimization," *International Journal for Numerical Methods in Engineering*, vol. 18, no. 5, pp. 661–673, 1982.

Index